AF361576

The Market in Birds

The Market in Birds

Commercial Hunting, Conservation, and the Origins of Wildlife Consumerism, 1850–1920

Andrea L. Smalley

With Henry M. Reeves

Johns Hopkins University Press
Baltimore

Johns Hopkins University Press
2715 North Charles Street
Baltimore, Maryland 21218-4363
www.press.jhu.edu

Library of Congress Cataloging-in-Publication Data
Names: Smalley, Andrea L., 1960– author. | Reeves, Henry M., author.
Title: The market in birds : commercial hunting, conservation, and the
    origins of wildlife consumerism, 1850–1920 / Andrea L. Smalley with
    Henry M. Reeves.
Description: Baltimore : Johns Hopkins University Press, 2022. |
    Includes bibliographical references and index.
Identifiers: LCCN 2021026358 | ISBN 9781421443409 (hardcover) |
    ISBN 9781421443416 (ebook)
Subjects: LCSH: Fowling—United States—History—19th century. |
    Fowling—United States—History—20th century. | Wildlife
    management—United States | Wildlife conservation—United States.
Classification: LCC SK313 .S525 2022 | DDC 799.2/40973—dc23
LC record available at https://lccn.loc.gov/2021026358

A catalog record for this book is available from the British Library.

*Special discounts are available for bulk purchases of this book. For more
information, please contact Special Sales at specialsales@jh.edu.*

Contents

Foreword

Henry M. Reeves, known by all as Milt, researched the connections among migratory birds, hunters, the market, and consumers. He was interested in the Gilded Age, when privileged diners, even at the White House, feasted on canvasback ducks, passenger pigeons, woodcocks, and a whole host of other migratory birds. Untold millions of these wild birds were trapped or shot by hunters and sold to merchants in city markets. Finally, after much avian bloodshed, controversy, and public outcry, Congress passed the 1918 Migratory Bird Treaty Act and other laws to halt the slaughter.

Milt was a wildlife biologist who believed the history of market hunting needed to be known. He began by collecting out-of-print hunting books, like *The Shadow of a Gun*, by Henry Clay Merritt. Milt wrote introductions for thirty-eight historically valuable out-of-print books and made them available through print on demand from Kessinger Publishing in Montana. He also became interested in the *Florentine Codex*, which describes the knowledge and use of migratory birds by the Aztec. Milt also assisted with the ultimate publication of the *Codex Canadensis* (François-Marc Gagnon and Nancy Senior, McGill Queen's University Press, Montreal, 2011), which received the prestigious 2012 Sir John A. Macdonald Prize from the Canadian Historical Association and the 2012 Governor General's History Award for nonfiction scholarly research.

Milt met Merilyn Bronson Reeves, his wife of sixty-two years, at Utah State Agricultural College (now Utah State University) in Logan, Utah, where Merilyn received her bachelor of arts in English and Milt received his master of science in wildlife management. His 1954 thesis, "Muskrat and Waterfowl Production and Harvest on Dingle Marsh, Idaho," provided the biological and ecological background that led to the eventual designation of the Bear Lake National Wildlife Refuge.

Later, as a federal game agent for the US Fish and Wildlife Service (FWS) in the Rio Grande Valley, Milt helped protect hundreds of thousands of wintering waterfowl along with the then plentiful white-winged doves. In one assignment, Milt joined about fifty agents, led by the undercover agent Tony Stefano, to break up an illegal duck sale operation, resulting in the conviction of fifty-three market hunters and sellers. At that time, this was the largest FWS market-hunting operation in terms of the most violators apprehended over the broadest area. This event is described in the epilogue. Milt noted in his autobiography that his role in the "East Texas Take Down" was "one of the most memorable of many experiences with the Fish and Wildlife Service."

Merilyn and Milt shared a love of nature and worked together in the field. For several summers, Merilyn joined Milt in Saskatchewan, Canada, where Milt was a crew leader banding fledgling waterfowl. Milt went on to work with respected wildlife management professionals in South Dakota and Minnesota, banding thousands of birds in the Central Flyway. While Milt began work in the Bird Banding Laboratory at the Patuxent Wildlife Research Center in Maryland, Merilyn became involved with the emerging environmental movement. Merilyn participated in a volunteer advisory capacity to promote the protection of the environment and public health. She served as a member of the National Board for the American Lung Association. As vice president of the National League of Women Voters, during the peak of the environmental movement, Merilyn represented and educated more than 125,000 league members in cleanup activities to protect the environment. She was also member of the University of California at Berkeley Natural Resources Advisory Committee, served the National Research Council on the selenium study at the Kesterson National Wildlife Refuge, and, for six years, was the first chair of the Hanford Advisory Board, a federal advisory committee. Merilyn received numerous awards for her work. Governor Harry Hughes of Maryland honored her with the prestigious award of Admiral of the Chesapeake Bay.

In 1983, Milt ended his thirty-year career with the US Fish and Wildlife Service as the branch chief of migratory bird management in Washington, DC. During his career and thirty-year retirement, Milt was a prolific writer, authoring or coauthoring fifty-six published articles, along with several books and book chapters. His book *A Contribution to an Annotated Bibliography of North American Cranes, Rails, Snipe, Doves, and Pigeons* (1975) remains a valuable resource. He was coeditor and contributor to *Flyways: Pioneering Waterfowl Management in North America* (Hawkins et al., 1984) and wrote three chapters in *Ecology and Management of the Mourning Dove* (Baskett et al., 1993), which was honored with The Wildlife Society's wildlife publication award for 1996.

Milt's papers are in the Utah State University Merrill-Cazier Library Special Collections and Archives and can be requested at https://archives.usu .edu/search.php. His professional obituary was published in *The Wildlife Professional* (Winter 2013), and his autobiographical *Recollections of My Career, My Colleagues, and My Retirement*, published in 2013, is available online.

During his final decade of life, Milt drafted seventeen chapters of a book he longed to see in print, provisionally titled *The Market Hunters*. The research was difficult and extensive: he filled many boxes and file cabinets. He used the historical expertise of many institutions, libraries, and individuals who assisted him, including the Valley Library at Oregon State University; the Bancroft Library at the University of California, Berkeley; the law library at Willamette University, Salem, Oregon; the Quinney Library at Utah State University, Logan; the Patuxent Wildlife Research Center Library in Laurel, Maryland; the Department of Agriculture Library in Beltsville, Maryland; the Department of the Interior Library in Washington, DC; Arthur S. Hawkins, FWS (retired); and Illinois Department of Natural Resources wildlife management professionals David Kenney, Larry David, Scott Jacoby, and Scott Simpson.

Merilyn was instrumental in the publication of this book. After Milt's death in 2013, Merilyn collected and organized his unfinished chapters. She reached out to Milt's former colleague Richard E. McCabe, who in turn helped to connect the Reeves family with historian Andrea Smalley, author of *Wild by Nature: North American Animals Confront Civilization* (2017). Dr. Smalley used Milt's manuscript and his abundant compilation of source material as a starting point for her further in-depth research, greatly expanding upon the historical aspects and implications, reorganizing the material, and skillfully developing it into this book.

The Reeves family is most grateful to Dr. Smalley for realizing Milt's ambition to tell this important, fascinating, and haunting story of the history of market hunting. Milt would have been impressed with this book and with Dr. Smalley's extensive insights, additions, and expertise. The Reeves family is also grateful to Dr. Clait E. Braun for his detailed review of Dr. Smalley's manuscript. The Reeves family also thanks Dick McCabe for encouraging the family's efforts throughout the process. The other members of the Reeves family appreciate daughter Julie's efforts in assisting her father in resolving many computer glitches in the early stages of this book and helping with format and organization.

The Henry M. Reeves Family

The Market in Birds

For the Birds

The partridge loves peas, but not those that go with her into the pot.
HENRY DAVID THOREAU, "WALKING"

If nostalgia were a force of nature, Henry Clay Merritt's reminiscences could return wildness to Illinois. Fond memories, though, were not enough to restore the prairies he first encountered as a young man in the 1850s. In the early years of the twentieth century, it took an old man's imagination to "reconstruct a map of this wonderful region where the hen cackled on every plain, the quail whistled on every hill, the snipe flew along the marshes, the woodcock whistled through the grove and the ducks swarmed in thousands in every conceivable bayou." Merritt pitied the "despairing huntsman" of this modern age, who would find that "the plough of the farmer has made the prairie, once vocal with wild birds, only a desolate solitude." Future generations would never see the avian abundance of the past, which had been "a sight grander than any temple and more ravishing to the hunter than the streams of Paradise."[1]

Merritt's lament for a paradise lost was a familiar refrain at the twentieth century's turn. Concerns about dwindling wildlife populations permeated popular discourse throughout the United States, provoking widespread worries that something uniquely American had been squandered. In Merritt's own Henry County, local historian Henry L. Kiner likewise mourned the wildlife loss that had occurred within his lifetime. The area had once "swarmed with wild pigeons . . . almost as plentiful as the leaves." Hunters had harvested wagonloads of prairie chickens and waterfowl. Now, in the early twentieth century, the birds were gone. "Seldom, in any walk of life," Kiner concluded, "has there been such wanton waste of the gifts of nature."[2] Decades earlier and several states away, the *Brooklyn Daily Eagle* similarly raised the specter of environmental loss. America had "suffered so much from reckless and indiscriminate slaughter" that even "the warble of song birds [was] scarcely heard in woods once vocal with their notes."[3] Parallel complaints came from California in the 1890s, where commentators warned that the unchecked extermination of ducks and quails would leave the region in "a condition of affairs equal to some of the Eastern States, which are now almost barren of any kind of game."[4] Repeated across the country, these predictions reinforced what many like Merritt had seen with their own eyes: the conspicuous diminishment of a formerly rich and abundant wildlife landscape.

Although the near extinction of the iconic bison provided a well-publicized example of humans' devastating effects on wild animals, it was the dramatic

Henry Clay Merritt (*farthest left*), in front of Butterwick Hardware Store in Kewanee, Illinois, early twentieth century. At times, Merritt stored wildfowl in Butterwick's basement. Courtesy of the Kewanee Historical Society, Kewanee, Illinois

changes in bird populations that most animated public awareness of wildlife issues in the late nineteenth and early twentieth centuries. Diverse and ubiquitous, birds were creatures far more familiar to both urban and rural Americans than were the exotic buffalo that roamed the distant western plains. Besides, the bison, along with other larger mammals, seemed like vestiges of a pioneering past; their disappearance was widely understood as the unavoidable consequence of taming the wilderness. "We hunted the Indians out of the country in the beginning," declared the *Brooklyn Daily Eagle* in 1871, and "then wild cats, deer, and rabbits."[5] To that list could have been added wolves, coyotes, squirrels, and other "vermin" species purposely exterminated by farmers and ranchers. Wild birds, on the other hand, did not appear to be so ill adapted to the modern world. Sheer numbers and mobility were thought to provide some degree of protection for avifauna, and, except for a few pest species such as crows and hawks, most were considered beneficial or at least inoffensive creatures. The visible diminishment of bird populations was more troubling because it presaged a future devoid of all wildlife. Modernity and wildness might ultimately be incompatible.

Merritt's doleful prophecy of a birdless environment was hardly singular in 1904. He was neither the first nor the best-known person to raise alarms. What made his assertion remarkable was that it came from a man who made his money killing birds. Merritt penned his observations at the end of a long and successful career as a market hunter and game dealer based in the small town of Kewanee, Illinois. Across the latter half of the nineteenth century, he had grown his business from a small-scale, mainly local concern into a far-flung interstate enterprise funneling enormous quantities of wildfowl from Illinois and points westward to urban markets in the Midwest and on the East Coast. He shared his profession with a significant number of other commercial hunters, processors, shippers, and sellers across North America. At its height in the last decades of the nineteenth century, the market in wild game brought a variety of animals—deer, elk, bears, rabbits, fish—to consumers. Wildfowl, however, constituted the overwhelming bulk of game sales in the period, making the turn-of-the-century game market predominately a market in birds.

Selling wild birds for their meat, feathers, skins, and eggs was certainly not a new phenomenon in the late nineteenth century. Wildfowl had long been articles of exchange in North America, traded by indigenous peoples, sold in colonial markets, and bartered in frontier settlements. After the Civil War, however, the market in birds changed dramatically. What had been a mostly limited, local, and seasonal trade was transformed into a modern enterprise with all the earmarks of industrial consumer capitalism: new technological innovations, finance capital, specialized divisions of labor, abstract commodification, segmented markets, advertising, and a system of prices and profits. As such, late nineteenth-century wildfowl commercialization was not merely an accelerated version of earlier forms of wildlife trading that had been common from at least the colonial era onward. It was, instead, a revolutionary shift in wildlife use, one that produced a new kind of interplay between biology and economy.

The modern market in birds was short-lived but broadly influential. At its height between 1870 and 1900, market hunting was a huge and lucrative business across much of North America, its rise and fall occurring simultaneously with the emerging Progressive Era conservation movement. This was no coincidence. Market hunting in its modern form became arguably the single most important issue for the initial generation of self-proclaimed wildlife protectors. Wildfowl commercialization forced these early conservationists to unravel the tangled connections between ecology, economy, culture, and law that the market had thrown into sharp relief. The first federal wildlife laws, passed in the opening decades of the twentieth century, were designed primarily to restrict this market in birds. Those laws dramatically expanded the national state's authority over public rights in a particular environmental resource.[6] By 1918, when the Migratory Bird Treaty Act (MBTA) finally brought the legal end to market hunting, American wildlife policy had reached a key turning point. Conservationists' signal achievement of the early twentieth century was the result of a newly nationalized public interest in wild animals that had been produced in large part by the expansive reach of the modern wildfowl market.

Market hunting not only triggered a fundamental transformation of wildlife law but also reshaped the cultural significance of wildness for American society. Late-nineteenth-century wildfowl commercialization assigned wildness a commodity value, linking dead birds' market value to their living wild natures. Wildness, however, had other influential proponents who attributed their own set of esoteric cultural values to wild creatures and defined new terms for wildlife consumption. The market crystallized these competing conceptions of wildlife's worth to Americans and provoked crucial debates about the place of wildlife in a twentieth-century mass consumer culture. The controversies over game commercialization at the turn of the century thus revealed that the issues underlying the Progressive Era conservation movement were not simply those of consumption versus preservation or wastage versus wise use. Rather, the market in commodified birds generated new and contested understandings about the intrinsic value of wildness—ideas that remain relevant and controversial in wildlife conservation policy today.

❧

Market hunting has not been ignored in either scholarly or popular histories of wildlife use in America, but its evolution, operation, and effects have yet to be fully analyzed. Some researchers in the first major wave of conservation studies, published between the 1960s and 1980s, recognized market hunting as a key factor in the origins of the Progressive Era conservation movement. Many of those often-celebratory studies credited the public relations campaign against market hunters, waged initially by activist sportsmen and outdoor publications, with spotlighting the devastating effects of wildlife exploitation and encouraging conservation causes. Contributing to this body of work, a few historians provided more detailed accounts of market hunting and the successes of the conservation crusade against the practice. Appraisals of wildlife law followed a similar progressive historical trajectory, focusing on early failures to protect wildlife from commercial overhunting and the eventual implementation of restrictive legislation.[7]

In the past two decades, environmental histories have complicated the mostly congratulatory picture drawn in early conservation historiography by exploring the ways power and privilege figured into the protection of wild resources. Marrying social history with environmental policy, these studies reveal how ordinary people resisted the modern state's efforts to assert its authority over nature. Class and racial conflicts about customary, common rights in nature explain turn-of-the-century battles over market hunting. Poorer subsistence hunters, middling-market gunners, and elite wildlife advocates each aimed to monopolize wildlife for their own purposes, scholars argue.[8] Wildlife legislation was the key to the redistribution of power over nature, and students of legal and policy history have carefully traced the transition from local rules to state game codes to federal action that took control of wild animal resources away from ordinary people and reassigned it to bureaucratic specialists. Current scholarship suggests that the shift to national regulation of wildlife first articulated in the Lacey Act (1900)—and bolstered later in the Weeks-McLean (1913) and Migratory Bird Treaty Acts—was caused by the mobility of wildlife coupled with inconsistent and poorly enforced state laws, making federal, and even international, legislation necessary.[9]

Environmental historians have generally accepted previous interpretations of market hunting's role in motivating public concern for wild animals and public policies to protect them—to the extent that they often take the expansion of late-nineteenth-century wildlife commercialization and its demise as givens. In these analyses, the game market was the predictable outcome of an early American doctrine of "free taking," which carried the seeds of its own inevitable destruction.[10] Historians cite widening criticism of market hunters' excesses at the turn of the century, along with the spread of restrictive game laws, wildlife activist organizations, and popular campaigns against the commercial exploitation of wild creatures in the millinery trade and live-bird trap shooting, as evidence of a slowly awakening conservation consciousness in the United States. Market hunting makes for a convenient marker to delineate a sharp break between the ecologically unenlightened past and modern sensibilities about nature.[11]

Missing from the conventional interpretations is a ground-level examination of the workings of wildlife commercialization. Striking images of buffalo hunters perched on enormous piles of hides or professional gunners festooned with strings of waterfowl obscure the far-reaching effects of the modernizing market in game. As a complex and changing enterprise involving a variety of human and nonhuman players, market hunting cannot be explained simply in numbers of animals killed. Its many accomplices cannot be subsumed under the single heading of "market hunter." Advocacy for federal intervention in wildlife regulation cannot be fully appreciated without attention to the shifting alliances and divisions in the conservation movement provoked by changing ideas about wildlife consumerism at the turn of the century. Above all, the history of wildlife conservation and law cannot be told without acknowledging that the market in birds is central to that story.

To date, market hunting has been most meticulously described in a small body of state- and regional-level studies as well as some popular histories devoted to the topic. Scouring little-used archival sources and interviews, authors of these accounts detail the legal and illegal activities of commercial gunners in the field, going well beyond the often spare, generalized depictions in more scholarly examinations. Often presented as rehabilitative portrayals, market hunting books and articles written for popular audiences mainly concern themselves with recording "that brief but colorful bit of American history" made by an extinct class of hunters.[12] Like much of the conservation historiography, these histories tell a tale of progress leading toward enlightened wildlife protection while commonly offering a sympathetic interpretation of professional hunters' practices as socially acceptable, even laudable, in historical time. Less attention is paid to the broader marketplace in these works or to the major legal confrontations that occurred there rather than in the field. Game consumers, dealers, shippers, processors, and other vital contributors to the market hunter's occupation are usually underrepresented in popular market hunting chronicles. Some state and regional studies of wildlife commercialization present a fuller account of game market operations, but their narrower geographic focus prevents them from situating both economic and legal changes within the wider contexts of interstate game markets and national wildlife statutes.[13] In short, these works are valuable for their fine-grained detail, but they often lack the analytical frameworks to explain economic, legal, and cultural transformations that produced modern wildlife consumerism.

Regional studies and popular histories as a body, however, have attended more than most other historical scholarship not only to the practices of professional hunters but also to the traits of the animals they hunted. Many of these accounts give wildfowl a prominent role in their stories since birds were the prey for most market gunners and harvesters. Conversely, broader analyses of hunting and conservation have tended to privilege larger, more charismatic mammals or treat wildlife as an undifferentiated category.[14] Small, more ubiquitous creatures have garnered less concentrated scholarly interest. Researchers have only recently taken up American avifauna as an object of historical study, with publications devoted to such topics as the development of ornithology, the organization of the Audubon Society, the creation of wildfowl refuges, and the popularization of bird watching.[15]

Both scholarly and popular accounts of wildlife use in the Progressive Era, however, share a common argumentative theme. Wide agreement exists on the point that a major shift in American wildlife value orientations occurred in this period. Dominionistic and utilitarian values that had shaped much of the early American approach to wild animals were modified by new aesthetic, ecologic, and moralistic sensibilities. Traditional historical interpretations posited a clear dichotomy of values in the conservation movement between utilitarian and preservationist impulses, usually exemplified by Gifford Pinchot, the first chief of the US Forest Service, and the eminent naturalist John Muir, respectively. Although scholars have since challenged this overly stark ideological divide in the conservation movement, the general notion remains that urbanization and industrialization reshaped Americans' relationships with animals at the twentieth century's turn and, in the process, reformed public attitudes about wildlife. The current literature explains popular support for new wildlife conservation policies as a direct result of these changing cultural values, coupled with a belief in the efficacy of scientific management and government intervention.[16]

Certainly, ample evidence demonstrates that shifting public valuations of animals shaped American political, legal, and cultural responses to diminishing wildlife populations at the turn of the twentieth century. Standard interpretations of that transformation, however, locate its causes in an ethical evolution driven by broad social forces of modernization. They do little to explicate specific mechanisms, processes, and personalities responsible for that change. Moreover, analysts' neat taxonomies of values dividing utilitarian from moralistic, or dominionistic from ecologic, attitudes obscure the extent to which those discrete categories are themselves historically constructed products of the same paradigmatic shifts they seek to explain. Sorting wildlife attributes into separate classes of competing, human-defined values is a modern phenomenon that warrants its own historical examination.[17]

In many ways, the market in birds and the debate it provoked produced that cultural phenomenon. Intensifying wildfowl commercialization in the latter half of the nineteenth century created the conditions for a broad, public reevaluation of wild creatures and set the cultural and legal terms for wildlife conservation efforts in the twentieth century. The game market assigned one measure of worth: an exchange value tangibly expressed in profits and prices. Of course, wildlife had been objects of trade and consumption long before, but their social utility was not defined solely by market prices. So long as America's natural abundance allowed for largely indiscriminate exploitation of animal resources, few bothered to rank wildlife uses into any hierarchy of worth. Instead, as John Smith's seventeenth-century descriptions of New World fauna indicated, wildlife afforded a multiplicity of human benefits: "not only chase sufficient, for any that delights in that kind of toil or pleasure; but such beasts to hunt, that besides the delicacy of their bodies for food, their skins are so rich, as may well recompense thy daily labor, with a Captain's pay."[18] Well into the nineteenth century, most Americans continued to believe that subsistence, recreation, profit, protection, and pleasure, pursued singly or in combination, were equally valid reasons to consume common wild species, so long as they were available. There was little need to make value distinctions in order to justify or condemn certain forms of wildlife use. More prominent in the limited legal regime governing human-wildlife interactions in early America were the issues of ensuring democratic access to beneficial creatures and encouraging the control of harmful ones.[19]

The modern market in birds fundamentally altered these views of wildlife use, and that crucial transformation is the focus of this book. The singular in its title—*The Market in Birds*—is deliberate, employed to distinguish the complex interstate and international system of game commercialization that developed in the late nineteenth century from the various kinds of wildlife exchanges common in early America. Wildfowl consumerism took on an industrial-capitalist form in this period, changing the ways in which many Americans encountered wildlife. As economic and industrial innovations broke through the natural barriers that had previously kept local wild game trades in check, the urban game market emerged as the locus of a public debate about the forms and limits of wildlife consumerism. That debate crystallized new, "pure" categories of economic and noneconomic motives for exploiting or preserving birds as it legitimized some uses of wildness over others.

The most conspicuous agent of those changes was the market hunter. A novel figure on the post–Civil War economic and cultural landscape, the market hunter quickly became an identifiable character in public discourse by the end of the nineteenth century, distinct from his deerskin-hunting and fur-trapping frontier predecessors. The modern market hunter was a new breed—part outdoorsman and part businessman—who was at times admired for his environmental skills and at others criticized for his environmental excesses. Although some observers grumbled about the lack of controls on commercial hunting, few in the nineteenth century considered the market hunter's profession to be wholly illegitimate. The popular image of the market hunter, however, obscured the actual variety commercial gunners, trappers, and collectors who harvested wildfowl meat, eggs, feathers, and skins for profit. Chapter 1 details this varied array of professional wildfowlers whose true identities and actions were often masked by the singular character of the "market hunter."

Even less visible was the broader cast of characters central to the modern, capitalist system responsible for moving wildlife from countryside to cities. Chapter 2 traces the economic and technological changes that spawned a new class of profit-seeking avian entrepreneurs in the late nineteenth century, including shippers, processors, commission agents, wholesale merchants, retailers, and restaurateurs. Game dealers stood at the apex of this system, positioned at the critical juncture between production and consumption. Much like the singular market hunter figure, "the dealer" represented varied conglomerations of workers and businessmen whose specialized labors transformed living creatures into marketable products. Game merchants and their allied industries connected what had previously been mostly local and regional game trades together into a modern, capitalist network designed to concentrate scattered wildlife resources and transport them to urban consumers. As the game business grew more complex, it came to be dominated by these so-called market men, leaving individual hunters subject to the whims of a far-flung economic system increasingly beyond their control.

Chapter 3 turns to the key characters in this drama: the hunted. Playing dual roles as biological creatures and economic commodities, wildfowl constituted the major portion of the modern game market and the central problematic in the turn-of-the-century commercial hunting debate. As the market's reach expanded across ecological zones in the late nineteenth century, a kind of economic migration redistributed avian populations as commodities to consumers. Yet even in birds' second lives as economic goods, their natural characteristics continued to shape the market. Modern American consumers valued birds for their wildness. Avian wildness, though, mattered to others in different ways. Foremost among those champions of wildness, at least if measured by the amount of their public activism, were sportsmen who sought to preserve the experience of wildness they found through hunting. Influential, organized sport hunters positioned themselves as uniquely qualified game protectors whose noneconomic interests in wildlife countered the market's exploitation. Despite all the organization and movement building, however, the "sportsman"— the subject of chapter 4—was no monolithic force when it came to regulating wildlife consumerism. Contentious disputes arose over what constituted proper sport and improper hunting excesses in the field.

Sportsmen found exploitation far easier to identify when it occurred in the market. Confronting the apparatus of wildfowl commercialization directly, amateur hunters found allies among scientists, bird lovers, and politicians who sought national prohibitions on avian marketing. Expanding legal restrictions on the game market near the end of the nineteenth century recast the market hunter and his accomplices from professionals to criminals. Chapter 5 explores how the conservation coalition came to criminalize "the man who hunts for the market."[20] Not everyone, though, agreed that the commercial hunter committed the greatest crimes against nature or that he was the only one motivated by wildfowl profits. Game wardens, state game commissioners, judges, and even activist sportsmen faced public charges of corruption, collusion, and profiteering from new forms of wildlife consumerism. The first federal wildlife statute, the Lacey Act, passed in 1900, added another level of surveillance, extra game law enforcement, and more opportunities for malfeasance and misuses of power. In the end, although wildfowl protectionists sought to label the market hunter as the "criminal" solely responsible for avian destruction, interpretations of criminality were more often a matter of perspective.

Those differences in perspective began to fracture the conservation coalition in the twentieth century, even as its members commenced work on their most ambitious legislative project to date: a federal migratory bird law designed to replace the patchwork of nineteenth-century state-level game codes with uniform, ecologically grounded wildfowl protections. Chapter 6 details the ways the "conservationist" sought to broaden public support for bird preservation beyond the narrow base provided by recreational hunters. By focusing on the spatial dimensions of the wildfowl market and ecological benefits of *migratory* birds, conservationists nationalized, and even internationalized, wildlife protection issues.[21] At the same time, activists advertised a variety of recreational, educational, and psychological contributions wildfowl made to the quality of human life. Thus, the value of living birds, many prominent conservationists insisted, could not be quantified simply in either dollars of profit or numbers of hunters.[22] This new emphasis in wildlife conservation rhetoric threatened to displace sportsmen from their long-held, preeminent positions in wildfowl protection and drove a wedge into the conservationist ranks. When the federal migratory bird protection laws once advocated by many recreational hunters faced critical constitutional challenges, they came not from commercial hunters or game dealers, but from organized sportsmen.

The stock characters in the usual story of wildlife conservation—the destructive market hunter, the profit-minded game dealer, the ethical sportsman, the criminal poacher, and the crusading conservationist—fit well with a teleological narrative of environmentalist progress. On closer examination, however, these caricatures dissolve into distinctive personalities: a business-minded San Joaquin Valley market hunter supplying game birds for San Francisco sportsmen's dinners; an outdoor writer branding both market hunters and sportsman with rank commercialism; a New York restaurateur trying to navigate a rapidly changing legal landscape policed by politically well-connected sport hunters; a successful St. Louis wildfowl merchant employing the most modern technologies profiled glowingly in the *Saturday Evening Post*; an embattled game warden seeking to reconcile Chicago's game dealers with local sportsmen's associations; and a Kewanee, Illinois, entrepreneur making a small fortune from an avian paradise while decrying its destruction. None could be deemed entirely representative of a class of historical actors, and no group championed a consistent set of principles. This complexity reveals the Progressive Era conservation debate to be no simple contest between culturally enlightened preservationist values articulated by wildlife protectors and antiquated utilitarian motives pursued by wildlife exploiters. These divisions had not yet been so clearly defined.

Paradoxically, the cultural terms on which the battle for wildlife conservation policies would be decided were forged in the market. Qualitatively different from earlier kinds of wildlife exchanges, the late-nineteenth-century, industrialized wildfowl business created a mass market for wildness. By repackaging far-flung wildness as an object of consumer desire, the market in birds transformed wildfowl from a localized economic resource into a nationalized cultural resource. Whether selling meat, eggs, specimens, or feathers, the interstate wildfowl market served as a mechanism, albeit an imperfect one, for reconstructing modern urbanites' attenuated connections to nature. Avian populations, meanwhile, suffered in the grasp of a rapacious consumer economy.

Sport hunters proffered a different vision for wildness in the modern age, turning the wildfowl market's terms into an argument for the recreational advantages of wildlife consumerism. While sportsmen-conservationists identified the market's profit motive as the main destructive force working against wildfowl, they calculated the "dollars-and-cents value of game birds" pursued by the recreational "shotgun exponent."[23] These monetary benefits, they insisted, far exceeded the returns generated by commercial sales—economic values that had been heretofore "not fully priced in the market."[24] Of even greater significance, though, were the many social advantages wildfowl shooting afforded by "supplying a means of healthful recreation in field and forest" to the consuming public.[25] Of course, wildfowl hunting was a "tidy business" of great "monetary importance," but its "moral value" was even greater.[26] Hunting was a "way to express in actions, if not in words, the love of nature" that continued, even in the modern age, to draw people "into the big and little wilds of old mother earth."[27] In order to distinguish legitimate sport from illegitimate market hunting, however, sportsmen needed to convince nonhunters—most of whom saw little difference between the two forms of killing—that the recreational hunter's active engagement with living wildness, his physical connection to the intangible qualities of nature, was morally superior to the consumer's purchase of dead, commodified wildness sold in the game market.

Henry Merritt disagreed. The old market hunter and game dealer passed judgment on "this exercise which is fancifully called sport" early in the twentieth century. "Is it sport or is it murder?" Merritt refused to say. But it was obvious that he doubted sportsmen's claims to the moral high ground. The "killing of birds for sport," he asserted, was "so brutal that Christian ethics should make it a thesis for future orations." Challenging the conventional wisdom about market hunting's destructiveness, Merritt added that "the man who shoots for a profit" could not be a factor in wildfowl depletion. When birds became scarce or profits were low, market calculations of efficiency, expenses, and returns necessarily limited the numbers harvested by the professional wildfowler. Once the numbers declined to an unprofitable level, economics compelled the professional gunner to move on to more productive grounds.[28]

Amateur hunters were not governed by these kinds of rational cost-benefit calculations. The actual "value of game killed" by sportsmen was "often less than the cost of killing," Merritt observed. While many previously "believed that the scarcer the game the less demand would be made upon it," the "rage for hunting" had nevertheless continued unabated into the twentieth century. It had grown "beyond all reasonable bounds" even as wildfowl populations dwindled. "Where men hunt for sport," Merritt claimed, the supply of game would continue to suffer "constant or yearly pursuit" as hunting enthusiasts chased after the seemingly limitless cultural, social, and psychic intangibles sport hunting advocates promised.[29] For them, the profits from hunting were not measured solely in the tallies of birds bagged or the efficiency with which they were killed. Wildfowl's value was gauged instead by the quality of the experience with wildness, the boundless pleasure in the pursuit. On this point, the game dealer would have found common cause with anti-hunting conservationists who came to see little difference between game market consumerism and sport hunting consumerism. Both were destructive of wildness. Advocating alternative, "nonconsumptive" uses of wildfowl, influential conservationists proffered their own visions of avian worth. This mounting "battle between bird-lovers and sportsmen" in the early twentieth century, warned ornithologist and cofounder of the Audubon Society, T. Gilbert Pearson, signaled "trouble brewing in the ranks of the so-called conservationists in this country and the extremists on both sides are increasing in numbers."[30]

The modern consumer-capitalist game market, though short-lived, brought about these crucial redefinitions of birds' legal status and cultural value—refinements that required ever-sharper distinctions between economic and noneconomic motives for wildlife consumption and preservation. Sportsmen's and conservationists' campaigns to expand wildfowl value categories to include more esoteric elements of the human experience with nature broadened popular support for wildlife preservation. The successes of federal migratory bird policies relied on this revolution in environmental values that isolated and elevated the aesthetic, hygienic, ecologic, and psychic qualities of experiences with living wildness over the market's calculations of worth. At the same time, however, these legal and cultural reevaluations transformed the experience of avian wildness itself into a consumer good and revealed the inherent contradictions of conservation in a mass consumer culture.[31]

The modern market in birds set these changes in motion. In a few short decades at the turn of the twentieth century, industrial and economic innovations loosened the environmental constraints that had controlled wildlife use and turned the game business into a capitalist enterprise. In the process, the wildfowl market made wildlife consumerism a national economic, legal, and cultural issue. For Merritt, this was the game market's greatest contribution: "What other department of business," he asked, "has so well represented a united country where the benefit of one State is the happy heritage of all?"[32] Although not everyone would have agreed with his cheery interpretation of the market, there was an emerging consensus that wildfowl commercialization had drastically altered Americans' relationships with wildness. The market in birds thrust the environmental challenge of wildlife consumerism onto the national stage, and it has never left.[33] It is high time we explain how it got there.

CHAPTER ONE

The Hunter

"I was born with a gun."

Henry Clay Merritt could think of no other way to describe his lifelong fascination with hunting and shooting, a fascination most of his family and neighbors did not share. To them, a gun "was like a bird of ill omen," and hunting was a distraction from the honest work of farming. A gun merely tempted indolence. The Merritt household's only firearm stood in the corner, behind the door, and was unlimbered only when the hawks and weasels got into the chickens. Yet Henry had always felt a strong, instinctual attachment to some long-forgotten frontier past when hunting was a way of life. A "taint of wildness," inherited from his dusty ancestors, he thought, must have infected him.[1]

He was no frontiersman, though. The bucolic village of Carmel, New York, where he was born in 1831, was no unsettled wilderness requiring firearms to survive. Henry spent his boyhood not far from the shores of Lake Mahopac, already a popular resort destination among antebellum vacationers. Nestled in the hills of Putnam County, the picturesque lake provided fine recreation for anglers, boaters, and hunters. An easy three hours' train ride from New York City, the lake and surrounding countryside were crowded in the summer with visitors seeking respite from their urban existences.[2] The farming families who made up the bulk of the region's full-time residents had little need of hunting prowess, and they looked askance at those who spent their time in pursuit of game.

Young Henry found his shooting role models not at home but in the hunters nearby: some of them visiting sportsmen, some of them locals who "made their living with a gun." Carmel's proximity to New York allowed those semiprofessional gunners to send "bushels and barrels of game" to the city during the colder months, and Henry imagined himself spending his time, like them, afield with gun and dog. He started hunting with his brother, using the old, unwieldy household gun to shoot beans at bobolinks and owls. It was a satisfying but cumbersome process that involved Henry lying on his stomach while his brother balanced the heavy gun across his back to shoot. When the opportunity arose to buy his own weapon from a local hunter, the aspiring shooter jumped on it without telling his father, causing "a rumpus" in the family when it was discovered. With gun in hand, he set out to be a sportsman, shooting birds on the wing. On other days, he thought he might "supply the market" with the bobolinks and woodcocks he killed. He tagged along with other hunters to practice their techniques for hunting game birds. By the time he was ready to leave for college, Merritt had already become known around the Lake Mahopac region as "an amateur hunter of some respectability" and a capable guide for visiting sportsmen.[3]

Entering Williams College in 1850, Merritt quickly learned to intersperse his studies of Latin and Livy with bouts of partridge hunting. When his shooting successes overstocked his landlady, Merritt shipped the excess to one of his sportsman friends, who paid for the birds based on values quoted by A. & E. Robbins, a premiere New York game merchant. This first taste of profitable wildfowling stuck with him after graduation as he returned home to spend his summer and fall hunting woodcocks and partridges. He sent them to the nearby market towns of Croton Falls and Brewster Station and, eventually, to New York, where his birds brought higher prices. When winter came, he ended his postgraduate "vacation" and took a respectable, though temporary, position teaching school in Madison, Connecticut. At the end of the spring term, however, he found himself out of a job and uncertain about his future.[4]

A chance conversation at the train station in Croton Falls suggested a path forward. Illinois, Merritt's informant told him, was ripe with prairie chickens. A skilled hunter would do well there. As it turned out, a couple of Merritt family relations had already gone to Illinois some years earlier, settling in the central part of the state. It seemed as good as any other plan Merritt could devise, so he left New York, arriving at the Illinois River town of Peru in June 1855. He was out hunting in the fields of Marshall County by the start of chicken season in August. "The birds we killed and the sport we had," he recalled years later, "would sound like a tale that is told, which nobody would believe." With a partner, Merritt could bring in upward of five hundred birds in one outing, but "there were no buyers or sellers at that early day," so they settled for supplying themselves, their families, and their neighbors with free meat.[5]

Merritt passed his first years in Illinois in this desultory way, working at various jobs and hunting whenever he got the urge. He taught school once or twice but decided he was not cut out for teaching as a lifelong profession. He reluctantly harrowed for a local farmer "for five long days," received five dollars for his trouble, and then promptly quit. He boarded with various relatives, but soon began to wear out his familial welcome. "I was practically an outcast," he said; "there was no one to take me in; there were plenty to throw me out." Though still uncertain about his future, Merritt wed Sarintha Moore in 1857, declaring that "I might have done better; she certainly could have done no worse." Now a supposedly respectable married man, he felt the need to seek some sort of gainful employment, or at least move out of his mother-in-law's house. In a quandary about his prospects, Merritt pondered his next move.[6]

The first clue to his future appeared as he was walking into town one day. Noticing the fence rows lined with prairie chickens, he "began to study what [he] could do with them." Golden plovers dotted the skies, but he could see "no way of turning them into cash." Birds were everywhere on the Illinois prairies. How could he make a living out of this avian abundance? With that thought in mind, Merritt threw a few grass plovers he had shot into a trunk as he and his new wife packed their bags for a honeymoon trip back to his home state early in 1857. The birds were understandably "a bit tender" by the time he reached New York, but Merritt proceeded to peddle them on the city streets anyway. When he was thronged with eager buyers at the Washington Market, an idea began to form in his mind. In the springtime, the meadows of Henry County were full of golden plover, the cornfields ripe with prairie chickens, and the sloughs crowded with ducks. "It did not seem a bad proposition," he conjectured, "to kill and market them if I could ship them safely to a market a thousand miles away."[7]

⊷

Merritt's idea, to make a living from killing and selling wildfowl, was not an entirely novel one. Harvesting wild creatures for the purpose of exchange had a very long history, and the trade of meat, furs, eggs, feathers, and skins was ubiquitous practice in ancient and premodern societies. Much of that trade was individual and local, but in some cases, wild animal products circulated in complex and far-flung systems of exchange. Hunters of the distant past certainly profited economically, politically, or personally from those exchanges.

But Merritt was no throwback. While his desire to hunt may have come from his pioneering ancestors, his plan to make his living with a gun owed less to the past than it did to a technologically and economically modernizing future. Merritt represented a new type of hunter—the modern market hunter—an ephemeral form that emerged in the decades after the Civil War, flourished at the turn of the twentieth century, and was forced out of existence by the 1920s. There was no singular version of the modern market hunter. Instead, a diverse group of hunters, trappers, netters, and collectors pursued wild creatures— particularly wildfowl—for profit. Some did so temporarily and sporadically. Others labored at it more consistently. In all cases, however, late-nineteenth-century market hunters were products of a modern, industrializing, consumer-capitalist economy.

Turn-of-the-century professional wildfowlers possessed knowledge of wild creatures and their habits, but to that biological knowledge the modern market hunter added a facility for business and a familiarity with costs, profits, prices, and economic cycles. The modern consumer-capitalist economy allowed market hunters to specialize, to develop long-distance trading networks, and to become commercialized. Market hunters like Merritt depended on modern technological innovations in weaponry, transportation, and food storage as well as commercial innovations in corporate organization, financing, and advertising. In sum, these were hunters operating in an economic environment quite different from that prior to the Civil War.

Modern market hunters differed from their predecessors not only in their economic functions but also in their social and cultural ones. Market hunters became identifiable characters on the late-nineteenth-century cultural landscape. Popular images, though often conflicting, distinguished between men who made their livelihoods from commercialized hunting and those who pursued game recreationally. These popular depictions separated market hunting from all other forms of hunting and classified market gunners as a different breed. These were men who made their livings by killing game.

Thoroughly Modern Market Hunters

To be sure, many had made their livings from killing game before. Much of the wildlife harvested in premodern and colonial America by indigenous peoples, Euro-Americans, and African Americans was traded or exchanged in some way. Colonizers amassed substantial capital from selling furs and skins. Public markets carried venison and wildfowl of all sorts, including ducks, geese, turkeys, and wild pigeons regularly supplied by hunters. Others who made their livings by hunting included those employed by military, exploration, and surveying expeditions to provide provisions, as well as hunters and trappers working for fur-trading companies. After the Civil War, professional hunters found employment feeding work crews on the transcontinental railroad.[8]

Were these market hunters? They all profited from hunting, trapping, or collecting meat, skins, feathers, or eggs. They sold, bartered, or otherwise exchanged wild animal products for monetary or nonmonetary gain. At times, they supplied markets or were employed by others to acquire wild animals or birds. In some cases, these hunters were key players in extensive, transatlantic economic networks, as in the case of the southeastern deerskin trade or the colonial fur trade that engaged Euro-Americans and Native Americans alike. Taking the broadest of definitions, all these hunters would seem to fit the basic criteria: they made their livings, at least in part, by hunting.

Yet the modern market hunter was of a different character than those professional and semiprofessional hunters who had come before. In fact, the term "market hunter" originated in the mid-nineteenth century, and its regular association with a class of professional hunters occurred only in the century's last decades.[9] Before the Civil War, hunters who killed wildlife for sale tended to be identified by the animals they pursued rather than by the forum in which they were exchanged: deerskin hunter, bear hunter, fur trapper. Census records show that while the term "hunter" was used in the early nineteenth century as an occupational category for Native Americans as well as hunters of European and African descent, by 1900 census takers used "market hunter" or "hunter for the market" to distinguish the predominately white commercial hunters from Indian hunters (who may also have been hunting commercially).[10]

Besides appellation, there were other characteristics that set modern market hunters apart from their predecessors. Many of these earlier hunters supplied mainly nearby markets within limited systems of exchange. Even well into the nineteenth century "there wasn't such a thing known as a market hunter," and most wildlife was consumed locally.[11] To be sure, some of the earliest, most professionalized hunters—deerskin hunters and fur trappers—participated in transatlantic trading networks, yet these enterprises were of a colonial, mercantilist nature. They were extractive ventures, designed to acquire lightly processed raw materials to be transported to colonial centers. Although hunting for the market, colonial deerskin hunters and fur trappers were not market hunters in late-nineteenth-century terms.

A nascent version of the modern market hunter appeared in the bison hide trade that flourished briefly in the western plains during the 1860s and 1870s. This business had many of the characteristics of commercial wildfowling. It was inextricably linked to modern industrial technologies: railroads, rifles, and new leather-making methods. The hunt was a specialized process, with organized teams of men carrying out specific roles. The long-distance trade in commodified hides was also specialized, and the marketing segmented. Yet buffalo hunters retained the characteristics common to earlier forms of wildlife exchange. In many ways, bison hunting still functioned as an extractive, colonizing venture, occurring ahead of white settlement on contested frontiers not fully incorporated into the US state. It was a venture carried out sometimes in cooperation and, more often, in competition with indigenous peoples similarly engaged in the business of producing hides for the market.[12] Although bison hunters were occasionally identified as market hunters in public discourse, they were more commonly known as buffalo or hide hunters. The business was also short-lived. By the time commentators began castigating bison hunters for driving the great beasts to the brink of extinction, the hide business was burning out. Supplanting the buffalo hunter as the market-hunting villain in conservation rhetoric was the turn-of-the-century commercial wildfowl gunner. The market hunter acquired his modern form in those men (and they were almost entirely men) who shot or otherwise captured wildfowl to sell for their meat, feathers, eggs, and skins.

The singular image of the "market hunter" popularized in the second half of the nineteenth century obscured the wide variety of gunners, netters, trappers, and collectors engaged in the game business at the turn of the century. This diversity makes identifying and quantifying modern market hunters difficult. Wildfowl commercialization cut a wide swath across American society, encompassing impoverished men duck hunting in the Chesapeake Bay, immigrant egg gatherers scouring islands off California's coast, and professional plume hunters working Louisiana's bayous. Some who sold wild birds would not have considered themselves to be market hunters. Others pursued the business more exclusively as professionals—at least for a time. In short, those who worked as market hunters often depended on season and circumstance. They can, however, be divided roughly into three main categories defined by the nature of their participation in the profession—opportunistic, independent, and salaried market hunters—while recognizing that distinctions between groups were not sharp and that any one individual might shift between categories depending on conditions.

Opportunistic market hunters took advantage of fortuitous chances to kill and sell game while spending the rest of their time in other pursuits. Farmers, youngsters, the rural poor, and sportsmen all might sell birds occasionally. Wildfowl hunting, if "approached as a business," wrote David Cartwright in 1875, offered "monetary success" to farmers during slack seasons.[13] Game sales also provided poor southern farmers with an income after the dislocations of the Civil War.[14] For youngsters, such as those working intermittent passenger pigeon nestings, market hunting represented a lucrative diversion. Boys could earn a penny or more for each squab they knocked out of a tree and brained with a stick.[15] "City sportsmen who like[d] profit and well as sport" could market excess birds to pay for future hunting trips.[16] Although contemporary observers disagreed about the extent to which such spasmodic hunting contributed to commercial wildfowl harvests overall, they agreed that amateur, opportunistic market hunters were ubiquitous in many locales frequented by wild game.[17]

Self-identified professionals, however, were the most visible of the modern market hunters: men (and rarely women) whose primary occupations, at least seasonally, involved the taking of wildlife for sale and consumption. Even within this category, however, there was great diversity. Many hunted independently, working by themselves or with partners, selling directly to customers or to wholesale buyers, commission agents, or retail markets. Phoebe Ann Moses worked as this kind of independent hunter, selling wildfowl and other small game locally to a grocery store in Greenville, Ohio, before she became famous as the Wild West's Annie Oakley.[18] Lon Main, another independent market hunter, shot birds as a solo act in northern Iowa to ship to farther-away markets in Des Moines, Minneapolis, and Chicago.[19] Most independent professional wildfowlers operated in a particular locale like John Cockrell's party of hunters based on the Sacramento River in 1850 or the four Kleinman brothers who harvested waterfowl at lakes south of Chicago.[20] But other independent hunters traveled extensively, migrating along with the birds they sought to exploit. Professional "pigeoneers," for example, were prepared to move at a moment's notice to any location where passenger pigeon nesting colonies were reported.

Another class of professional hunters worked as salaried employees for larger commercial concerns. Hotels or restaurants sometimes hired hunters to provide wild birds for their menus.[21] Other employers included game dealers who built ice plants near waterfowl-rich wetlands and recruited local gunners to hunt exclusively for them, paying them by the day, week, month, or season. The Winters Brothers operated a large freezer near Spirit Lake, Iowa, and employed a number of professional hunters. These men were reportedly well compensated, earning seventy-five dollars a month in the 1890s plus their room and board.[22] The Winters paid for ammunition but expected that their hunters would "bring in at least one duck for every two shells."[23] Landowners, too, sometimes hired shooters for as little as a dollar a day, furnishing meals, blinds, and live decoys in return for the harvested birds.[24] When circumstances allowed, however, salaried market hunters might work for themselves.[25]

The variety of commercial hunter specializations also complicates any easy generalizations about market hunting. The professional gunner who shot game for sale in established food markets is surely the best known, but other market hunters worked in less familiar ways: "plumers" acquired feathers and skins for milliners; "eggers," predictably, gathered wildfowl eggs; collectors sought out bird, nest, and egg specimens to sell; and trappers captured live birds for use as hunting decoys, targets, or pets. These specialized exploiters sometimes shifted from one occupation to another depending on opportunity, the season, or the vagaries of the market: egg stealing could include shooting or otherwise taking the breeding parents for their plumage while gunners might collect "cripples" to sell as decoys. In all, as the *Saturday Evening Post* pointed out early in 1903, modern commercial hunting was not simply a process of game being "shot here and there by straggling hunters and sent to the nearest market." While some opportunistic and independent shooters still operated in that "haphazard and incidental way," new varieties of professionals had emerged. Armed with specialized methods and imbedded in an extensive commercial system, the many types of modern market hunters were obvious evidence of an expanding, consumer-driven market in birds.[26]

Given this diversity, it is almost impossible to know how many market hunters were active at any one time in the locality, region, or nation. Scattered anecdotal accounts offer only clues. One commentator suggested in 1872 that more than a thousand men in Louisiana earned "a fair living during four months of the year, shooting [waterfowl] for the New Orleans market."[27] Henry B. Roney, a crusader for passenger pigeon protection, claimed that about five thousand men in the United States worked annually as pigeoneers, while biologist T. Gilbert Pearson reported that between 1903 and 1910 some four hundred duck hunters were working in Currituck County, North Carolina, who marketed no less than $100,000's worth of game annually.[28] Iowa game warden E. C. Hinshaw claimed that "eighty per cent of the farmers in this whole territory supported themselves and families partially from market hunting" at the century's turn.[29] In many cases, though, observers did not distinguish between professional and recreational hunters in their counts. When sporting author John Mortimer Murphy estimated that "over five thousand persons were engaged in wild-fowling along the Maryland side" during the winter and early spring seasons of the 1890s, he failed to note how many were shooting for the market.[30]

Census records provide only fragmentary evidence for market hunter numbers. The occupational category of "market hunter" appears only in the 1900 and 1910 decennial censuses, with a limited number of market hunters or "hunters for the market" listed in the returns.[31] But these numbers do not begin to account for all the persons engaged in the different kinds of professional hunting. In part, this problem reflects the varying terminology used by census takers, who listed "game hunter," "hunting wild game," "hunter and trapper," or even simply "hunter" interchangeably as an occupation or industry. One isolated set of census statistics from 1890 reported 1,516 commercial plumers working in Florida, Louisiana, and Texas who collected feathers worth nearly fifty thousand dollars that year.[32]

The seasonal and intermittent nature of market hunting also meant that many self-proclaimed commercial hunters were not identified as such by census takers. For example, Walter Raleigh Zeneer, a Missouri market shooter, was classified in the 1900 census, at the height of his market-hunting career, as a farmer.[33] Market hunting watermen of the Atlantic coast who sold waterfowl

and other birds seasonally resorted to related endeavors such as fishing, crab-
bing, oystering, clamming, and trapping as well as boat building, net mend-
ing, and decoy carving. Daniel T. "Squealer Dan" Whealton, who hunted ducks
in the Chincoteague area at the turn of the century, was listed as a sailor and
captain in the 1900, 1910, and 1920 censuses.[34] Furthermore, since market hunting was often a young man's occupation, any person whose career spanned
less than ten or twenty years might appear as a hunter in only one census or
in none. Take the records of John E. Ades of Marquette, Wisconsin, as an example. Ades was reported to have bagged more 45,000 birds during his more
than fifteen–year career as a commercial duck hunter. Identified as a "game
hunter" in the 1880 census, Ades was listed as a "produce dealer" in 1900.[35]
In all, while census records provide only suggestions about the numbers of professional hunters, they clearly reveal the fluidity and variability of modern
market hunting and those involved in it.

Modern market hunters shared some common characteristics, however.
The successful market hunter intimately understood how his game-taking
weapons—guns, nets, traps, or other devices—functioned, as well as when
and where he might use them most effectively. For commercial shooters,
marksmanship skills were obvious everyday necessities. Many market gunners
became superb marksmen, while a few attained an admired, even revered status because of their shooting skills.[36] Commercial shooters also needed to be
knowledgeable about the properties of firearms, powder, and ammunition.
Professional gunners like the Kleinman brothers often loaded their own shells
at the end of each tiring day on the Lake Calumet marshes—a chore they hated.
Spending an evening "reloading these confounded shells" was necessary,
though, because the cost of disposable, preloaded paper shells cut too far into
their potential profits.[37]

Other outdoor knowledge, skills, strengths, and dispositions were equally
important. Market hunting often required long, uncomfortable days afield,
far from shelter, fighting insects, mud, briars, heat, cold, or rain. Some duck
hunters spent their nights in boats in order to be on the water in the best
hunting spots by daybreak.[38] They had to withstand the vagaries of weather,
from broiling, sultry heat to rainy, windy, and frigid cold. Coastal gunners
needed to understand the risks associated with raging waves and tides. Those
hunting aquatic environments had to be proficient with watercraft, whether
rowed, paddled, sculled, poled, sailed, or engine powered. Professional wildfowlers had to be familiar with the outdoors in all its elements and know-

ledgeable about how to carry on their pursuits efficiently—or when it was best to just quit. Market hunters needed to be in reasonably good physical condition as well. Alexander Hunter, a late-nineteenth-century sportsman-writer described the "typical" market hunter of the North Carolina coast as "slab-sided and always thin; I never met a fat one yet."[39] More upscale market hunters operating at Tulare Lake in the San Joaquin Valley of California in 1892 were characterized by another writer as "the best natured men in the universe." The market hunter's "exciting outdoor life" gave him "health and a good appetite." Even more, his lifestyle made him "better morally" than those imprisoned in "the filth and dirt of the big city."[40]

Besides possessing knowledge about terrain, weather, and weapons, the market hunter needed a deep understanding of wildfowl behavior. Professional hunters understood how environmental changes affected wildlife presence, abundance, or usage of habitats—how an abundant year for beech nuts increased passenger pigeon presence, or how wild rice growth influenced sora rail availability. Market hunters needed to be familiar with the seasonal migration and daily movement patterns of each species to determine when prey might be available. The relationship between physical topography or terrain and bird movement, such as the concentration of some species on peninsulas or points or the major migration route along the Mississippi River flyway, was another piece of the market hunter's knowledge base.[41] The commercial wildfowler was a naturalist, whether he considered himself one or not, as the outdoor magazine *Forest and Stream* argued in 1889. The successes of the men "who shoot for the market," wrote "Hoodoo," depended on "their intimate knowledge of the habits of the game they seek" as much as "their dexterity with the gun."[42] Beyond that, the professional required a sense of detachment from wild creatures and their sufferings. It was not work for the weak, squeamish, or softhearted.

Of course, these were attributes of all hunters, but modern commercial hunters needed something more: knowledge of and liking for the business.[43] At the very least, the commercial hunter had to ensure that the market value of his quarry did not exceed his costs. These could be substantial, considering the prices of guns, powder, shot, and accessories such as boats or other means of transportation. Hunting dogs, bait, ice, game storage, and rail transport were additional expenses. Most active professional hunters, like Edward Alders of the San Joaquin Valley, kept a close watch on their accounts. Alders meticulously recorded his turn-of-the-century shipments of waterfowl to San Francisco markets, along with those of his hunting partners, noting numbers of each species sent daily and money earned from each shipment. At the end of the season, he totaled his weekly receipts and subtracted his costs in ammunition, powder, wads, expressage, and other cash expenses to arrive at his net gain, which in 1901, amounted to almost four hundred dollars.[44]

In addition to monitoring their own economic positions, profit-seeking market hunters had to have some understanding of supply and demand—a task made more difficult by the distance and time involved in sending game to far-off urban markets. This was a lesson Merritt learned early on. Upon shipping some of his first prairie chickens to New York in 1858, he found that "in a week or two we forwarded more birds than the market could use." When cold weather returned, more orders came in than he could fill. In this way, Merritt's new business seesawed from one day to the next—"now a surplus and then a dearth"—until the end of the season.[45] Market demand, consumer desires, and prospective prices influenced professionals' decisions about what species to hunt or whether to hunt at all. Economic calculations determined the methods hunters used to store, process, and transport game. These were especially important considerations prior to the widespread availability of refrigeration and dependable swift transportation.

For independent and opportunistic market hunters, profit was the primary goal and their main incentive for wildfowling. Goose hunters could make as much as one hundred dollars a day or sometimes even more, wrote author John Mortimer Murphy in 1882, since "the birds [were] valued at from fifty cents to a dollar each, and from ten to fifty [were] bagged at every discharge" of the huge "scatter cannon" used to hunt them.[46] Snipe hunters on the prairies of south-central Illinois, wrote Franklin Satterthwaite in an 1887 issue of the magazine *Outing*, did very well for themselves. "There have been millions [made] in the snipe trade for the profits were enormous," he explained, causing many to go "snipe-mad." The birds were "exchangeable everywhere in trade" and were even "the groundwork for many promissory notes. People owed other people snipe."[47] Passenger pigeons brought some hunters enormous profits, at least according to conservationists eager to end the practice. Pigeon trappers could bring in a daily income of ten to forty dollars in the late 1870s.[48] One professional pigeoneer was reported to have made $60,000 in the business.[49] *Forest and Stream* editor George Bird Grinnell told of a market hunter who killed over two hundred dollars' worth of ducks in one day in 1899.[50] Another market shooter in Texas shipped enough ducks to earn ten to twelve thousand dollars a year.[51] Waterfowl hunting profits that Edward Alders earned in San Francisco markets after six years of gunning bought him some ranchland and a vineyard and transformed him into one of the "most prominent and influential" men in the Farmington District of San Joaquin County.[52] Altogether, even deducting for expenses, the reported profits from market hunting seemed lucrative for a seasonal occupation.

Salaried market hunters could do well for themselves, too. The Winter Brothers paid Fred "Dude" Gilbert and other hired hunters five dollars a day plus room and board.[53] Richard Harker, who worked for the Winter Brothers in the late nineteenth century, was paid seventy-five dollars a month while being furnished with everything except his guns.[54] These professional shooters probably would have agreed with Merritt's assertion that "hunting was more profitable than farm work, and was as good as a trade," especially in an era when a workman's or farmhand's wage was just one or two dollars a day.[55] In fact, Iowa market hunter Tom Ditto, convinced that he could make more money shooting ducks than farming, paid a hired hand sixteen dollars a month each spring to do the chores while Ditto went hunting.[56]

While many testified to the profitability of market hunting, other commercial hunters countered with somewhat less extravagant estimates of the money to be made. E. T. Martin, a Chicago game dealer, contradicted the conservationists' claims about the profits made by pigeoneers. Martin argued that only "under very favorable circumstances a good netter . . . would have made from $100 to $200, but by far the larger portion would not reach $100 over expenses." The average daily catch for netters was no more than five to twenty dozen birds.[57] A Chesapeake Bay shooter, looking back at the so-called golden age of market hunting, insisted that "the entire picture has been exaggerated . . . you only hear of the good days." The reality was that "anyone clearing a thousand dollars a season was lucky."[58] Weather, wildlife movements, and hunter fatigue limited the number of hunting days and the size of hunting profits. Unstable markets also decreased the chances for financial success. Bird hunter J. Kemp Bartlett Jr. agreed: "I don't believe the [Chesapeake] market hunters thrived financially. I knew a number of them. They lived in small houses and their valuable assets were their guns, boats, and decoys."[59] Illinois market hunter William Brown put it even more starkly: "There were damn few market shooters that made any money at it."[60]

For many commercial hunters, however, the incentives for taking up the business were more than economic. Surely the prospect of making money and a decent living probably was a paramount consideration, but few market hunters became wealthy from hunting alone. Other inducements encouraged wildfowl hunters to turn professional. Simply being afield in some very special, spectacular locales held its own attraction. Hunting trips offered the excuse to travel to new places and to spend time in the out-of-doors. The freedom that market hunting provided in a modernizing world was another incentive. Commercial gunners, particularly those independent market hunters, enjoyed an autonomy that was unknown to the closely supervised wageworker of the time. Additional satisfaction in being a productive market hunter could be found in the challenge of outwitting their often-wary quarry. Masculine competition with other local market hunters, typically measured by birds killed and sold, motivated some. In later years, outwitting law enforcement officers added a new form of competition and new appeal to the market hunting experience.

Market hunting could also become a family tradition, offering the incentives of a home-grown business and shared familial purpose. The Kleinman family of Illinois was not unusual in this regard. Henry, Abe, and George Kleinman all hunted waterfowl in the Lake Calumet marshes south of Chicago.[61] In Louisiana, Anthony Tragle and his two teenaged sons worked as hunters for the market.[62] Merritt's first introduction to the world of commercial hunting came via the "Gay boys . . . who made their living with a gun." Later, he became acquainted with the Joles family on the Mississippi River. Nelson Joles and his boys "were all desperately fond of hunting as soon as they could carry a gun," so much so that "they soon forgot their little farm" and turned toward "prosecut[ing] this new branch of industry with ardor and success." Mrs. Joles was a key part of this family enterprise, assisting "very much in the game business" and earning the title "The Gamey Woman" from Merritt.[63]

Offsetting market hunting's benefits were the dangers inherent in the business. The possibilities for death or injury accompanied professional wildfowlers afield daily. The greatest risk came from firearms, and not a few commercial hunters bore lifetime scars from gun mishaps. In coastal villages and towns, so-called Big Gunners could often be identified by disfigured faces caused by the cannon-like guns they used. A market hunter on Smith Island, Maryland, had his big gun explode in his face on one bitterly cold night. Blinded and injured, he miraculously paddled his disabled boat six miles to shore. Others less fortunate went down with their burst guns or shattered ves-

"The Dewey Brothers, market hunters and crack shots, Fergus Falls." Market hunting was often a family business, as this circa 1910 photograph from Minnesota shows. Courtesy of the Minnesota Historical Society, St. Paul

sels.[64] Percy Cushing's market-hunting father laconically recorded some of these accidents in his 1860s diary: "Delancey Hallet fell out of his boat and got drowned trying to pick up a cripple off Fire Island. . . . Delancey helped hunt for Timothy's brother Job when he got froze to death in his ice dingy in Whig Inlet. I got a bad cold."[65] Cramped quarters in sinkboxes and boats, along with gunners' usual practice of placing the gun barrel muzzle on the stern alongside their feet, meant that an accidental discharge could result in the loss of a foot or leg, coupled with possible drowning in the shattered craft. Concocted, homemade weaponry of questionable safety doomed some unlucky men. The gun fashioned by two brothers at Horicon Lake in Wisconsin in the early 1900s, for example, used four- to six-inch galvanized pipes "mounted on a boat and filled with black powder, nails, bolts, and any other scrap metal that would slaughter the geese." They used the weapon until "it finally blew up and killed one brother and blew an arm off the other."[66]

Although the hazards of bird specimen and egg collection might seem few, significant dangers could accompany this kind of work, as well. Some specimen hunters paid dearly for their efforts, like John Feilner, who was killed by Sioux in 1864 while collecting birds for the Smithsonian Museum in the Dakota Territory.[67] Eggers faced hazards from shipwrecks and being marooned on rocky island coasts, to falls from the perilous cliffs where seabirds laid their eggs.[68] The egging experiences of one diarist included a near shipwreck during weather "bad enough to discourage us landsman" and a too-close encounter with a dislodged rock that could have hurled his egging partner "into the sea that boiled below."[69] Egger James Foster was not so lucky. While gathering gull's eggs in 1858, he fell "over the edge of a precipice" and was "dashed to pieces on the rocks below."[70] Less serious falls could end with a shirt full of broken eggs. Conflicts over the rights to prolific egging islands sometimes resulted in violence between egg pickers, hijackings of transport boats, and forcible evictions by officials. Greek and Italian eggers on the government-owned Farallon Islands were "themselves half pirates, trespassing on Uncle Sam's islands," according to newspaper accounts.[71] A well-publicized "egg war" on the Farallons in the 1850s and 1860s amounted to a series of "many bitter and fatal encounter[s] between large parties of rival claimants to exclusive rights in the *muir*-egg 'placers,'" culminating in the 1863 death of one man and the wounding of several others in a pitched gun battle.[72] In all, the work "to rob gull's nests, and smuggle eggs into market" was "anything but fun."[73]

Few market hunters, though, made much ado about the hazards their work involved, choosing instead to see the risks they faced in the field as tests of their masculinity. Because many market hunters operated alone, distant from the nearest possible help should it be needed, self-reliance was crucial. A long-lived market hunter had to practice his profession safely if he were to survive its many perils, such as gun injuries, drowning, falls, and hypothermia. Perhaps as important, he had to possess a high degree of mental toughness that inured him to such conditions. As the professional gunner Adam Bogardus explained, those dangers were part and parcel of wildfowl shooting. "Overcoming obstacles makes the man."[74]

Obstacles did not dissuade Merritt. To him, the idea that he could turn wildfowling into a business seemed an incredible prospect—one that he almost forfeited while on vacation in New York. Hunting alone near his family home in Carmel, he tripped and fell, causing his gun to accidentally fire. The shot tore through his right arm "in a frightful manner" and lit his clothes on fire. Merritt

"Eggers trolley to Sea Island Islet." Egging could be a hazardous occupation for those who collected eggs from cliff-dwelling seabird nests. Falls were not uncommon. This photo from 1896 shows a man being conveyed by a rope-drawn trolley to the cliffs on South Farallon Island. Image from Arthur L. Bolton Family Papers © California Academy of Sciences, San Francisco

managed to throw himself into a nearby stream to extinguish the flames and then tried to drag himself toward help, but he fell from the loss of blood. His cries alerted others who came to his rescue. Although the accident resulted in permanent damage to his right arm—his shooting arm—he refused to abandon his lifelong passion.[75]

When Merritt returned to Illinois in the fall of 1857, he was determined to find some way, despite his injury, to turn wildfowl into cash. The locals of Henry County made little use of the droves of prairie chickens, plovers, snipes, and woodcocks that seemed to multiply every day. Muzzle shotguns were few. Farmers in the region used their rifles for bird shooting and perforated their prey with rifle balls almost to the point of inedibility. Some birds were sold, but no one thought of sending them any farther than the local market or maybe

Chicago, and even then, only during the colder months. The coming of the Rock Island Railroad to Kewanee in the late 1850s had changed little along those lines. The trains were infrequent and irregular. Express shipping was cost-prohibitive and unreliable. Refrigeration was unheard of, and ice was expensive and available only in the cities and larger towns. Faced with these obstacles, "nobody seemed to think there was any market for any game anywhere." But Merritt did.[76]

The Hunt

Merritt shipped his first box of grouse, packed in ice, to New York in 1858. Thereafter his shooting became more purposeful as he tried to make a business out of hunting. He went out alone or, more often, with a few partners, scouring the region for birds. In spring, Merritt hunted the nearby Annawan marshes with his partners, Isaac Bice and W. K. Porter, shooting ducks, geese, and some snipe in the late afternoons and then camping out overnight on one of the small river islands. Up early for the next morning's flight, the team would return to Kewanee by late afternoon. The birds were plentiful, but he struggled to kill very many—some forty golden plover and snipe out of the hundreds the team might have gotten—because his crippled right hand was an impediment. He was still learning how to shoot left-handed. Summer found Merritt hunting woodcocks along the Green and Rock rivers with George Cutmore until July. By October, he was in Geneseo, going after quail with a group of five or six other gunners. The outbreak of the Civil War hardly registered in Merritt's consciousness, fixated as he was on shooting. After 1861, he took to the Mississippi River, hunting there throughout the spring and summer months. Other times, he headed out to the Iowa prairies by horse and buggy to harvest prairie chickens. On a few occasions, he went after pigeons in farmers' late-summer wheat fields, shipping hundreds to Chicago taken in one afternoon's shooting. A trip to Minnesota to see his brother was another chance to hunt, and Merritt brought two partners along, taking blue-winged teal "with good success" when he could turn the others away from snipe shooting.[77]

Merritt's market hunting had no single method. He went out alone and with others. He hunted on foot, in a horse and buggy, and by boat. Upland, wetland, forestland, and field could all be exploited for game in their own season. He shot nearly every avian species available: plovers, snipes, woodcocks, quail, passenger pigeons, grouse, ducks, and geese. Among market hunters, Merritt was hardly unique. The goal of every professional hunter was to kill as much marketable game as possible, in the shortest time, and by the most efficient and economical means possible. While some market hunters specialized in a particular species or method, for many of those seeking to make a living at the game business, "the hunt" necessarily involved an assortment of methods for capturing, killing, and transporting a variety of birds from the field to the market.[78] Recreational hunters employed many if not all the same techniques, and their practices usually appeared little different from those of the professionals. While there were some "enlightened" sportsmen who decried certain kinds of wildfowl taking—firing at sitting birds or spring shooting, for example—just as many others saw no harm in emulating the market hunter's ways. The main differences between sport and commercial hunting lay not in the methods used but rather in the frequency and intensity of the hunt.

The market hunter's first task was simply locating his prey. Access to wild birds, particularly migratory species, varied by season, thus making commercial hunting a largely seasonal occupation. The work of market hunters in northern states usually commenced in late summer to early fall with the arrival of the initial migrants headed south for winter. Hunting activity resumed in intensity with the return of wintering birds and tapered off with their departure to breeding areas farther north. The slackest time was from about April through August, though where resident woodcocks and certain shorebirds were present, commercial shooting as well as sport hunting of these species started as early as July. In the Northeast, sora and reedbird hunting ushered in the fall gunning season starting in September. The most intense period for market shooting occurred in the colder parts of the year, when migratory birds were most plentiful, in the best condition, and the most marketable.

Plume hunters worked southern Atlantic and Gulf coast islands and mangrove swamps during the brief winter breeding season, when the birds' nuptial plumage was most colorful. Feather hunting along the northern Atlantic coast where market hunters shot snowy egrets, seagulls, and even starlings to sell to local hat manufacturers, however, was somewhat less seasonally dependent.[79] By the late nineteenth century, as plume hunting on the East Coast waned, activity increased in the western marshes along the Pacific coast and on Pacific islands where terns, grebes, white pelicans, and albatrosses nested. From the early 1890s until about 1906, market hunters took tens of thousands of western grebes during summers from breeding colonies on Lower Klamath Lake, Oregon, and other western marshes.[80] Passenger pigeon hunting carried an added element of temporal and geographic uncertainty since the birds' migration routes changed from year to year. They appeared erratically, at times nesting where none had been seen before. Consequently, the birds were often targeted by opportunistic farmer-hunters as well as professional pigeoneers. The great Petoskey, Michigan, nesting of 1878 brought out "every homesteader in the country" to harvest the unexpected avian bounty.[81] At a major pigeon nesting in Wisconsin, one game dealer noted that of the six hundred pigeoneers in his register, nearly all were famers, their wives, and daughters.[82]

In addition to season, time of day mattered for locating birds. The most productive times for a market hunter to be afield varied, depending on the hunting method, weather, and species hunted. Since prime wildfowling seasons were often short, market hunters had to maximize the days available. Commercial duck shooting, according to Abe Kleinman, meant working seven days a week from before dawn until it was too dark to see. "Best shooting is in the morning and evening," Kleinman explained, but if the wind blew, it could "be good all day." Warm, still days meant no shooting.[83] Hunting opportunities were best when birds were habitually active, such as on morning and evening feeding flights or, in the case of doves and pigeons, flights to drinking water or mass flights from nesting colonies. For colonial nesters such as passenger pigeons and many plume bird species, night hunting by torchlight could be particularly productive.[84] One such New England hunt in 1865, led by market hunter Forrest Bentley, bagged one thousand passenger pigeons in a single night. "Sometimes but few or none would fall, at other times twenty or more would come rattling down through the limbs and bury in the snow," an observer reported.[85] Waterfowl, too, could be hunted at night by illuminating them while they were settled, or "rafted," on the water's surface. Such nighttime gunning of resting birds usually resulted in massive kill numbers.

Once the market hunter determined the right time to go afield, he had to find a way to get there. Transportation to wildfowling environments was a necessary consideration and a key component of the hunt. Walking was the predominant mode of transport for many professional hunters, especially beginners. Jack Miner and his brother, who started market hunting in the early 1880s, walked "miles away in morning" and then hunted back toward home so that they would not have to carry their heavy loads of birds too far.[86] Horse and buggy, or wagon, was another way to get to the birds, especially upland species such as snipes and woodcocks. Boats were a requirement for some hunts—so much so that Merritt wrote floridly about their value to the market hunters he knew. Professional shooters working around Henry County would not have been able to "clear the ducks out of the whole marsh of Annawan," or to "have exterminated a hundred ducks at one discharge," or to have "swept the woodcock infesting the shores of the Mississippi and soaring above the grave of Dubuque" without utilizing everything from canoes to steamboats. "Without a boat, alas, how could Charon transport us poor mortals across the Styx into the land of the immortals?" Merritt wondered.[87] Modern transportation technologies revolutionized the hunt. Railroads opened new fields to commercial hunters, offering them faster and quicker transport to wildfowl habitats farther away. When the Des Moines River flooded in the late nineteenth century, attracting an unusually vast concentration of ducks, the Rock Island Railroad brought hundreds of gunners to the area within a matter of days, leading to what became known as "The Big Slaughter."[88] In the twentieth century, automobiles brought even greater mobility to both commercial and recreational hunters, raising concerns among some that there would be no place left where wildfowl could remain unmolested.[89]

Finding places to hunt was usually not a problem, since bays, estuaries, navigable rivers, and other waterways were generally open to public use and market hunters freely trespassed on private uplands. Ownership boundaries were often unmarked, and, with a relative plethora of game, some landowners had little interest in protecting it for their own use.[90] In fact, many farmers welcomed hunters into their fields to shoot avian stealers of grain.[91] For his part, Merritt was unconcerned about property rights. Most times, no one bothered him. An exception happened on a hunting trip to Cassville, Wisconsin, when he ran afoul of ex-governor Nelson Dewey, who complained about the market hunters "killing those birds of his and shipping them to the nabobs of New York to feast upon."[92] Other market hunters had more contentious encounters with property owners. Richard Harker was chased by Minnesota farmers wielding "pitch forks and monkey wrenches and anything they could get a hold of."[93] Southwestern Ohio farmers suffered from "a severe attack of the 'posting fever,'" reported *Forest and Stream* in 1878, "swearing vengeance on anyone hunting on their premises"—market hunter and sportsman alike.[94] This "'ungentlemanly' habit" of posting property received very little respect from professional shooters like Iowa's W. W. Sawyer. To him, a no-trespassing notice was simply an advertisement for good hunting on an underexploited parcel.[95]

The next task for the market hunter was to get within range of his prey. This could be done by stalking the quarry or by attracting it to the hunter by decoys, bait, or calling. Some birds were relatively sedentary while feeding or resting, so the hunter had to seek them out on foot, by horse and buggy, with stalking livestock, or other strategies. On the water, various kinds of boats—oared, sculled, or sailed—brought gunners within shooting range. Buggy hunting was a method employed on the plains to bring shooters closer to wary birds since wildfowl appeared oblivious to larger vehicles approaching on a tangential course. Buggy hunters on the western prairies were sometimes able to get so close to unsuspecting birds that their heads could be snapped off by hand.[96] Trained domestic livestock could also be used as a kind of moveable animal blind to allow gunners to approach feeding birds. Stalking horses or cows were guided by the hunter and learned not stampede when the firearm was discharged over its back or below its neck. In mid-century California, "bull-hunting" was a common technique employed by hunters who had the wherewithal to pay the three hundred dollars for the specially trained animals.[97] L. A. Sischo, a well-known California hunter who supplied San Francisco markets, used this method to great effect in the Los Baños region of the San Joaquin Valley. Sischo and his two assistants reportedly killed four hundred ducks in only six shots while using trained steers for stalking. When the practice was outlawed early in the twentieth century, Sischo continued to use it, substituting horses hitched to a cart that he could unhitch for hunting and reharness quickly if a game warden approached.[98]

Perhaps the most valued animal tool used by both market and sport hunters alike were dogs that could find, flush, and retrieve birds.[99] In many cases, dogs were much more than aids—they were partners and companions. Forrest Bentley's team of professional gunners, for example, had several canine members, each named for a Civil War general. General Grant in particular, was a "noble dog," Bentley observed, whose field detail was "to take care of the killed and wounded" birds.[100] For Merritt, dogs functioned as "the better half in all warm weather shooting." He owned four dogs—all pointers—over the course of his hunting career, and he was uncharacteristically sentimental about them. Merritt recalled with sorrow the final resting place of each dog: the first "ended his life on the banks of the Green River in Massachusetts"; the second died at the mouth of the Apple River near Savannah, Illinois. Merritt's third pointer, named Sancho, was his most prized companion. He was the "best dog in Kewanee" and "knew everything there that could be known about a woodcock."

Tragically, Merritt accidentally shot and killed his beloved Sancho, hitting him broadside while taking aim at a bird. His "painful . . . and somewhat dramatic" death near Cassville, Wisconsin, left Merritt bereft.[101]

Hunters did not always go after the birds; sometimes they waited for the birds to come to them. Disguising techniques were thus necessary parts of the professional's toolkit, like the method used by California market hunter Claude Kagee. He shot flying waterfowl from a small pit along the edge of a field, camouflaged with cornstalks, sticks, or leaves.[102] On the water, a sinkbox functioned in a similar way. A common sight along the Atlantic seaboard, the sinkbox was a low, decked-over craft with just enough space for a seated or reclining gunner. It was like "a tin coffin sunk beneath the water."[103] Surrounded by decoys, it was usually occupied by one or two shooters and a boatman who picked up the birds killed.[104] Market hunters in the Chesapeake, such as brothers Jesse and John Poplar, shot hundreds of ducks daily from their sunken stations.[105] Though quite productive, sinkbox shooting was cold, wet, uncomfortable, and tiring work and required considerable skill. It was also likely to be extremely dangerous. Rough water or gun mishaps could sink the craft or cause bodily harm. Robert H. Gilbert found that out when he accidently shot himself in the foot while stepping into his cramped sinkbox.[106]

All the efforts to get to the birds meant nothing if market hunters could not come away with enough birds, eggs, or feathers to make their outing profitable. Capturing prey quickly, easily, and in large numbers was key to making hunting a vocation rather than an avocation. Methods for acquiring birds or bird products varied among the different market hunting specialties. Some techniques employed by market hunters were time-tested tools and weapons relatively unchanged since the premodern era. Others were modern killing machines, innovations that made commercial hunting more efficient. While a few of the taking devices employed by market hunters were specific to the business, most were also used by amateurs. The difference lay in the professional's ability to deploy those devices with skill and persistence.

Collecting was the least technologically dependent of commercial wildfowling methods. Egg hunters worked by hand, robbing wild bird nests, particularly those of colonial nesting species, for eggs to be used for food or other products. Eggers worked intensively during the relatively brief egg-laying period, causing great disturbances among breeding populations. To ensure that the eggs they were gathering were not partially incubated, and therefore worthless, eggers made a practice of initially destroying all the eggs in a nest

"Duck Shooting from Sink Box, Havre de Grace, Maryland." This early-twentieth-century postcard shows Maryland market hunter Jesse Poplar shooting from a sinkbox surrounded by decoys. Poplar was dubbed "the greatest sink-boat duck shot in America." Image from Dr. Harry Walsh Collection (accession no. 0213-0005), courtesy of Chesapeake Bay Maritime Museum, St. Michaels, Maryland

and then returning immediately once replacement clutches had been laid.[107] In addition to eggs, some collectors robbed nests of shed feathers and down to sell. Eider duck down was especially valued for pillows, cushions, and bedding, and, as Audubon noted at midcentury, eggers in some locales "collect[ed] it in considerable quantity" to sell.[108]

By the late nineteenth century, "game eggs" were "always marketable" and the egg hunter had "always a profitable, if perilous business."[109] Along the Texas shore, egg hunting was "a yearly custom." Every sort of watercraft headed out to nearby islands when gulls, terns, and herons nested, making away with thousands of eggs to be "exposed for sale in towns along the coast."[110] In Chicago markets, mud hen eggs gathered by the thousands in the marshes of central Illinois, found ready buyers at six cents a dozen and netted egg hunters a cool profit in the process.[111] Egging off the San Francisco coast was a big business in May, as small boats headed out to the Farallon Islands, where seabirds such as puffins and various auk species nested. Crews, "mostly composed of Greeks and Italians, with a liberal sprinkling of water-front castaways," scoured the rocky cliffs, wearing specially designed shirts with open pouches sewn in the front for storing eggs. "A good hand," reported the *San Francisco Chronicle* in 1891, "will pick up and get away with 200 eggs a day."[112]

Specimen hunting was another field in which commercial egg hunters operated, supplying taxidermists, professional naturalists, and amateur collectors. A popular "egg-collecting mania" at the turn of the century gave rise to a class of commercial collectors who went to great lengths to find unusual eggs to sell. To supply the demand, countless professional and amateur collectors rumbled through the forests, prairie uplands, marshes, and seashores to gather interesting eggs as well as other avian collectibles: feathers, skins, and nests. Samuel C. Clarke "supplied Eastern collectors with specimens" of "the rare and beautiful bird known as Wilson's phalarope," which bred in Chicago's marshes.[113] Scaling the rocky cliffs off the California coast, "a few adventurous spirits" made "a business of trying to secure" eagle eggs for sale to private collectors.[114]

Museums and research institutions sponsored numerous collecting expeditions as scientific endeavors and became significant consumers of the avian specimens provided by amateur and professional collectors.[115] This kind of collecting, when practiced by amateurs or scientists, usually provoked less criticism from observers than when done by professional marketers. Unlike the market hunter who sought avian products for sale as food, the "humble collector" was "irreproachable," according to some commentators.[116] Yet the "humble collector" could wreak havoc on nesting colonies and bird populations, especially once avian specimens became a sought-after consumer product. For some, the sale of specimens for scientific study made for a lucrative sideline.[117] One New Jersey contributor to the magazine *Ornithologist and Oologist* reported in 1883 that he had observed a collector "obtain 400 to 500 [marsh wren] eggs in a day," while the correspondent himself had "added several hundred to the stock of eggs for exchange."[118] Ornithologist Elliott Coues's 1874 field guide encouraged bird collecting with almost no limits. "How many of the same kind of birds do you want?" he asked rhetorically. "All you can get." Fifty to one hundred was not unreasonable, in his view. "Birdskins are capital," Coues explained; they were "a medium of exchange among ornithologists," having both scientific and commercial value.[119]

Market hunters employed an even wider variety of methods than collectors to capture wild birds. Nets, placed in or suspended above water or on land, could be used effectively to entangle or enclose birds while flying, swimming, diving, or walking. Netting offered the advantage of taking many birds at a time and was a commonly used technique for capturing passenger pigeons. Decoy "stool pigeons" tied to the ground lured birds into baited areas where nets suspended from trees could be dropped onto the feeding birds. In other cases, pigeoneers strung nets between trees to catch the birds in mid-flight. Some were left alive and sold as shooting targets; others were killed by crushing the birds' necks or skulls. With a double net, a thousand birds could be taken in a single throw, and a market hunter could sometimes acquire hundreds of dozens of birds in a single day's trapping.[120]

Besides passenger pigeons, waterfowl and small birds such as bobolinks were sometimes caught by trapping, though it was never the predominant technique for commercial wildfowl taking. Trapping had many advantages over other methods of capture, particularly shooting, because it did not require the hunter's presence except to bait the trap and remove the birds. Duck traps were small, inexpensive to make, and could be used unobtrusively in marshes. Baited with kernel corn, they often caught surprising numbers of ducks. Trapping was also virtually noiseless, nearly undetectable, and could be utilized during the night when waterfowl fed. Finally, trapping avoided damage to the carcass caused by shooting.[121]

Although wild birds for the market were taken by many means, probably most were shot by firearms. These weapons varied among hunters and across time as vast improvements in guns and ammunition occurred over the modern market hunting era. Firearm innovations in this period included improved steels, the internal placement of detonating primers, the development of the breechloader, the creation of self-enclosed cartridges, improvements in the quality and hardness of shot, the invention of smokeless powder, the development of shell extractors, and the advent of the choke, which improved the pattern of shot leaving the gun barrel. The last major advance at the turn of the twentieth century was the invention of repeating firearms that could shoot multiple shots before reloading.[122] Efficiency was further increased by the arrival of preloaded cartridges that eliminated time wasted in reloading in the field and allowed professional gunners to reload their own customized cartridges at the end of the day. Such shells were considerably cheaper than commercially loaded shells, but the nightly task of reloading hundreds of them was an unwelcome chore for exhausted market hunters like Adam Bogardus. He hated his evenings spent "reloading these confounded shells."[123] The task was necessary, though, since the savings over disposable paper shells was crucial to the market hunter's bottom line. These kinds of innovations helped make hunting a viable commercial enterprise.

Not all wildfowlers immediately accepted the major advances in guns and ammunition when they became available. Improvements often required one to two decades to find general use while controversies raged among hunters over the relative merits and disadvantages of each new advance.[124] For the market gunner, however, the key determinant for change came down to economic efficiency. Early in his career, Merritt preferred using a double-barreled, muzzle-loading shotgun, but he transitioned by the early 1870s to a breechloader—a gun that could be discharged four times faster than his old muzzle-loading fowling piece.[125] One of the market hunting Kleinman clan in Illinois swore by the "new ten-pound, ten-gauge breech loading Lefever guns" since they did not "get out of order." Other market hunters preferred the British-made breech-loading Greeners, but Abe Kleinman found that gun far too costly a drain on his duck sales to be worth it.[126] Most market hunters used double-barreled breechloaders until repeating shotguns appeared in the 1890s and became the mainstay weapon of commercial wildfowl shooters. In 1897, Winchester introduced the highly popular slide-activated pump shotgun, which was capable of firing five shots. Even more efficient and murderous repeating shotguns were employed during the wildfowling industry's latter years. Browning patented the first mass-produced semiautomatic shotgun in 1900, and manufacturers Winchester and Remington soon followed with their own versions.[127] These weapons could be fired rapidly, resulting in high bird mortality. Market hunter Atley Lankford of Elliott Island, Virginia, for example, claimed to have killed more than thirty-five thousand ducks using a large-gauge semiautomatic, single-barreled shotgun.[128]

Big guns were another firearm innovation that made commercial waterfowl hunting even more deadly. Called "punt guns" in England where most were custom manufactured, these monstrous water cannons had to be secured in special small boats that were paddled into shooting range of rafted waterfowl.[129] Even a modest-size punt gun could weigh from forty to seventy pounds and measure seven to ten feet in length, with a one-inch diameter or larger barrel. A large one might weigh well over a hundred pounds and require two men to handle. Because the big gun had to be fixed firmly in the skiff by ropes and chocks, it could be aimed only by shifting the direction of the boat. The reclining gunner leaned over the gun stock and when the boat was properly aligned, he squeezed the trigger. Most punt gunning occurred at night, using torchlight to freeze the birds in place, but under favorable conditions a camouflaged skiff might be effective even during daylight hours. In either case, the Big Gunner fired directly into the resting, sleeping, or feeding birds. Most big guns were muzzle-loaders, so their boats had to be taken into shallow water or ashore for reloading. Swivel guns were versions of big guns that were firmly mounted into a sturdy gimbal that moved both horizontally and vertically. The hunter could then adjust his gun aim if his boat got off course. Many of these guns were family heirlooms handed down through many generations. Indeed, they became almost objects of affection and were sometimes given names such as "Big Tom" or "Dutch Mike." Others were identified by their first owners' family names: Cheseldine or Bundick, for example.[130]

The big gun was the ultimate market hunter's weapon, even though much more wildfowl overall was taken by conventional firearms. Often only a single shot per outing was possible—though under fortuitous circumstances more firings could be made; but given that forty or fifty ducks or geese could be killed in a single discharge, these cannons were highly productive and could bring professional shooters substantial profits. W. N. Smith recalled his uncle slaughtering 180 birds at one shot on the Chesapeake Bay.[131] Another pair of hunters harvested more than a thousand ducks in one night's outing. "The water was covered with dead ducks," one of the hunters reported. They brought $3.50 a pair in Baltimore, making it "the best night's work we had ever done."[132] Some Chesapeake Bay men used what was commonly called a "battery."[133] This device was composed of several (usually three to seven) conventional gun barrels or pipes that were welded together fanwise and permanently secured onto a heavy wooden block firmly fixed to the bow of a small, specialized skiff. All the barrels in the battery were fired simultaneously by lighting a trough of black powder. Firing in a ten-foot-wide spray, these guns inflicted great mortality and crippling. Batteries were usually employed at night, often with a lantern and mirror that confused resting ducks and allowed for the gunner's close approach.[134]

Whether using big guns, batteries, or conventional firearms, most professionals shot massed birds on the ground, rather than lone, flying targets. This was called "shooting sitting" in contrast to "shooting flying" or "wing shooting," which was considered the more sportsmanlike method. Densely concentrated birds made for extremely large kills per shot. Market hunters also sought out the

"Battery Gun and Skiff." Market hunters devised efficient and particularly dangerous weaponry like this seven-barreled, 12-gauge battery gun, also called an armada. Using spotlights to freeze resting waterfowl in place, commercial gunners could slaughter large numbers of birds with a single firing. Image from the Dr. Harry Walsh Collection (accession no. 0214-0002), courtesy of Chesapeake Bay Maritime Museum, St. Michaels, Maryland

best environments for numbers of wildfowl and shot in the same area until it became unproductive. They felt little to no compunction about shooting birds that had already begun their reproduction cycle. If birds were available, it was hunting season.[135] The main consideration was always to kill or capture the most birds with the least effort for the best prices. Consequently, market hunting accounts routinely listed the amount killed in a single shot, a single day, or in a single season. A sense of competitive boasting sometimes accompanied the reports, many claiming to record the largest bags ever made. "Big shots" were the lowest denominator of shooting success, measured by the number of wildfowl killed by a single firing of gun, such as Charley Chichester's muzzle-loader shot from a sinkbox on Great South Bay, off Long Island, which netted thirty-five redheads.[136] Big guns brought down even greater numbers, with hundreds of waterfowl taken in a single firing. A big gun operated by Peter Marshall, a Chesapeake Bay market hunter, killed twelve geese, sixty-seven pintails, and more than one hundred brant in one shot.[137]

Another way to gauge success was to total up the birds harvested in a day's shooting. Before bag limits were established by law, market hunters (and not a few sport hunters) gunned down as many birds as they could before their arms got too tired, their guns too hot, or their ammunition too low. Few could match market hunter William Dobson's tally of 509 ducks killed on the opening day of the 1879 season in Maryland. Dobson bested himself, however, shooting 529 ducks in 1884.[138] In Illinois, Henry Kleinman recalled that shooting on the Calumet River brought him a one-day bag of more than two hundred teal. It took him no more than two hours and only sixty shells to make that record.[139] Smaller songbird species that were commercially hunted, such as robins, bobolinks, and blackbirds, could also be harvested in impressive one-day bags. The Sam Armstrong family at the head of Delaware Bay reportedly would shoot between twenty-five and thirty dozen bobolinks in flight each day. Alone, Sam Armstrong could shoot as many as five hundred blackbirds on a good day.[140]

The seasonal and the lifetime bag were other frequently used units for recording hunters' game harvests. For commercial shooters, seasonal takes were the most important economic measure, since those numbers ultimately determined the profitability of their enterprises. Commercial wildfowlers paid close attention to their seasonal totals, often tallying them against their expenses at the end of the wildfowling term. The more than seven thousand geese W. W. Levy bagged during the 1847 and 1848 seasons in the Chesapeake garnered a nice gain in the Baltimore markets, while a single California hunter reaped the profits from the 6,380 geese, 5,956 ducks, 367 sandhill cranes, and 60 swans he harvested across the 1877–78 season.[141] An Iowa market hunter achieved a personal record bag of 2,000 prairie chickens, 100 woodcock, and two or three wagonloads of ducks, among other game, in 1873.[142] For Reese Knapp of Browning, Illinois, the 1,200 ducks he shot in the spring of 1879 with his 4-gauge, single bore, sixteen-pound gun represented the limit he could harvest, "only when prices were good and the weather cold enough to ship."[143] Market hunters kept records of lifetime bags less often than sportsmen, among whom it was a common habit, though some chroniclers estimated fabulous kill levels based on market shooters' occasionally hazy recollections. Reports claimed that Atley Lankford shot nearly a half a million ducks during his early-twentieth-century career.[144] These kinds of personal shooting records, almost tedious in their repetitions, often became the stuff of local legends—stories that marked commercial gunners as uniquely skilled professionals.[145]

In fact, quite a few commercial gunners achieved a surprising level of notoriety from their wildfowling exploits. Havre de Grace local Jesse Poplar was known as the "all-time great" and was revered as "the greatest wing shot" in the Susquehanna flats.[146] Stories of Jesse—"Maryland's crack duck shot"—and his three-generation, duck-hunting family reverberated far beyond Maryland in widely circulated newspaper and magazine accounts.[147] Other market hunters turned to exhibition shooting, displaying their marksmanship in public displays and competitive matches. In one 1872 "money match," Illinois market hunter Fred Kimble bagged a whopping 128 ducks to Reese Knapp's humiliating 37.[148] For Captain Adam Bogardus, professional trapshooting replaced market hunting as his primary occupation. His skill as a marksman gained him national and even international fame as "the champion wing shot of the world."[149] For his part, Bogardus, who also hunted the Illinois marshes, recognized locally famous Henry and Abe Kleinman as the "greatest duck shooters of the world"—high praise from the accomplished gunner.[150] In Iowa, Fred "Dude" Gilbert was known as the "Wizard of Spirit Lake" because of his exceptional marksmanship. Previously a salaried gunner for the Winter Brothers, Gilbert turned professional after winning the DuPont Cup for trapshooting in 1895. Thereafter, "Dude" represented the DuPont Powder Company as a champion wing shooter, competing in matches all over the country and abroad while advertising for his sponsor.[151]

For some, the fame they achieved through professional, competitive shooting led to new business endeavors. Bogardus invented the first practical glass ball trap in 1877 and patented glass target balls, both of which were widely advertised.[152] His 1874 book, *Field, Cover, and Trap Shooting*, quickly became the trapshooting bible, bringing him notoriety among sportsmen.[153] Popular audiences encountered Bogardus during his stint with Buffalo Bill's Wild West in the early 1880s until he was replaced by former market hunter Annie Oakley.[154] Kimble was credited by many with the invention of the first choke-bore shotgun, and he dabbled in the production of clay shooting targets as well, creating the "Peoria Blackbird." He also marketed his own mallard duck call.[155]

Jesse Poplar Jr., the son of famed market hunter Jesse Poplar, also gained widespread notoriety as a shooting prodigy. At the age of seven, the younger Poplar toured trapshooting competitions around the country, putting on exhibitions with his 28-gauge shotgun and promising to kill twenty-five pigeons at each event. From *Baltimore Sun*, 29 June 1902, 6. Image courtesy of Newspapers.com.

Gilbert and Bogardus both endorsed certain gun and ammunition manufacturers, working as spokesmen for their products.[156] Market hunting was thus an entrée for some particularly successful—or, at least, particularly accurate—shooters into a modern, consumer capitalist world of corporations, advertising, and entertainment.

The Man Who Kills for Profit

Competitive bird shooting was just one element in the multifaceted world of modern commercial wildfowling. Modern market hunters were a diverse lot themselves. There was no single type of market hunter, and professional wildfowlers followed no typical path into the developing commercial market in birds. Nevertheless, a singular, though conflicted, image of the "market hunter" emerged in public discourse during the latter half of the nineteenth century. No throwback to an earlier pioneering era, the market hunter was a modern invention, a product of industrializing economic forces and technological advancements. In the relatively brief period during which commercial egging, pluming, and bird hunting flourished, both positive and negative popular images of professional wildfowlers emerged, eventually coalescing into a single, recognizable form that came to dominate public discussions about legitimate and illegitimate wildlife use.

A pair of children's stories illustrates just how familiar the market hunter character became by the late nineteenth century and how different portrayals cast the professional wildfowler in contrasting roles. In 1884, the popular children's magazine *St. Nicholas* published "Marvin and His Boy Hunters," a serialized feature that was part adventure tale and part gun safety manual. The story centered on "Marvin the Market-Hunter," who took it upon himself to train two boys in the correct ways to pursue game. Along the way, he regaled his young listeners with tales of his own extraordinary hunting adventures. According to the story's author, Mr. Marvin knew "everything that [was] worth knowing about guns and dogs and the habits of wild game." His business was confined to wealthy epicures, high-class restaurants, and private clubs in Chicago, Cincinnati, and New York, and he was frequently employed by the Smithsonian Institution to acquire bird, egg, and nest specimens for its natural history collections. He hunted nearly year-round, shooting game in the North and West during summer and fall and moving south to Georgia and Florida in winter where Marvin killed herons and roseate spoonbills for their feathers and collected specimens. He "was a very well-known and highly esteemed man among the best class of sportsmen and naturalists" in the country, so it was little wonder that the boys dreamed of becoming market hunters, just like Mr. Marvin.[157]

A year later, the juvenile novel *The Young Wild-Fowlers* told a very different tale about market hunters. Written by the pseudonymous Harry Castlemon, *The Young Wild-Fowlers* followed the exploits of two boys on a hunting trip to the Chesapeake Bay. The book's major storyline involved a market shooter by the name of Mr. Barr, whose illicit use of his massive "duck-gun" kept him under almost constant surveillance by the local game warden. Market hunters were "a desperate lot," according to one boy, as proved by Mr. Barr's unyielding attempts to conceal his illegal activities by threatening, imprisoning, and kidnapping one youngster who had happened upon the well-hidden weapon.[158]

Though separated by only one year, the two depictions were poles apart. "Marvin the Market-Hunter" was a friend to youngsters, offering lessons in woodcraft and conscientious sportsmanship. Mr. Barr, the market hunter in *The Young Wild-Fowlers*, was the dangerous enemy of young boys, responsible for destroying their chances at valuable outdoor sport and capable of harming those who crossed him. Mr. Marvin was a respected, hardworking, and vigorous man. In contrast, Mr. Barr was a "desperate" criminal who carried out indiscriminate slaughter of wild creatures solely for the profit they would bring.[159] Mr. Marvin appreciated all wild creatures. He drew clear distinctions between the right and the wrong ways to hunt, telling the boys, for example, that there was "a vast difference between a market-hunter and a song-bird-trapper."[160] Simply put, "Marvin, the Market-Hunter" was a hero; Mr. Barr, the market hunter, was a villain.

Of the two, the villain image would come to dominate popular discourse across the late the nineteenth century. Market hunters rejected this negative characterization, identifying the distinction between professionals and "sports" as mainly one of skill. In the 1880s, recalled W. W. Sawyer, many people "spoke derisively of market hunters. They were called Pot Hunters, Market Hogs and the like by sportsmen, amateur hunters and other who envied them."[161] Richard Harker agreed, noting that some sportsmen raged against invasions of market hunters because "we could shoot and they couldn't hit anything."[162] In many cases, however, sport hunters readily acknowledged the professional's wildfowling expertise, and some took advantage of it by going into the field with market hunters as their guides. A massive 1902 waterfowl hunt in California, for instance, led by market gunners, netted a group of four Sacramento sportsman 783 birds in two days of shooting. This was not "pot hunting," one of the participants claimed. Rather, it was "wild goose shooting as practiced . . . by experts."[163]

For recreational hunters, the critical difference between proper and improper wildlife use was profit seeking rather than expertise. Sportsmen made it clear that they never sold any game. Saloonkeeper George Keenhold, for example, refused to sell quail to a Chicago businessman for a game dinner, shipping the birds gratis instead.[164] Billy Wiggins was another prolific hunter who gave away the hundreds of birds he shot to local hotels in Spirit Lake, Iowa, rather than selling them. Wiggins qualified as a sportsman, according to his friend, since he shot all his birds in a sportsmanlike way—"on the wing."[165] Hoyt Elbert recalled that he and his sport hunting friends tried "to educate" one local market hunter out of the business "because we didn't believe in it."[166] Increasingly, most self-professed sportsmen came to subscribe to the definitions proffered by Robert B. Roosevelt in 1866: "The man who shoots with a view of selling his game is a market-gunner; and he who kills that he may eat is a pot-hunter."[167] Although both were to be scorned, it was the market hunter's pursuit of profit that excluded him from the ranks of sportsmen.

Without the element of pecuniary gain, the actions of market hunters could appear, on the surface, little different from those of self-described sportsmen or amateur naturalist-collectors. All took wildfowl, often in large quantities. Many exchanged the spoils of their outdoor forays if not for money, then for other rewards, including the goodwill or gratitude of their neighbors, friends, and families. Amateurs used most, if not all, of the same methods for killing game as did market hunters. The crucial distinction was that the market hunter was "the man who hunts for profit."[168] This one trait subsumed the diverse and fluid ranks of wildfowl eggers, plumers, netters, trappers, and gunners into the singular category of the "market hunter." Over the course of the late nineteenth and early twentieth centuries, the market hunter's pursuit of profit became an ever-brighter line separating legitimate from illegitimate wildlife killing.

Profit was very much on Merritt's mind as he became a professional hunter.[169] The Civil War had depressed prices for game, with prairie chickens selling for as low twenty-five cents each, but Merritt continued hunting anyway, waiting for the war's end to revive prices. He was rewarded for his persistence and cleared over a thousand dollars in 1866. The next year, he hunted farther north, going up the Mississippi from the Galena River and beyond La Crosse, Wisconsin. Merritt and his partners found a wealth of birds, and even more mosquitoes, in the Potosi and Kickapoo river bottoms.[170] On the prowl for new areas to exploit, Merritt conceived of the idea of getting his own small steamboat that would allow his hunting team to "penetrate all the back places along the river where the birds were most plenty." An added advantage was that such a boat would allow Merritt to convey his birds to the nearest landing himself, rather than relying on others for transshipment. A Davenport, Iowa, boat builder constructed the craft in 1868—a shallow-draft, forty-foot-long, steam-powered, side paddlewheel vessel that Merritt named *The Firefly*.[171]

Taking aboard nine additional hunters, Merritt set out on an extended hunting trip, this time accompanied by his wife and two young sons, Herbert and Clarence. His hunters were "a motley crew, with no bonds to bind them besides the profit they expected to make and the novelty and fun of the enterprise."[172] They headed out from Le Claire, Iowa, and up the Mississippi, past Dubuque and into the sloughs near Guttenberg. The group shot birds up and down the river until October, learning how to handle their new craft and shipping over three thousand woodcocks as well as boxes of partridges and other birds from depots along the way.[173]

In 1871, Merritt took a new crew up the river in *The Firefly*. Although the shooting was consistently good and the prices for birds relatively high, each of the hunters fared quite differently. Some, like John Barton, worked hard and shot well "in the roughest of cover," bagging more birds than any other gunner. Barton's vices ashore—drinking and gambling—ate up his hunting profits within a month, however. Others, like William Morris, lacked the drive to turn a profit. Morris "exhibited no weariness," Merritt complained, "for he did not work hard enough to induce it." He was as skilled a marksman as Barton, but Morris failed to clear fifty dollars in the season when Barton accumulated five times as much. The hot-tempered Marion Ogden learned all he could about habits of game birds and the techniques for capturing them, never failing to make a large bag in the process. Quite different was his brother Billy Ogden, whose willingness "to lay in the sun while the shooting was fine" affronted Merritt's business sensibilities.[174]

Merritt did well for himself regardless. After paying off his team, he was left with five hundred dollars stuffed down his bootleg and "quite a few birds left to market." He shipped four hundred dozen prairie chickens and between twenty and twenty-five hundred woodcocks to market that summer, losing only seven birds in the process. Every bird brought a dollar or more. This was "splendid success" for the season.[175] The money in his bootleg, the steam launch in the dock, and the boxes of birds waiting to be sold proved just how much had changed in a short time for the market hunter.

CHAPTER TWO

The Dealer

If I had the last pair of game birds on earth, I'd ring their necks.

E. S. BOND, GAME DEALER

Henry Clay Merritt's successful 1871 season came after more than a decade of trials, missteps, and innovations. Marketing his growing stock of wildfowl required a great deal of experimentation. Years earlier, in the summer of 1858, Merritt sent his first grouse to New York in boxes so loaded with ice that the express shipping costs outstripped his profits. Although he cooled the birds prior to transit, the shipments still arrived somewhat worse for wear due to the heat. He shipped his first carload of game to the East Coast in 1859. After some failed attempts to get birds to the city in good order, Merritt tried various types of insulated shipping boxes and different recipes of ice mixed with sawdust to keep the game fresh for longer. Eventually he hit upon a successful process that involved thoroughly cooling the birds on ice before packing them tightly into wooden shoe boxes, doubled to protect against melting ice on the inside. Unopened, the birds stayed fresh for two or three days when packed this way. Birds packed in ice invariably arrived wet, but customers seemed not too concerned about the condition of the plumage. City merchants simply extracted the birds from the sawdust-filled containers and doused them in cold water, "by which means they were restored as good as ever" and, after drying and fluffing the feathers, they "could hardly be distinguished from those that were fresh killed."[1]

In 1866, Merritt and his hunting partner Nate Tompkins cleared a profit of one thousand dollars for birds sent to markets mainly in Chicago and New York. Alone, he could make as much as five dollars a day, but he no longer relied solely on his own hunting efforts. Instead, he hired a staff of salaried gunners to bring in more birds while buying game from other independent hunters, skimming off a nice profit in the process. In one instance, Merritt was able to buy snipe at $1.25 per dozen only to resell them in the Chicago markets at $3.00 or more.[2]

By early in the 1870s, Merritt and his business partner had found ways "for packing and marketing all kinds of game," shipping in every type of box, barrel, or can that could hold birds. It was becoming apparent, however, that "no man could carry on the game business with success where his market was a thousand miles or more distant" without some way of keeping wildfowl fresh for longer periods of time. By trial and error, he tested, rejected, or retained various methods of long-term storage of game. In 1870, Merritt acquired a newly patented butcher's freezing box for $150.00. He bought two more in Chicago that same year and started construction of his own subterranean freezing rooms in Kewanee. The next year, he built a freezer in the nearby town of Atkinson. Packed with snipe in the spring and summer, Merritt was able to keep birds frozen and in somewhat saleable shape until September, when they sold for a dollar a pair in New York.[3]

Getting his birds to markets "a thousand miles or more distant" also required reliable transportation, and this, too, was becoming more accessible by the 1870s. Illinois's early railroads "did little to foster trade," in Merritt's estimation. Express trains were "infrequent, their time little better for passengers or perishable goods than present freight trains." Charges for express shipping of perishable goods were "high and often prohibitory," sometimes double the fees for regular freight. In the late 1850s, each hundredweight of game destined for New York cost six dollars to ship, a sum that diminished the market hunter's profit to almost nothing. Consequently, "trade languished in all but the necessities of life and goods in which slow time was unavoidable." As railroads spread across the country, however, their speed and reliability improved. When the Great Central Route of the Blue Line first seamlessly connected Chicago with the eastern seaboard in the 1870s, Merritt's business was plugged into a vast "net work of moving trains" capable of "distribut[ing] goods to all parts of the country."[4]

Despite the technological progress, Merritt's evolution from market hunter to marketman was fraught with failure. Early efforts to increase his wildfowl stocks led to some notably ill-timed purchases from hunters at prices higher than the market would ultimately bear. Advancing ammunition to shooters in exchange for birds proved a bad proposition when "some outsiders" refused to pay up. Freight transportation, though improving, remained slow, and even express shipping by rail could be unreliable, leaving game to spoil en route, especially during warmer weather. Many of Merritt's experiments with cold storage were also spectacular failures. In the 1860s, he repurposed a beer keg to store his game until flies came in through the spout and damaged some hundred dozen snipe, depriving him "of a large commission." His first freezing rooms often lacked proper ventilation and adequate supervision. After having to discard nearly half the birds in his Atkinson freezer in 1871, Merritt learned that his storage facilities needed to be periodically inspected for rotting game. Purchasing enough ice to keep freezers cold enough was also a challenge—one that resulted in the loss of several dozen snipe in one instance.[5]

More business failures accompanied Merritt's hunting successes in the field. Expanding his operations, Merritt built a new freezer in Cresco, Iowa, in 1872, a year he long remembered as "fertile with disaster." Once the storeroom was completed, he commenced to buy all the prairie chickens he could. Immediately swamped with a flood of birds, Merritt was unable to freeze them all, forcing him to ship the excess game during the heat of summer. "When I got reports of the first shipment they went through in bad order," he recalled, "not bringing one-half of what I paid for them, and I stopped shipping and attempted to hold them in the freezer" in Cresco. Unfortunately, the rooms had too little ice and were too poorly supervised by the employee charged with overseeing the facility. "Here was our fatal mistake," Merritt admitted. Buying birds instead of ice caused massive losses. His overstocked freezer could not keep the birds frozen, and his man in charge seemed unaware that the stock was rotting. Meanwhile, the New York market was being flooded with birds coming in from all along the rail lines from Iowa to the East Coast, and Merritt's stock gained a bad reputation among merchants. Shipping low-quality birds that were soft or spoiled "threw a cloud on all frozen stock, so that for a year or more such goods could not be disposed of for any reasonable price." After that disastrous season, Merritt dumped all the birds left in his Cresco plant out on the prairie, dismantled the facility, and had it sent back to

Kewanee, where it was reerected to serve as his main cold storage warehouse. By that time, Merritt was hunting only fitfully, convinced it was "better business only to buy and not to hunt regularly." The financially shrewd thing to do, he thought, was to purchase more domestic poultry, employ more help in packing, and find more market hunter-suppliers farther west.[6]

In all, though, Merritt's successes outweighed the setbacks. The "game business was getting brisk and the purchases were so large, so often and so continuous" that he sometimes struggled to keep up with the distant demand. He took in all kinds of saleable goods—"game, furs, poultry, and sometimes hides"—and sold most at a profit.[7] He was able to hold and ship game birds, cooled or dry chilled, as far away as New York easily in the right weather. And, by the 1870s, he was branching into frozen stock that could be held for longer periods and shipped during any season.[8] While there were still some kinks to be worked out, Merritt's game business was developing into a true capitalist enterprise. In the process, Merritt reinvented himself. The market hunter was becoming the game dealer.

Merritt's transition from hunter to dealer was one path into the modern game business. There were others that involved the participation of a variety of players. Market hunters worked hand in hand with game dealers, processors, packers, shippers, freezer owners, and retail merchants, so it was not surprising that some independent hunters would become marketmen themselves in due course. One such entrepreneur was Englishman Fred Dudley, who purchased ten thousand acres of cheap Louisiana marshland at the turn of the twentieth century. There, he set up what was reputed to be "the most efficient market hunting operation ever." Dudley's "camps" provided hunters and their families with food, lodging, and even tutors for their children. Each hunter was assigned a section of marsh to harvest. With his army of shooters, Dudley was able to send two thousand birds to the New Orleans game market daily.[9] Nat Wetzel, another major game merchant, started out as an independent hunter like Merritt and eventually became "the pioneer of organized game hunting," according to a *Saturday Evening Post* article. His wholesale businesses and retail markets in St. Louis and Beaumont, Texas, handled all sorts of game brought in by "a force of almost one hundred 'crack' hunters," dispatched by telegraph to meet flocks of migratory wildfowl wherever they appeared. In the fall, they provided ducks, geese, and turkeys, and in the spring, quail and prairie chickens, some of which were shipped as far away as

England. Wetzel's methods, explained the *Post*, represented "the most advanced practices in the industry of supplying game dinners for the tables of American epicures" and were akin to the organization of the modern steel industry or department store. Wetzel was a "commercial game 'magnate.'"[10]

Dudley, Wetzel, and Merritt were all variations on the turn-of-the-century game dealer, a distinctive and key component in the larger process of late-nineteenth-century wildfowl commercialization. Technological and economic innovations in this period transformed market hunting from an intermittent local or regional trade into a coordinated, large-scale business with all the earmarks of a modern, industrial-capitalist enterprise: a division of labor, a price system, market segmentation, and a new class of profit-seeking avian entrepreneurs. The market hunter became but one part of a broader structure working to transform wild birds into commodities. Participants in the game business included those who might never have picked up a gun or tramped the fields in search of wild prey. They were businessmen, financiers, and merchants engaged in the accumulation, commodification, and capitalization of dead bird bodies. In its heyday at the turn of the twentieth century, commercial hunting and marketing of wildlife was a huge, often lucrative business over much of North America—one that drew the countryside into closer connection and communication with cities. As it grew ever more complex, the game business came to be dominated by the so-called marketmen, leaving individual market hunters caught in a far-flung economic system increasingly beyond their control.

"The Modern Nimrod Who Hunts by Wire and Slays by Companies"

Although wild game had long been an article of commerce, the second half of the nineteenth century saw the sale of game, especially wildfowl, expand dramatically. Where geography and biology had previously circumscribed the market, industry and technology intervened to ease, if not eliminate, natural limits. Late-nineteenth-century innovations in refrigeration, transportation, and communication allowed for wildfowl to be efficiently processed, preserved, and transported in astronomical numbers.[11] Merritt believed that the game business "stood at the foundation" of these industrializing improvements. While his somewhat overblown claim might have confused cause with effect, he was nevertheless correct about the extent to which modern technologies "eliminated the great obstacles which had so long opposed progress" in wildfowl commercialization.[12]

Prior to the mid-nineteenth century, most wild meats and eggs were marketed locally since early preservation methods limited the distance game could travel to market. In earlier days, various techniques had been used to preserve wildfowl, including drying, smoking, salting, curing, pickling, canning, cooling, and freezing. During his antebellum childhood, Iowa market hunter Orin Sabin recalled, passenger "pigeons along with pork were stored in the oats bin and covered with oats."[13] Occasionally birds were "potted," meaning they were put, cooked or raw, into a crock or tin and then sealed with an airtight layer of oil or grease.[14] Eggs were commonly preserved by chilling, pickling, or immersion in oil.[15] Early preservation methods for meat usually involved natural cooling, using caves, springs, ice cellars, or hand-dug wells for longer-term storage. Masonry springhouses, packed with ice blocks cut in winter from lakes and ponds, eventually came into wider use in some areas.[16]

Preservation of wildfowl meat was initially of less concern, reflecting the widespread belief that game birds should be aged to improve taste. Well into the nineteenth century, many connoisseurs argued that game birds needed to be hung for several days before consumption. Others disagreed.[17] *Chambers's Encyclopedia* noted in 1889 that game was "generally eaten in this [putrefied] condition," and though it was "readily digested and admirable in flavor," such "high" meat was also "apt to disagree with many people." In worst cases, its ingestion could result in "fatal consequences."[18] Writer Henry William Herbert, who was better known in sporting circles by his pen name Frank Forester, explained that some heavy-bodied domestic and game birds were "better for keeping," but waterfowl were "ruined by it."[19] Elisha Lewis went further in his 1855 book, *The American Sportsman*, to declare that "no kind of Game-Bird improves by keeping over two or three days." Although snipe, woodcock, and other upland birds could be kept without ill effects for several days, Lewis was contemptuous of the notion that keeping game birds hung until the meat was "high" was in any way desirable. "We have no fancy for putting into our stomach half-decomposed substances of any kind; much less have we so distorted, so depraved a taste as to desire to partake of tainted, par-rotten game."[20]

While aging improved the taste of meat to some palates, most agreed that the quality of game bird flesh was best maintained by immediately cooling the carcass. Since the modern business of market hunting was chiefly based on the sale of fresh meat, natural deterioration presented special problems to hunters and game dealers and limited the extent of the trade. Market hunters and sportsmen air-cooled their birds in order to delay decomposition.[21] Early photographs of killed waterfowl show them suspended from nails or pegs driven into houses and barns as well as from houseboats and railroad cars used as hunting quarters. Frank Hale, who hunted the Suisun marshes near San Francisco Bay, California, described how "they hoisted the ducks into the rigging of their boat to cool at night. Once a week, they loaded well-cured birds on a steamer bound for San Francisco."[22] Efficient air cooling and preservation, however, were possible only during colder weather, and thus the early game market was limited by season and the biology of decomposition. Unlike livestock, which could be sent on the hoof to towns with no need for refrigeration, wildfowl destined for consumers' tables had to be shipped dead. Therefore, the market for birds was mostly a short-distance, cold-season business. Hot weather presented "a burden" to dealers in game.[23]

In the mid-nineteenth century, game dealers began to experiment with ways of preserving game for longer periods that would allow increased time

Wisconsin market duck hunters standing atop a houseboat covered in dead ducks, probably between 1900 and 1905. Courtesy of the Oshkosh Public Museum, Oshkosh, Wisconsin (P2299.1)

for shipment and sale. As Merritt discovered, the expenses and uncertainties associated with ice tended to eat up potential profits. Newly patented mechanisms for freezing first appeared in the 1870s, and game dealers immediately grasped the revolutionizing possibilities of these innovative designs.[24] Some even devised their own "preserving apparatus" for keeping their stocks frozen from one year to the next.[25] Game dealers quickly shifted to frozen game, using ice refrigeration at first and later mechanical refrigeration for both longer-term storage and longer-distance transportation. In the process, they liberated the wildfowl market from climate and season.[26]

By the early 1880s, every town of any size had an ice or cold storage house that held commodities, cooled or frozen, on behalf of dealers, shippers, retail merchants, and restaurateurs. In wildfowl-rich, rural areas, freezer storage facilities run by wholesale game dealers kept birds supplied by local hunters. The Winter Brothers, for example, operated houses in Lakefield, Minnesota, and Spirit Lake, Iowa, to store game destined for the Minneapolis, Chicago, and New York markets. Their earliest freezer was "about fourteen feet square" with walls eighteen inches thick and insulated with sawdust. Three or four hundred pounds of ice and plenty of salt filled twenty galvanized pipes placed in the freezer to keep the stock cool. A duck could be frozen "in there in one night as hard as a bullet."[27]

After "freezers came into fashion" in the 1880s, the wild-bird business became highly profitable, allowing wholesale dealers to stock up on birds during flush seasons and hold them until market prices were favorable.[28] To counteract the high cost of ice, many dealers and shippers developed their own nearby sources and sometimes sold it as a sideline.[29] Cold game storage buildings sprang up near railroad depots or along rail lines for quicker and easier transshipment to eastern cities.[30] The demand for frozen game birds rose as "one successful packing made the way for another." States west of the Mississippi River saw the wildfowl industry prosper in this period with "some places packing only snipe and plover, others packing prairie chickens and quails, and still others dow birds [Eskimo curlew] for the short season they tarried, and some packed only ducks." Game dealers that were located in the East where wild bird populations were declining shipped in what they could not supply locally. Hunters, packers, shippers, and dealers purportedly all "made good profits" in the process.[31] As new freezers went up in many places, it seemed that all dealers had to do "was to freeze up the birds and wait quietly for the profit."[32]

This "great expansion of the game market" in cities, opined *Harper's Weekly* in 1890, was due entirely to "the development of the refrigerated car and cold storage," which had truly "revolutionized the business."[33]

At the turn of the twentieth century, improvements in mechanical refrigeration using compressed gases such as ammonia or sulfur dioxide freed game dealers from reliance on natural ice and removed one more seasonal barrier to the market in birds. Many mechanical refrigeration plants were constructed in larger cities with active game markets and urban retail dealers readily availed themselves of these facilities, thus partially offsetting the advantages of rural wholesale game suppliers who had their own large freezers. Many substantial marketmen in Chicago and New York had not only retail stalls in city centers but also huge storage facilities where they could keep enormous quantities of stock and ship in bulk to other retailers. The great refrigerator of the A. & E. Robbins Company in New York City could hold up to fifty tons of game from which supply it could make large shipments to markets as far away as Europe.[34] Between 1869 and 1909, the number of commercial refrigeration facilities in the United States rose from four to more than two thousand.[35]

While mechanical refrigeration represented a technological advancement, it had its drawbacks, mainly in expense and upkeep. Only larger businesses engaged in preserving other types of meat or produce could make mechanical refrigeration profitable.[36] Older methods of ice-based cold storage, on the other hand, were simple, making it possible for one person to "maintain the efficiency of a large plant." Costs for operation were "trifling," especially for rural dealers who had access to their own ice resources.[37] Yet even these storage facilities required constant care to replace melting ice and to inspect stored game for signs of deterioration.

Mechanization in transportation had an even greater transformative effect on the modern game business. Before the Civil War and the expansion of railroad networks, it simply was not possible to quickly ship substantial quantities of game to distant markets. Modes of transportation employed by early market hunters were usually human or horse powered.[38] Even after rail transport became more widely available, horse-drawn wagons were still used to get shooters to their hunting grounds and to get bulky quantities of game, such as passenger pigeons, to depots or markets. Along the eastern and western seaboards and in major inland waters, market hunters relied on various types of watercraft, especially for waterfowl hunting. Everything from rowboats to skiffs and sailboats to steamers were used by commercial gunners for hunting as well as for transporting their spoils. Where sinkbox hunting prevailed, a larger tender boat fitted with a boom was required to transport, launch, and retrieve the heavy, awkward shooting box, its aprons, and large numbers of decoys.[39]

As game markets expanded in the late nineteenth century, hunters transported more and more birds to urban markets on larger ships. On Virginia's Eastern Shore, the market hunting Cobb family would load dressed ducks, packed in barrels, onto "one of the sailing sloops they built and run them out the inlet to the sea." They would then "hail the first northbound ship beating up the coast and dicker with the captain as to freight rates" to ship their birds to New York.[40] Paddlewheel steamboats plied the navigable water routes of the mid-Atlantic coast and many, if not all, supported the market hunting business in some way, carrying untold millions of ducks and other wildfowl to seaboard cities over the decades.[41] In the Delta region of central California, riverboat companies plied the Sacramento River throughout the nineteenth century, connecting Sacramento, Stockton, and other smaller towns of the Delta with the game markets of San Francisco.[42]

While water transport was invaluable to market hunters for providing access to waterfowl habitats, untapped hunting grounds, and remote ports, it was a relatively slow method for moving game to markets. Game dealers realized the full potential of wildlife commercialization only with the advent of faster rail transportation. Initially, though, railroads made only limited differences in the game market, modestly increasing the distances wildfowl and other game could be carried before they spoiled. The rapid growth of rail networks across the country and the linkage of developing western railroad hubs to eastern commercial capital centers, however, brought vast northern and western wildfowl hinterlands within the reach of urban markets from Chicago and St. Louis to New York and even abroad.[43] In Illinois, for example, the Rock Island line reached Henry County, where Merritt's market hunting business was located, only in the mid-1850s, thereby connecting the western county to Chicago's developing rail network. The first railroad in Iowa quickly followed in 1855. By 1860, eastern Iowa was well serviced by rail, and, after a temporary slowing of railroad construction during the Civil War, rail lines spread across western Iowa and into Nebraska and Kansas by the decade's end. The completion of the transcontinental railroad in 1869 was followed by a period of accelerated expansion, with some forty thousand miles of new track laid in the 1880s alone.[44] The effects of improved transportation on game markets, especially in the numbers of migratory birds available, were substantial. In Louis-ville, Kentucky, for example, "daily arrivals by railroads and steamers" of wildfowl from distant points "account[ed] for the plentifulness of the game market" in the mid-1870s.[45]

Though somewhat unreliable at first, railroads became much better for shipping perishable freight over time, especially after the midcentury introduction of refrigerated boxcars. Early on, these rolling iceboxes were considered one of the "freaks of railroad transportation," according to an 1842 article outlining the Western Railroad's plans for adding refrigerated cars to its trains running west from Worcester, Massachusetts. This new technology would allow shippers to keep "fresh beef, pork, veal, poultry, pigeons, venison, wild game and other fresh meat . . . in perfect order in the heat of summer" thus adding greatly to the market value of such products. Rail industry journals approved of the development, noting that the Western Railroad's directors clearly understood "the art of converting the whole country into a garden for our great cities."[46] One could have added that the railroad converted the entire country into a hunting ground for cities, as well. As early as 1844, the New York and Erie Railroad was carrying one thousand tons of game annually.[47]

For hunters, both commercial and recreational, the advent of rail transportation provided increasingly easier access to fresh wildfowl hunting grounds in more-distant locales.[48] The pseudonymous Frank Forester recognized the modern innovation's appeal to sportsmen in 1848 when he predicted that plans for an Oregon railroad might bring a day when New Yorkers could "shoot Cocks of the Plains ourselves, and bring them home the next day to dinner at Delmonico's."[49] The linkage between the growing network of railroads and the rising popularity of sport hunting following the Civil War was reflected in the vast number of hunting and fishing guidebooks that cataloged popular shooting locales and rail access to them.[50] William C. Harris's 1888 *The Sportsmen's Guide* listed more than two hundred transportation companies, mostly local or regional railroads. Harris noted that, in many cases, "the best shooting grounds" were within one or two miles of a station.[51] One late-nineteenth-century bibliography of sporting books listed numerous railroad-produced outdoor guides, including the Philadelphia, Wilmington and Baltimore Railroad's 1883 publication, *A Paradise for Hunters and Anglers*; the Southern Railway's *The Land of Bob White*; and the St. Louis–San Francisco Railroad's 1898 *Feathers and Fins on the "Frisco."*[52]

Some local trains running through prime waterfowl areas, informally called "duck trains," specialized in transporting hunters in season. One well-known route of the Northern and Western Railway Company carried hunters from Des Moines, Iowa, to Spirit Lake, a renowned ducking area. Hunters on the "Duck Special" could stop the train and be let off whenever they saw ducks along the way. The engineer would give a blast of the train's whistle, hoping to flush the birds for waiting shooters.[53] A locally famous duck train ran through the western end of the Suisun marshes east of San Francisco Bay, dropping off and recovering gunners at various stops named aptly for different ducks such as Teal, Cygnus, and Pintail. A hunter in the Bay Area could easily make a one-day round trip by rail, as opposed to a several-day trip by boat.[54]

The appeal of railroads to all sorts of hunters led professional wildfowlers to routinely blame them for the invasion of sportsmen who came west after the Civil War. Despite their protestations otherwise, such "sports" sometimes marketed their excess game in direct competition to bona fide market hunters. Yet market shooters benefited even more than sportsmen from late-nineteenth-century railway expansion. Trains brought market shooters to new hunting areas, but, more importantly, they became the primary method used to ship wildfowl to urban markets. Hordes of plume hunters and pigeoneers destined for great nesting colonies made extensive use of the rails. "Hardly a train arrives that does not bring hunters or trappers," reported a local newspaper during the 1871 passenger pigeon nesting near Kilbourn, Wisconsin. The birds killed were then "shipped to all places on the railroad" and as far away as "Philadelphia, New York, and Boston."[55] Some large Chicago merchants even chartered whole trains to deliver birds directly to market. Commercial gunners working in California's Central Valley sent game to San Francisco via the rail, flagging down trains in order to load sackfuls of ducks.[56]

By midcentury, the impact of rail transportation on the game market was already obvious. The reason that vast numbers of game birds could found in urban markets was not because wildfowl populations were increasing anywhere in the United States, insisted zoologist Daniel Elliot in 1864, "but because the facilities for transporting them have increased, and now birds killed in Minnesota can be brought to New York perfectly fresh and fit for the trade in winter months, and if packed in ice, in summer also."[57] Chicago's game market stood poised to benefit the most from the opening of the Pacific railroad, reported the *Memphis Daily Appeal* in 1868, with "parties in Cheyenne making arrangements with some of the Chicago game dealers for consignment of regular supplies from that locality" in the near future.[58] In the East, the Blue Line cars ran "with regularity IN FOUR OR FIVE DAYS to and from New York and Boston," according to an 1871 advertisement in the *Railroad Gazette*, without any excess delays required for changing cars between railroads, promising "Safe and Quick Transport" of "all Perishable Freight."[59]

Safe, quick transport was crucial to commercial hunters and game shippers since unforeseen delays and losses during shipment were ever-present and costly risks. In the 1860s, before railroads had reached some areas, game had to be transported by wagon or steamboat to the nearest railroad depot—a time-consuming journey that often resulted in spoilage, especially in warm weather.[60] Water-shipped game could be unsaleable before it ever reached the market, like the barrels of rotten red knots (a medium-sized wading bird) that George Mackay observed being thrown into Boston Harbor in 1893.[61] Other lots of birds were lost because the shipper unwisely chose to use cheaper freight shipment rather than express rail. Passenger pigeons were unusually susceptible to waste during storage or transport simply because of the large volumes involved and the fact that many were gathered, trapped, or shot during warmer months. Live pigeons, trapped for shooting contests, were more valuable but also more likely to die during transport. Detroit game dealer Henry T. Phillips lost three thousand birds in one day, at a cost of $250, simply because a railroad car for shipping was not available at the time promised.[62] In the 1880s, of a shipment of sixty thousand live pigeons from Missouri, forty thousand died for want of want of water before they reached market. The loss of live birds could be high even when railroad attendants watered and aired the birds at every railroad stop.[63]

Game shippers thus had an incentive to achieve greater efficiencies in packing and preserving their game for shipment in order to reduce both the risk of spoilage and the weight of shipped birds.[64] When market prices were depressed, the sale proceeds sometimes failed to cover shipping costs. Misjudgments in the quantity of game packed per container, the means of shipping (regular versus express), quantities of ice and insulation needed, and the type of container could well result in game arriving in a deteriorated condition, which would command lower prices, or worse, a total loss. Even worse, a dealer's reputation could suffer from too many spoiled shipments. Some suppliers had to defer shipments during warm summer months when the expense of ice packing and transport became greater than the anticipated price paid by urban retailers and consumers.[65]

Since dealers absorbed the risk of losses, rail shipment of game was a lucrative business for railroads and express companies, particularly when large volumes or weights were involved. Express charges on passenger pigeons sent to New York from the Great Lakes region in the 1870s could run from six to eight dollars per barrel. Live birds were shipped for three dollars per hundredweight.[66] As suppliers developed new methods for cold storage and packing in the latter half of the nineteenth century, railroads carried increasing quantities of game. Minnesota railroad records show that wildfowl made up the largest share by weight of wild game transported in 1876, with 25,810 pounds of birds shipped during November and December alone. For comparison, 4,750 pounds of venison hams and 3,200 pounds of rabbits were shipped during the same months.[67] In Iowa, a whopping 23,114 tons of poultry, fish, and game were transported by rail during the 1889–90 fiscal year.[68] These figures clearly demonstrated that "the development of cheap and rapid transit and of cold storage" marked the origins of the "unrestricted trade in game."[69]

Alongside railroads and refrigeration, the telegraph proved vital to game buyers and sellers engaged in transforming market hunting into a modern, capitalist enterprise. This innovation not only enabled wildfowl dealers to reach agreements on quantities and times for shipment and delivery, but also allowed them to track distant marketing trends. Marketmen could use the telegraph to bypass middlemen and sell straight to retailers, and they could hold their birds until a price was agreed upon beforehand.[70] Telegraphed orders to hunters and packers at rural stations allowed more retail dealers direct access to game suppliers, expanding the market while fostering competitive pricing. In the field, news of wildfowl migrations, such as erratic passenger pigeon nestings, could be rapidly disseminated to hunters, netters, and dealers.[71] Game dealers like Nat Wetzel used dispatchers to wire weather conditions and shooting locales to his mobile squads of salaried hunters while messengers carrying killed game to railroad depots for shipment would telegraph requisitions back to headquarters for ammunition, ice, decoys, and other supplies. With the telegraph at his disposal, the modern game dealer thus worked "in precisely the same manner as the train dispatcher," gathering and disseminating information while directing the interconnecting components of an extensive enterprise.[72]

Avian Entrepreneurs

The dealer stood at the forefront of wildfowl commercialization, an increasingly modern and complex endeavor characteristic of late-nineteenth-century industrial capitalism. Commodified birds were valued according to a system of prices and exchanged in segmented markets for the profits they would bring entrepreneurial game dealers. An arrangement of specialized labor extracted avian bodies from their environments and transformed them into salable products, in the process spawning a diverse class of game marketers and their allies. Many in the game business held multiple roles, with some dealers operating in a combination of capacities. Many were innovators themselves, developing new storage, transportation, or marketing methods to enhance the wildfowl supply and whet customers' appetites. With new refrigeration and transportation technologies available to help smooth the seasonal cycles of supply and demand, the marketmen had a reliable, salable, and profitable commodity to offer consumers—at least for a time.

In this emerging consumer market for wildlife, dealers found that wildfowl held a comparative advantage over other game animals. Birds' smaller sizes and weights rendered them more easily stored and transported. They were also more widely salable. A single prairie chicken, Merritt claimed, could make a meal for a small family; two birds could feed a large one.[73] Another advantage was that wildfowl required minimal preprocessing before sale. The variety of migratory birds available at different seasons across a range of geographic zones also contributed to wildfowl marketability. A dealer could buy and sell an assortment of species as each came into season. By 1890, New York City markets offered epicures as many as "eighty to one hundred varieties of game" at various times throughout the year.[74] This variety and the relative ease with which wild birds could be stored and shipped made them the mainstays of the modern game market.

Like any capitalist enterprise, commercial hunting was all about prices. For the market hunter and game dealer, the prices birds could bring and the ultimate profits to be made were their main incentives for pursuing the business. Early game market prices varied widely because they were tied to local conditions, local supplies, and local preferences. Hunters of passenger pigeons, with their easily exploited, massive nesting colonies, for example, often produced a glut on nearby markets. At such times they were offered for as little as a penny apiece or twenty-five cents a bushel.[75] In 1860, Lafayette, Indiana, was awash with prairie chickens that sold for a little more than $1 a dozen.[76] About four years later, local scarcity drove Chicago prairie chicken prices up to $3.75 a dozen.[77] High prices for wildfowl in Baltimore early in 1859 were also blamed on a scarcity of local wild game.[78] In California, before the Civil War, sandhill cranes seasonally brought up to $20 apiece in San Francisco markets, while wild ducks and geese were so plentiful that they depressed poultry prices at times.[79]

Although the introduction of freezers and refrigerated shipping substantially increased the market supply, wildfowl prices generally rose in the latter half of the nineteenth century, in part because the game dinner had become fashionable among the urban upper classes. "No first-class hotel or restaurant was considered worthy of the name" if it did not have wild game on the menu, Merritt insisted. Elites who wanted wildfowl for their tables were quite willing to pay the expense for express shipping and costly refrigerated storage, to the extent that from the 1880s onward there was an "active demand for game birds of every description" and all involved in the business "made good profits." As wildfowl populations declined around urban centers, "demand on the remaining fields of supply" increased further.[80] In the maturing Chicago market, a growing population and refinement in local tastes rapidly raised the price of canvasback ducks in a few years from about forty cents to five dollars a brace in 1889. "Such is the effect of educating the western palate," concluded the *American Poultry Journal.*[81]

With refrigerated storage and speedy transport, faraway marketers and consumers increasingly determined the value of wildfowl from the hinterlands. Consequently, information about current prices was a necessity for all those involved in the business. Trade publications, commercial newspapers, and even sporting publications routinely reported urban market prices in the latter half of the nineteenth century, noting when periods of oversupply or shortages affected game costs.[82] Individual market gunners, however, more often had to rely on dealers to provide relevant price information.[83] The prices Edward Alders received from San Francisco dealers for the waterfowl he shot in the San Joaquin Valley varied widely, with sprig (northern pintail ducks) bringing anywhere from $1.50 to $4.50 each in the early twentieth century.[84] Merritt likewise found that he had little control over the prices he could ask for the birds he sold to urban dealers and commission men. In the 1876 season, he tried to get $4.50 a dozen for fall snipe—the price he received from New York game dealer A. & E. Robbins the year before—but city dealers refused to pay more than $4.00. No snipe orders came in the next year, and when Merritt made inquiries, he found that a new supplier in Erie, Illinois, was sending all that the New York market could handle. Since this man "knew little or nothing of possible values," he sold his birds to commission agents at $2.75 a dozen, effectively undercutting Merritt's business.[85]

A system of market prices was just one of the capitalist characteristics of the modern game business. Wildfowl commercialization also involved a range of specialized labor, including game buyers, commission men, retailers, restaurateurs, egg collectors, milliners, and biological specimen dealers. They were supported by a host of secondary associates employed in processing, storing, refrigerating, and transporting game, as well as those who communicated information about wildfowl availability and prices. Without these facilitators, modern market hunting could never have become any more than a limited local or regional trade.

The business of market hunting began with the hunter. But even here, as discussed earlier, the labor was sometimes segmented as hunters and collectors specialized in particular avian products. In the field, market hunting was often a team effort. On the Chesapeake Bay, where the use of sinkboxes prevailed, it took a team to support one or two gunners. The sinkboxer needed a "mother" boat for transportation to the gunning site and to serve as lodging during the trip. A tender boat was required to transport large quantities of game back to shore, and one or two small "pick-up" boats retrieved the downed birds. Thus, a hunting crew might consist of the gunner, the captain of the tender boat, one or two deckhands to man the retrieving boats, and probably a cook.[86] Merritt employed a rotating coterie of hunters and maintained a daily record of the birds they supplied, paying them a lump sum for the entire hunt at the end of the trip.[87] Massive passenger pigeon hunts could involve thousands of participants, from netters and trappers to the itinerant "buyers, shippers, packers, Indians, and boys" who were all "engaged in the traffic" at a single nesting. In the 1878 Petoskey nesting, observers estimated the number of netters alone at seven hundred, with an additional fifty teamsters working to haul birds to the nearest railroad station.[88]

Capturing birds was only the first step. Even though wildfowl generally needed relatively little processing in comparison to other game, a certain level of postkilling preparation was usually required before sale. Early in the nineteenth century, birds were often sold with their entrails intact, and many hunters sent their birds to market without plucking, skinning, or gutting them beforehand. In fact, according to one author, the best market prices came from game birds that were packed entire and "in the natural state."[89] Some consumers preferred to eat small upland birds whole, as Audubon noticed early in the century. Snipe and woodcock, he wrote, were eaten by "many epicures . . . with all their viscera, worms and insects to boot."[90] Lewis considered this a "disgusting and filthy habit" practiced only by those "fond of half-cooked leeches, partly digested ground-worms, tough viscera, and ugly insects of all kinds."[91]

Gutting before sale became more common later in the nineteenth century. This practice not only reduced the weight of the birds for transport but also allowed for better air circulation to keep the meat cool. Even then, few market shooters gutted birds in the field or at the conclusion of the hunt. Iowa market hunters "didn't touch the birds at all when we put them in the cooler. We didn't cut off any part of them."[92] Those who did their own processing found quick and efficient ways of doing the job, such as the California market hunter who gutted ducks by "slicing open their abdomens and then expertly swinging them by the neck with a whip-like motion."[93] Plucking was rarely required for meat birds destined for the market. More often, wildfowl bound for shipment were just eviscerated, especially in cold weather, since well-chilled birds were kept clean, insulated, and preserved by their plumage.[94] Besides, buyers found the attractive plumage pleasing. Commercial waterfowl hunters often did not need to do any processing or preserving at all, because in some places "duck speculators came right out to the boat and bought them" as they were unloaded.[95]

The time-consuming chores of eviscerating, plucking, and skinning were distractions that the market hunter could ill afford. Instead, the limited processing required was often done by someone else. In areas of high waterfowl concentration, picking was often a cottage industry with local men, women, and children plucking birds for either a small fee or in exchange for the feathers and down.[96] One New York market hunter in the 1860s "paid the neighbors a penny apiece for the ducks they plucked for him." He traded the feathers for shot, powder, and food at the local store, which had its own "feather bin" for such exchanges.[97] Paul Marshall, who was born near the tail end of the market hunting era, recalled that nearly all the residents on Smith Island in the Chesapeake Bay spent the winter picking birds. At the local store "there would be men settin' and talkin' and big piles of fowl just heaped up everywhere, and people pickin'" at night. One of the pickers, nicknamed

"The Christmas Season—Game Stand in Fulton Market, New York." In this illustration by A. B. Frost originally published in *Harper's Weekly* on January 5, 1878, many ducks, a swan, and probably upland game birds such as prairie chickens (in baskets) are readily identifiable, still clad in their plumage. Unplucked birds were easier to preserve, and customers usually desired them still feathered. Image from Miriam and Ira D. Wallach Division of Art, Prints, and Photographs: Picture Collection, courtesy of The New York Public Library

"John L. Lewis for the labor boss," demanded thirty-five cents per bird plucked. "He finally got the price up some," according to Marshall, "but not that much."[98]

Feathers and down were additional marketable products that some commercial hunters traded locally or sold to any number of feather marketers operating in the nineteenth century.[99] A variety of waterfowl including eider ducks, scoters, loons, swans, and geese were stripped of their feathers and down for use in pillows, cushions, and bedding. Canada geese feathers and down could net a seller in Philadelphia an extra thirty cents per bird.[100] Plume birds such as egrets and herons supplied feathers for the millinery industry, which employed numerous individuals, including many women and children, who worked at cleaning, curling, and sewing feathers.[101] Millinery plumes and bird skins highly sought by merchants in the late nineteenth century went to plumassiers both in the United States and abroad. According to the *Millinery Trade Review*, there were nearly one hundred persons in London alone engaged in the feather and bird skin trade in the mid-1870s, "either as manufacturers, merchants, dealers, purifiers, plumassiers, naturalists," or others.[102]

Although feather and down production involved divisions of specialized labor, the storage of wildfowl meat was an even larger industry with more participants. Cold storage warehouse owners in both rural and urban locales became indispensable components of the modern bird business. Freezers owned by local merchants and produce dealers in prime bird-hunting areas either bought birds from area hunters or held them on commission until they could be shipped and sold. The expansion of the game trade brought even more of these "enterprising proprietors" of cold storage warehouses into nearly every section of the country.[103] In cities, large cold storage facilities held perishable stock for multiple merchants, while smaller freezers operated by retailers and restaurateurs kept supplies for their businesses. Warehouse owners like A. I. Dexter, president of the immense Milwaukee Cold Storage Plant (and inventor of the "Dexter System of Cold Storage"), profited from the growing "refrigerating business," which included the commercial storage of wild game of all sorts.[104]

Another set of players in the game trade consisted of financiers—bankers, brokers, stockholders, and insurers who advanced dealers necessary operating capital. For a brokerage fee, rural game dealers like Merritt could write drafts drawn from local banks on a percentage of their future shipments.[105] Large urban wholesalers with cold storage facilities could raise money from bankers using their warehouse receipts for security, much in the same way as grain dealers could use elevator receipts. Before the invention of freezing technologies, the perishable nature of game made such transactions impossible, but by the 1890s, bankers would "accept storage certificates for collateral as readily as they would United States bonds." The value of frozen game and the minimal hazards to this kind of preserved collateral meant that "the cold storage warehouse [was] considered an excellent insurance risk."[106]

Getting marketable wildfowl to buyers engaged this cast of players, but the dealer was the crucial link between the field and the market. While market hunters could sell birds to consumers themselves, only those who hunted near metropolitan areas were able to deal directly with restaurants, hotels, or the public. Few commercial wildfowlers had the time, access, or inclination to be marketers. That job fell to the game buyer or dealer. Even among this class of avian entrepreneurs, however, there was great variety—no homogenous group of marketmen existed. Some retail dealers bought birds directly from hunters and marketed them to consumers. Other dealers worked as wholesalers, buying wildfowl meat, eggs, or feathers from shooters or collectors and re-selling to retailers. In some instances, dealers acted as importers or exporters, bringing foreign-raised game birds from abroad or sending American birds to overseas markets. Commission agents received birds on credit from market hunters and sold them to wholesalers for a fee. Game dealers quite commonly operated in multiple capacities as wholesalers, retailers, or commission agents as the situation allowed. In all, though, wildfowl marketing increasingly became the domain of these specialized business practitioners whose efforts in the latter half of the nineteenth century extended the reach and complexity of the market in birds.

Market hunters commonly sold to a dealer who paid cash on the spot and then remarketed the birds either to other retailers or to consumers. Small-town meat, poultry, and produce dealers had long been game buyers in the season, retailing birds locally. The expansion of the wildfowl market after the Civil War, however, allowed many to add wholesaling to their businesses, collecting game from area gunners and selling to more distant urban game merchants. The self-proclaimed "Boss Game Dealer," W. R. Savage, made a name for himself in southeastern Kansas in the 1870s and 1880s, where he "built up a big trade." He paid hunters "the highest market prices for game of all kinds," the *Sumner County Press* reported, and thus provided "employment for many persons and support to their families."[107] By offering top cash prices to commercially inclined gunners, Savage made Independence, Kansas, "a better game market for Prairie Chickens and Quaile, than Kansas City or St. Louis," according to local advertisements.[108] In Gibson City, Illinois, White's Game Market promised "the highest market prices in cash to shooters."[109] So many birds, both wild and domestic, were shipped from that area in the 1880s that the *Gibson City Courier* proclaimed the small town "one of the best markets for poultry and game in Central Ill."[110]

Itinerant dealers took to the field during hunting seasons, seeking areas with high concentrations of wildfowl and market hunters. Ed Johnson, a store proprietor in North Carolina, became a traveling waterfowl buyer when hunters were active. In the late weekday afternoons, he cruised through Currituck Sound in his cargo boat. Market hunters knew his schedule and met him with birds they had killed that day. Johnson paid them in cash or, as an option, credited them for ammunition or future purchases at his store. Then he rushed the birds to the mainland in order to meet the evening train.[111] G. W. Williams bought ducks at Back Bay, Virginia, for the Norfolk storage warehouse of R. L. Dyer.[112] He "would sail his boat to the different gunners who were operating in sink-boxes, batteries or blinds and buy their ducks" on site. Packing the ducks into barrels, he sent as many as one thousand at a time to Norfolk by wagon. Williams had "pretty much a monopoly on the duck shipping business in Back Bay" from the late 1880s well into the second decade of the twentieth century. Passenger pigeon dealers flocked to nesting sites, "anxious to make rapid purchases" of freshly killed birds. Bringing their own barrels and ice, these buyers followed the netters and shooters into the forest "in order to be first in the market."[113]

Along with itinerant bird buyers, the commission merchant was a product of the modern game market. Commission agents functioned as middlemen, funneling game birds from hunters, freezer owners, and small-town game dealers to urban wholesalers and retailers. Their profits came from extracting anywhere from 5 to 10 percent of the sale price.[114] These "special brokers" operated "over a radius of several hundred miles," collecting game from hunters to ship to city dealers.[115] Without storage facilities of their own, smaller commission agents sold game "without delay to the retailer," seldom moving the birds from the wharf or depot to which they were shipped.[116] The "local buyer" on commission had become "a necessity" in many places, noted a turn-of-the-century *Farmers' Bulletin* from the US Department of Agriculture. As the shipper's agent, the commission man acted in the larger markets on the game supplier's behalf "at far less cost to the shipper than would be attendant on the latter's coming to the city to market his own wares." Though commission agents were usually willing to handle any kind of "country produce," most preferred consignments in their own specialty lines: "butter, cheese, and egg handlers; fruit and vegetable dealers; or poultry and game handlers."[117]

Smaller agents competed with a growing number of larger, urban commission and wholesale businesses that maintained their own facilities for game storage.[118] Midwestern newspapers based along wildfowl migration routes were filled with notices from such firms soliciting game consignments from hunters. Just one page of an 1874 issue of the *Daily Iowa State Register* in Des Moines contained more than twenty-five dealer advertisements for commission houses and wholesalers located in cities such as Baltimore, New York, Philadelphia, and Boston. Many of those identified game as their specialty.[119] On the West Coast, large commission houses dealing in game were especially prominent in San Francisco, where prices were high and profits for middlemen were easy to find.[120] Early in the twentieth century, several California commission men banded together to form game transfer companies that transported game as well as livestock, poultry, and agricultural products for others. These corporations aimed to circumvent restrictions on game marketing and encourage continued commercial wildfowl harvests.[121]

Many urban game dealers did it all: took birds on commission, accepted shipments from rural wholesalers, bought game directly from nearby hunters, transshipped to other domestic and foreign wholesalers, sold to restaurants and hotels, and operated retail market stalls. Railroad hubs such as Chicago and St. Louis attracted many game dealers who amassed wildfowl harvested from points farther north and west not only to retail in their growing metropolises but also to sell wholesale to East Coast merchants. To offset the seasonal fluctuations in the game market, few dealers limited themselves to selling wildfowl alone. In smaller burgs, the game merchant or cold storage proprietor usually sold a variety of merchandise locally—poultry, meat, produce, or dry goods—while wholesaling bulk shipments of wildfowl to city markets when they were available. Urban dealers also usually handled other kinds of stock besides wildfowl and other game. The most common combination was domestic poultry and game, but other product mixes, including fish, dairy products, honey, hides, fruits, or vegetables, were not unusual.[122] Along the East and Gulf coasts, game dealers often included oysters and terrapin among their wares.[123]

Although most avian entrepreneurs usually dealt in other types of products, some were specially recognized for their wildfowl business, and by the end of the nineteenth century, the market in birds had become segmented into meat, feather, egg, and specimen submarkets. Among those, wildfowl meat dealers were the most prominent. In New York City, Knapp & Van Nostrand and A. & E. Robbins (later, A. & M. Robbins) were renowned for handling considerable quantities of wild birds and other game meats.[124] The latter firm, *Harper's Weekly* boasted in 1878, could store up to fifty tons of game in its own "great refrigerator."[125] An 1885 publication claimed that there was "no superior" game dealer in Chicago than F. M. Smith, who "receive[d] his supplies direct from the leading game producing sections of the country." His "superior facilities" included "coolers and all other appliances for the effective conduct of his business."[126] In 1899, the Chicago-based firm of George H. Sloan was considered by the *Inter Ocean* as "one of the largest handlers of game" there.[127] The "principal game dealers" in San Francisco, wrote John S. Hittell in his 1882 economic summary of the Pacific coast, included Richard D. Mowrey & Company, Hart & Goodman, and F. N. Woods & Company, all game and poultry dealers in the city's California Market.[128]

The divisions in the market allowed some sellers to specialize in certain species or in types of avian products. W. W. Judy, for example, made his reputation in passenger pigeons, becoming "the ruling game dealer" in St. Louis. His company handled "more dead and live pigeons than any other firm in the country."[129] Brothers Joseph and Isaac Allen of Manchester, Michigan were another of the better-known passenger pigeon wholesalers.[130] In Detroit, grocer and provision dealer Henry T. Phillips was the passenger pigeon king.[131] Live pigeons for trapshooting were a particular dealer specialty, and some marketmen established their own facilities for keeping enormous quantities of avian livestock, such as game dealers Bond and Ellsworth of Chicago, who had cages for holding 12,000 live pigeons. Tom Stagg, also of Chicago, made a business of the trap trade, keeping thousands of birds at a time in his big barn on the city's north side.[132] A single New York supplier provided 20,000 pigeons for one 1881 contest at Coney Island. At a little over a dollar a dozen, live pigeon dealers had to work in these kinds of bulk numbers in order to clear a reasonable profit.[133]

Egg dealers, though not nearly as common as those dealing in meat birds, served a geographically limited market mainly along the coasts. Perhaps the most commercialized of the wild bird egg producers operated off California's shores in island seabird colonies.[134] Between 1850 and 1856, an estimated three to four million eggs from the Farallon Islands made their way to San Francisco markets. During the laying season from April to July, boatloads of eggs came to city wholesale produce dealers who sold them to retail merchants and bakers for twenty cents a dozen or less.[135] While supplies remained high, the annual value of the trade was significant, much of it monopolized by the Farallon Egg Company.[136] Competing eggers, often commercial fishermen, contested the Farallon Egg Company's exclusive control of the islands and its crop of gull and murre eggs which it had claimed since 1851, provoking the short-lived "egg war" that broke out in 1863.[137] Commercial egger businesses also operated off the Texas Gulf coast, on Hawaii's Laysan Island, and on the Pribilof Islands off the Alaskan coast.[138]

Feather merchants and bird skin sellers represented still another segment of the bird market. Joseph Rosenthal in New York, for example, identified himself specifically as an "Importer and Dealer in Raw Feather Stock Birds, Bird Skins, Wings, Aigrettes, and Plumage."[139] Feather dealers advertised directly to distant suppliers, asking for the plumes and skins from the most desirable species. One 1899 advertising postcard sent by Louis Stern, a New York commission merchant, to the Gulf states requested laughing gulls, royal terns, barn owls, sea swallows, white sea pigeons, and grebes, among others. Stern also noted the birds he did not want: "Small Blue Crane, Night Heron, Brown Egret, Surf Snipes, Water Turkey, and Grosbeak."[140] Taxidermists provided bird skins and whole-bird mounts for some of the most elaborate hat designs. One New York taxidermist in 1886 "had in his shop thirty thousand bird-skins . . . made up expressly for millinery purposes."[141] Some dealers went to the source, as did one Long Island merchant who prepared more than eleven thousand skins on a single trip to South Carolina.[142] The milliner's agent was almost everywhere at the turn of the century. An Indianapolis taxidermist reported that some "five thousand bird-skins collected in the Ohio Valley" had been shipped in 1895 alone.[143]

Specimen dealers occupied another slice of the bird market, serving collectors, amateur naturalists, sportsmen, and scientific institutions. The specimen-collecting fad was widespread in the nineteenth century, as evidenced by the more than five thousand individuals listed in an 1884 directory of collector-naturalists. Prominent American specimen specialty shops were found in New York City, Boston, and Philadelphia, as well as many smaller cities across the country. Entrepreneur Henry Augustus Ward's natural history specimen supply house, established in 1862, was one of the most important. Based in Rochester, New York, Ward's firm became a major seller to both individual and institutional customers, providing over $700,000 of specimens to one hundred American museums by the end of the nineteenth century. Ward's was also a center for teaching taxidermy (one of the firm's more notable students was William T. Hornaday, director of the New York Zoo and national conservation leader). By the end of the nineteenth century, more than a hundred specimen dealers, large and small, were scattered across the United States.[144]

The development of the specimen dealer's submarket was advanced by new advertising opportunities provided in popular, scientific journals such as *The Oologist* (1884), *Natural Science News* (1895), and *Bird Lore* (1899). Interested readers could find scores of dealers and taxidermists offering a wide array of bird skins, bird eggs, and nests for sale. F. T. Corless in Los Gatos, California, for example, offered twelve varieties of bird skins for one dollar each in 1895.

Mounted birds could be had for fifty cents.[145] Ernest H. Short sold everything from a "very fine" pintail duck skin at one dollar, to chickadee skins for ten cents apiece. Mounted birds, including wood ducks and California quails in a fancy case, ranged from seventy-five cents to over five dollars.[146] Though never as extensive as the market in meat birds, wildfowl specimen dealing was clearly commercialized in the late nineteenth century. The differentiated avian products offered by specimen sellers, along with those sold by feather merchants and game dealers, revealed the segmentation of the modern market in birds.

In sum, the dealer had no singular identity or role. The market in birds involved different kinds of avian entrepreneurs performing an array of market functions from processing, storing, and shipping to wholesaling, consignment selling, and retailing. For some, wildfowl was a seasonal sideline; for others, it was a mainstay. Some sellers specialized and marketed their stock to a targeted segment of consumers. Others sold whatever they could. As the game business matured during the latter half of the nineteenth century, an increasing number of market-hunting accomplices divided the transformative work required to turn living creatures into saleable products. In other words, wild birds passed through numerous hands on their journeys from the field to the consumer. This complex network of players, of which the market hunter was but one part, drove the evolution of the modern, industrialized market in birds.

❧

The last year Merritt hunted was 1873. He put down his gun, surprisingly with no regrets. He had hunted for fifteen long years and accomplished what he had set out to do, and more. He had not only made his living by hunting but built a thriving commercial wildfowl enterprise as well, learning the lessons of supply and demand along the way. He understood how to secure operating capital and negotiate advance payments from commission men and game dealers. He had experimented, innovated, and adopted new technologies in transportation and refrigeration. Now, Merritt thought, it was time to turn all of his entrepreneurial attention to expanding his game business.[147]

To increase the efficiency of his freezer facilities in Kewanee and Atkinson, which were already up and running, Merritt added ice production to his commercial repertoire. He purchased the defunct brewery on Kewanee's south side and converted it into an icehouse. Damming a slough of the Spoon River that flowed through the property created a small pond from which he could harvest his own ice supplies in the winter. In the 1880s, he enlarged the

The Providence, Rhode Island, natural history supply establishment of Southwick & Jencks published this guide in 1883. Natural history pamphlets such as these mainly functioned as advertisements for specimen companies and as catalogs for collectors. Public domain via Wikimedia Commons pond with a newly built, two-hundred-foot brick dam and, in the process, turned Merritt's Pond into the town's most popular swimming hole and skating rink. Merritt and his business partner Henry Parker operated a retail store in Kewanee, selling dry goods, clothing, and groceries in addition to game. Parker & Merritt employed ten people in 1876, including a bookkeeper, three clerks, and five salesmen. Independently, Merritt worked as a wholesaler, packer, and shipper of game and as an ice supplier. When Parker left Kewanee for Kansas City in 1880, Merritt carried on with his business concerns alone. He bought the old Kewanee House Hotel and replaced it with an extravagant brick building that one wit would later dub "an architectural joke."[148] The "New Kewanee" featured over its threshold a stained-glass window inscribed with a quotation from Sir Walter Scott's *Lady of the Lake*—a detail reflecting Merritt's literary fancies. In addition to the hotel, Merritt owned an entire block of buildings in Kewanee, as well as properties elsewhere.[149]

Wildfowl, however, remained Merritt's specialty. All of his diverse business interests took a back seat to his growing commercial network for moving birds from countryside to city. No longer hunting himself, Merritt devoted his time to the market in birds, building relationships with commission agents and urban game dealers. Even though he was a game merchant himself, his opinion of marketmen was generally low; they were too often "proverbially

Henry Clay Merritt's evolution from market hunter to game dealer and diversified businessman is evident in this late-nineteenth-century engraving depicting his retail store in Kewanee, Illinois. Courtesy of the Kewanee Historical Society, Kewanee, Illinois

foolish and filthy talkers."[150] A few were to be admired, however. Merritt had a long history with Amos Robbins, general manager of the A. & E. Robbins firm, whom he characterized as the "Caesar" of the game dealers in New York and "a colossus of finance." Robbins's opinions about game held sway among restaurant and hotel managers in New York, according to Merritt, and years of observation had proved Robbins to be an able capitalist, speculating in the market and always getting the maximum profits.[151] Commission men were also a bad lot, in Merritt's view, more interested in short-term profits than cultivating enduring business relationships, but John A. Lyon was worthy of praise. The New York–based agent was "Nature's nobleman," a man "absolutely void of cheat or deceit," in Merritt's eyes.[152]

By the 1880s, Merritt was selling more birds than ever. New York was his primary market, but he was also supplying Chicago, Boston, and other cities. He took time to rebuild his freezer, instituting a better-insulated design that allowed him to keep his stock frozen and in good shape for years at a time, even when outdoor temperatures rose. His first batch of frozen canvasbacks and redheads stored this way kept from 1883 to 1885 and still sold for a good profit. When he finally disposed of his thirty-six barrels of frozen ducks, he cleared almost three thousand dollars.[153]

Merritt's decision to give up hunting was not just personal, it was pragmatic. By the late 1880s, "Henry County was practically denuded of game." Formerly abundant hunting grounds along the Mississippi were likewise emptied of wildfowl. But farther west the great expanse of the Great Plains contained birds "of every description." Woodcocks and partridges could still be found northward in Wisconsin, Minnesota, and into the Dakotas and Montana. This bird-rich region promised years of continuing business for Merritt, and toward that end, he sent his eldest son, Clarence, out west—not to hunt but to buy from the market hunters and rural freezer owners popping up along the rail routes west. Situated at the juncture between the urban East and the rural West, within striking distance of the great metropolis and railway hub of Chicago and tied with commercial and personal connections to eastern game dealers, Merritt stood poised to make an even bigger killing in birds.[154]

The Market

Writing from his East Coast vantage point in 1848, Henry William Herbert predicted there was "little doubt" that "the western species" would soon be commonly "exposed for sale in our markets."[155] Within in a few years, Herbert's prescient assertion was being borne out in cities and towns across the country. Technological and economic innovations allowed urban game markets to draw their merchandise from an ever-widening wildfowl hinterland. Where once it had been a rare occurrence for an urban merchant to have a "brace of plover or snipe" to hold "on exhibition to attract the eyes of customers," from the 1850s onward it became routine to see a wide variety of wild birds for sale in big-city markets.[156] The growth of Chicago's game market epitomized the change. "No other city is the centre of a game-producing region so vast and inexhaustible," wrote the *Chicago Tribune* in 1872. "From Southern Texas to British Columbia, and from the Calumet River to the Rocky Mountains,— this is Chicago's larder"[157] New York City, though, was hardly lacking in this regard. The "variety, quantity, and quality of wild-fowl and birds" available in public markets was "not surpassed in any other city in the world," proclaimed one observer. "The prairies of the West, the forest-regions of the North, the gulfs and coasts of the Northern and Southern States, and even European cities" contributed, year-round, to the city's burgeoning wildfowl market. Improvements in transportation continually brought "new and rare varieties of game" to New York's consumers.[158] On the West Coast, band-tailed pigeons taken by the thousands in Oregon's Coast Range and sandhill cranes from California's interior grasslands were "always for sale in the markets of San Francisco," where they were "highly esteemed."[159]

This late-nineteenth-century expansion of urban game markets reshaped the economic geography of commercial wildfowling. Local sales had once been the norm, but increasing demand from cities—and the technological means to deliver on those demands—redirected birds from nearby markets to distant ones. For market hunters, this meant sending the better, more sought-after species to dealers and commission houses in the big cities while selling less desirable birds locally, and on the cheap. "The only time when ducks were eaten around our house," recalled commercial gunner Atley Lankford, "was when there was no market."[160] For dealers, the game trade's changing geography freed them from the limitations imposed by local conditions, both natural and human-made. New York's markets, for example, were well supplied at the turn of the century even though few birds were "killed in near-by places." Mallards from Delaware, grouse and quail from Iowa, and partridges from points west and south, reported the *New York Evening Post*, were available and abundant, nonetheless.[161]

This commercial migration of wildfowl from countryside to city extended beyond the borders of the United States. By midcentury, dealers were exporting more wild game overseas. Rapid transit, reported *Scribner's Magazine* in 1877, had made it possible for "our trans-Atlantic cousins" to enjoy the Chesapeake Bay's wild produce, including the prized canvasback duck. "There have recently been dinners in England and Paris," wrote W. Mackay Laffan, "at which every article of food upon the table came from America."[162] Game wholesalers shipped western wildfowl, including quail, prairie chickens, and wild turkeys, from Chicago and St. Louis to appreciative foreign markets. By the early 1880s, the export trade in wildfowl "had assumed considerable proportions," amounting to nearly a million and a half dollars in trade for Chicago dealers in 1881 alone. St. Louis's game exporting likewise brought in more than one million dollars a year. A single St. Louis merchant reportedly supplied the London market with enough wildfowl to ensure that a "great many English families will have their Christmas and holiday dinners from American game."[163] American plume birds were another significant part of the export market in wildfowl, with feather merchants sending much of their fancy plumage to European millinery centers, especially in France.[164]

Many active game exporters insisted this was just the start since "the wild game trade [was] yet in its infancy." International trade in birds, these dealers predicted, could grow to at least ten million dollars annually.[165] Others were not so sure that avian populations could withstand the drain. As early as 1879, the New York game dealers Drohan & Powell complained that American canvasback ducks had become so "dear and scarce" that the firm had sent no more than one hundred of them overseas. The previous year they had delivered more than one thousand.[166] Ornithologist Daniel G. Elliot conjectured that the canvasback's dwindling numbers in the 1890s were caused by the twinned problems of high market prices and foreign demand. With "the thousands that are shipped to Europe," he wrote, "it is surprising that there are many left."[167]

As demand for wildfowl increased and bird populations declined, cheaper competing game from Canada, Europe, Mexico, and South America infiltrated US game markets. Southwestern Ontario supported a "thriving market hunting industry" in the late nineteenth century. Imports shipped by fishing boats and Great Lakes freighters were not uncommon in northern cities such as Cleveland, Toledo, Detroit, and New York.[168] As early as 1866, Canadian sportsmen worried that too many Canadian birds went to "the Yankee deal-ers."[169] The South American tinamou (a ground-dwelling, chicken-like bird) was "shipped from Argentina to London" in quantities, and then imported to be "sold in the markets of Washington D.C."[170] From Europe came increasing numbers of estate-raised game birds, creating a veritable "invasion of the American game market by foreign game." Commercially raised birds from private preserves were "not only more plentiful and cheaper in Europe than in American markets," but were also "sold at a lower price in the United States than corresponding American game." In contrast, American birds were harvested far from urban markets and came from "wild and practically unregulated stock."[171] As a result, US game merchants had to search farther afield to expand their domestic supplies.

For a time, they were quite successful. Across the latter half of the nineteenth century, dealers' efforts brought astonishing amounts of avian capital to urban markets, and thriving game markets expanded in many major and not-so-major cities. San Francisco was the preeminent West Coast game market city. The city possessed its own abundant supply of wildfowl, which was supplemented by shipments of birds from the already-legendary waterfowl hunting grounds of the Suisun marshes to the northwest and the vast wetlands of the Sacramento–San Joaquin river delta to the east.[172] San Francisco newspapers in the 1850s compared the city's game market to New York's, claiming that although New York drew its supplies from vast interior hinterlands from Georgia to Illinois and beyond, game in Gotham remained "much dearer and rarer" than in San Francisco.[173] Boosters claimed that "few markets in the world are better supplied with an abundance and variety of game at low prices than the market of San Francisco."[174] Across the latter half of the century, the city's game business exploded. In the 1890s, city directories contained listings for nearly fifty dealers in poultry and game, almost double the number of game merchants working there in 1869.[175]

Other major cities clamored for the title of greatest game market in the world. St. Louis game wholesalers and commission merchants shipped "the largest quantity of game of all kinds in the United States," averred Missouri's state fish commissioners in 1884, with most of it going to "points outside the State, principally eastward."[176] Seventeen game and poultry dealers appeared in the St. Louis city directory in 1880, most located in Union Market, near the railroad tracks.[177] As a transportation hub, the city facilitated the migration of wild commodities from west to east, standing at the bioeconomic juncture between western supply and eastern demand for birds. As a result, the *St. Louis Post-Dispatch* reported in 1881, the city was peculiarly well situated to "be the great Western center for the distribution of this class of comestibles."[178] Wildfowl brought into St. Louis traveled eastward to New York, Philadelphia, and Washington and westward as far as Denver.[179]

St. Louis's position as the "central game market of the United States" was hotly contested by Chicago's wildfowl dealers.[180] Possessing many of the same geographic advantages as St. Louis, Chicago could boast of an oversized economic role in the modern interstate game market that allowed dealers there to monopolize much of the trade. Commission houses lined South Water Street "as far as the eye could see," where "long strings of ducks, prairie chickens, quail, and so on" hung from the awnings.[181] The business expanded dramatically through the late nineteenth century. The number of game merchants in operation grew from twelve in 1871 to more than forty in a little over a decade.[182] The *Chicago Tribune* proclaimed the city's game market in 1889 as "the greatest one in the civilized world." All the spoils of "the pot hunters, snarers, and 'sportsmen for revenue' of the west and northwest finds its way to this city," the newspaper declared, making it "the exclusive supply market" for some of "the rarest birds" and many other "delicacies that must be packed in ice."[183] Wildfowl was shipped to dealers in bulk from western suppliers and then reshipped eastward to New York, Boston, Philadelphia, and overseas, making the season for wild game in Chicago never ending.[184]

Although Chicago's central location gave dealers a commanding position, it was the demands of the game markets in New York City that drove much of the avian trade. The city's markets had long been famous for their wild bounty. In 1828, James Fenimore Cooper wrote glowingly that it was "difficult to name fish or fowl, or beast, that is not . . . to be obtained in the markets of New York," making the city a veritable "empire of gastronomy."[185] Midcentury canal and railroad construction brought even greater access to wild resources from farther away. The rapid change in game availability was noticeable. At the end of the Civil War, one longtime merchant observed, "a brace of plover or snipe" to be "held on exhibition to attract the eyes of customers" was still "quite a find for the market-man."[186] Within a few years, though, steam transportation had reached into "the most distant uncultivated States and Territories." Thereafter, not only the "variety of game" imported into the city increased, but also the amounts.[187] "There was little edible that could not be found in the New York markets," journalist Junius Henri Browne wrote in 1869. Fish,

F. M. Smith & Company was just one of the many game merchants in Chicago. This photo of the store's interior shows several bunches of ducks, possibly mallards, hanging from the center display table. Active wildfowl wholesalers and retailers, many located along South Water Street, made Chicago a major game market city in the late nineteenth century. Postcard for F. M. Smith and Co., Detroit Publishing Co., ca. 1895–1910, courtesy of Library of Congress, Prints and Photographs Division, Washington, DC

fowl, and all sorts of game came to the city from coastal rivers, mountain streams, seas, lakes, marshes, forests, and prairies—"everything that can appeal to and gratify the epicurean sense."[188] By the end of the nineteenth century, New York City's extensive game market rivaled Chicago's for its diversity and numbers of game merchants. Where in 1860 there had been less than fifteen poultry and game dealers in the city's Washington Market, by the 1880s dozens of smaller commission houses dealt in birds and other game alongside some particularly large wildfowl merchants.[189] With the increased efficiency of modern shipping and refrigeration, merchants could bring in even more. In the 1890s, the markets offered "from eighty to one hundred varieties of game"—enough to more than equal Chicago's "extensive choice of game."[190]

New York's game market was "a specialty," controlled almost entirely by the wholesale dealers in and around Washington and Fulton markets.[191] These dealers kept display and retail stands in major market centers while holding much of their wholesale supply nearby in cold storage until needed. The firm of Drohan & Powell, operating from Washington Market, was "the most extensive and most influential game firm" in New York, claimed *Frank Leslie's Illustrated Newspaper* in 1879.[192] Merritt's preferred dealer, Amos Robbins, dominated Fulton Market, where his opinions set the fashion and prices in game.[193] These and the many other large game vendors in the city monopolized the market by the century's end and aggregated a "trade running close to a million of dollars" in game meats brought in mostly from points westward.[194]

Consequently, control of the market in birds rested mainly in the hands of these large urban dealers, whose specialized trade publications, such as the *American Produce Review* (1895–1939) and the *Millinery Trade Review* (1876–1938), provided the most up-to-date information on current trends, prices, and sources of supply.[195] Professional associations formalized and strengthened merchants' control in the late nineteenth century, allowing them to regulate prices and manage competition. Such organizations aimed to police the market, at times provoking charges of monopolization from critics. In 1876, about thirty San Francisco and Oakland dealers vowed not to buy from middlemen who sold "poultry or game to any parties other than the members of the said Poultry and Game Dealers' Protective Association." The goal of the group, reported the *Placer Herald*, was "to take the business out of the hands of the commission men, and wholly control the same through their organization."[196] The *New York Times* attributed the high cost of canvasback ducks in 1880s not to scarcity but to a southern "canvas-back duck trust," formed by Baltimore-based dealers who held back supplies to artificially raise prices.[197]

Organized dealers could wield considerable influence over not only the economics of the market but also its legalities. New York's Marketmen's and Game Dealer's Protective Association was instrumental in crafting a new state game bill in 1887, the result of its cooperating with fish and game preservation organizations to devise laws amenable to both groups. Concessions to dealers' interests were necessary, argued *Forest and Stream*, since the proposed statutes repealed provisions for state game wardens and relied entirely upon voluntary self-enforcement instead.[198] In Chicago, sportsmen and game dealers came together in 1882 to form their own organization devoted to the passage of uniform game legislation and the creation of a national association of like-minded sport hunters and dealers. Leading the effort were Dr. N. Rowe, of the Chicago-based sportsmen's magazine *American Field*, and Colonel E. S. Bond of Bond & Ellsworth, one of the major South Water Street game wholesalers. Many other large Chicago game dealers were represented at the inaugural meeting, along with the presidents of several shooting clubs and the local Audubon Society.[199] Little more than a year later, however, the association disbanded. Upon resigning as president, Bond declared that the group had "outlived its usefulness" since it had failed to arouse members' interests in reforming wildlife legislation.[200]

Dealers more often found common cause with other businessmen than they did with the sporting sect, though. Organized dealers, commission houses, hoteliers, and restaurateurs in urban centers joined forces at the turn of the century to oppose any legal restrictions on the wildfowl market. In the feather trade, designers, manufacturers, and dealers formed the Millinery Merchants Protective Association to respond to the increasing pressure from bird protection groups like the Audubon Society to end the use of wild plumage.[201] Commission merchants, game dealers, and retailers organized in St. Louis, Cincinnati, Chicago, San Francisco, New York, and other places to combat any attempts to curtail free trade in meat birds.[202] Such associations were necessary, claimed the marketmen. There was no other way of "properly protecting the rights of dealers" to ensure that they were not saddled with "absurd" game laws or "unjustly dealt with."[203]

The rights of market hunters in this system were of secondary concern, at best. Operating far from the urban markets they supplied, commercial hunters had little control over the prices dealers offered or the conditions of the sales. They ultimately depended on middlemen, including shippers, commission agents, and wholesalers. Only the commission house or retailer truly knew the precise figure at which a hunter's game ultimately sold. Indeed, if the commission agent "wrote back and told you your ducks were all spoiled . . . you had no way of finding out whether that was true or not."[204] The same was true for rural wholesalers who sent their stock to urban sellers, as Merritt found out in the spring of 1860 when he sent a shipment of more than a thousand snipe and plover to A. & E. Robbins. After a long delay, he received a check for about thirty dollars, which paid for the fraction of the stock Robbins had been able to sell. The rest of the birds "had been thrown away."[205] Only when a buyer accepted the market hunter's birds was the seller's responsibility for its quality absolved. If birds were refused, the supplier had to pay the shipping costs for their return or allow them to be destroyed.[206] Even if a hunter's birds were sold, each of the marketmen along the way skimmed a bit of the profits. In the 1870s, passenger pigeon squabs, for example, bought from hunters at twelve cents per dozen were sold in Chicago markets for sixty to seventy cents, "a possible profit range of about 400 percent." Live pigeons brought even more.[207] In San Francisco, commission agents charged exorbitantly high rates to handle game that they, in turn, sold "without delay to the retailer." The city's retailers also commonly made "25 to 50 per cent profit" on game in the 1870s. Market prices thus reflected the multiple divisions of labor and specialization in modern wildfowl commerce—the growing distance between production and consumption.[208] As the US commissioner of agriculture reported in 1871, the commercialization of wildfowl represented, the widespread "evils . . . of a bad market system."[209]

Market hunters had to accept on faith the urban dealer's word as to the receipt of game, its condition, and its sale price. Consequently, the marketman's prospects for profit and opportunities for deceit were high. Merritt learned that "the character of the persons to whom you ship" was of utmost importance. Dealing with an unscrupulous commission man or city dealer could lead to goods being "sold at a ruinous price" with "no chance of recovery for you."[210] Transporting, storing, and selling a perishable commodity—one that was highly variable and subject to the whims of fashion and taste—to consumers in faraway places enmeshed commercial hunters in an economic structure that had quickly grown beyond their control. By the end of the nineteenth century, the market in birds had become a complex enterprise with many participants. The market hunter was only one component. Avian entrepreneurs of all sorts participated in this transformation of wildfowl into salable products. At the apex of this newfangled, industrialized, and capitalized enterprise, standing between the market hunter's birds and his profits, between biology and commodities, was the dealer.

CHAPTER THREE

The Hunted

"The snipe is a pansy, the woodcock the rose of summer," Merritt proclaimed as he waxed poetic about the American woodcock's charm. When in flight, the bird was "the most perfect picture of beauty . . . the flower of the air." His rapid, twisting maneuvers while airborne suggested an exquisite "rhythm of movement" expressed in the "cadence of flapping wings." On the ground, his track was "as smooth as if pressed in molten glass" and as straight as if drawn with a rule on paper. The woodcock was a master of concealment, able to remain hidden in dense thickets until the hunter was nearly upon him, only to burst out explosively, almost underfoot, with whistling wings. He was the "most sagacious" of birds, in Merritt's estimation, endowed with a "system of telepathy" that allowed him to "divine the intentions" of the hunter, "whether they were good or bad." To the lifelong wildfowler, the woodcock was the epitome of all game birds—"the cock of the forest"—at once both an adversary to test the hunter's skill and a muse to awaken the watcher's imagination. "You want him," averred Merritt, "you are not debating how you might get him nor what he is worth, but you want him, and so the plot thickens."[1]

As a hunter, Merritt could prize the woodcock for its nature. Merritt the businessman, however, could afford no such prejudices, since market hunters

American woodcock. Currier & Ives lithograph, ca. 1871. Courtesy of Library of Congress Prints and Photographs Division, Washington, DC (LC-DIG-pga-10172).

and game dealers had to make more-rational calculations balancing production and profit if they wanted to succeed. On the production side of the equation were a multitude of biophysical factors that determined what could be hunted or harvested: the variety and availability of avifauna, the accessibility of their habitats, the physiologies and behaviors of different species, the vulnerability of wildfowl to various hunting methods, the seasonality of avian movements, the variability of the weather, and other environmental realities. On the profit side of the balance, consumers' desires weighed most heavily. Buyers could be fickle and, as Merritt conceded, "fashion" ultimately "set the prices for all game birds."[2]

Luckily, in the late nineteenth century, consumers' desires for wildfowl grew exponentially, especially among urbanites. "The business was very profitable" in those decades, Merritt insisted, mainly because of "the active demand for game birds of every description." Except for periods of commercial disasters between 1893 and 1896, game dealers easily found buyers for their stocks. "Game dinners" were regular fare in every "first class hotel or restaurant worth the name." As market hunters exhausted wildfowl populations in one region, commercial pressure increased "on the remaining field of supply." In "nearly all the states west of the Mississippi the industry flourished," Merritt observed. The market in birds converted wildfowl into "a cash article" that "could be sold for cash as readily as government bonds."[3]

Simple laws of economics guided Merritt's business decisions. Seeking the nexus between supply and demand, he sought out the best market conditions for his stock by tracking prices in distant cities, monitoring his game-dealing competitors, judiciously choosing shipping times, and limiting quantities to avoid glutting urban markets. His choice of prey was determined by economic considerations, as well. He found small- to medium-sized game birds such as woodcock, snipe, and prairie chickens to be the most remunerative. Consumers preferred portion-sized meat birds, and these found ready sale in dealers' stalls and restaurants. Geese and large ducks were less attractive, according to Merritt, because they were "generally unprofitable," with the mallard being "the cheapest duck in the market."[4] Keeping an eye on the vagaries of the business and exercising keen economic acumen, the Kewanee game dealer ensured the profits that made up his entrepreneurial bonus.

Yet biology figured equally into Merritt's bottom line. Merritt's dismissal of geese and ducks as unprofitable prey, for example, was due not only to economic considerations but also to the biological facts that these larger birds were heavier to carry and more likely to spoil in warmer weather.[5] His choice of avian specialties, such as his favored woodcock, were as much a calculation about the availability, accessibility, and anatomies of these birds as they were about the laws of market supply and demand. Seasonality influenced Merritt's business as well. Since meat birds harvested in the fall to midwinter were generally in better condition than those killed in the spring, he worked to maximize his stocks of fall birds. At times, Merritt tried to pass off spring-shot fowl to unsuspecting buyers, not always successfully. In one instance, six barrels of canvasback ducks he sold to dealers in New York were refused by the city's up-scale Delmonico's restaurant, whose owners recognized the order as inferior "Spring birds."[6]

The balance Merritt tried to strike between environmental and economic realities was replicated many times over in the market in birds. Clearly, avian physiology and behavior, distribution and migration patterns, and species variability shaped the contours of wildlife consumerism in the late nineteenth century. But economic logic, along with biophysical dynamics, governed the wildfowl market. Avian entrepreneurs, armed with cutting-edge technological and financial tools, were able to transform avian biological diversity into economic commodities, creating, in the process, new patterns of human-wildlife interaction that were mediated by the market. That market in birds refashioned wildness into new forms suited to a modern American consumer culture.

The Nature of the Market

Merritt's preference for wildfowl was typical among turn-of-the-century market hunters and game dealers. Although professional hunters pursued mammals such as deer, bear, squirrels, and rabbits, wild birds made up the overwhelming bulk of commercial game sales in the period. In many ways, wildfowl were biologically fitted for this kind of extensive exploitation. Avian distribution, diversity, physiology, behavior, mobility, and other natural characteristics combined to make wildfowl widely available, particularly vulnerable to hunters, and easily marketed by dealers. In short, birds' natures shaped the nature of the market.

Sheer numbers and varieties of birds were perhaps the most obvious biological factors facilitating wildfowl commercialization in the late nineteenth century. Long before the market hunting era, the richness of the New World's avian environment drew the fascination of observers from Columbus to Audubon and, in the process, produced a distinctive North American ornithology.[7] Early European explorers, describing the immense flocks of birds that "obscured the heavens" in North America, concluded that "in all the world the like abundance is not to be found."[8] The "wild Pidgeons" in seventeenth-century Virginia migrated in flocks so large that one "could neither discern beginning or ending, the length nor breadth of the Millions of Millions."[9] English naturalist John Lawson identified fifty-six types of land birds and fifty-three kinds of waterfowl in his 1709 book, *A New Voyage to Carolina*, claiming these were birds "more beautiful than in Europe."[10] A few decades later, Mark Catesby's two-volume, lavishly illustrated natural history of the Carolinas, Florida, and the Bahamas listed one hundred bird varieties that Catesby believed encompassed the extent of American avifauna.[11] The publication of Alexander Wilson's multivolume *American Ornithology* in the early nineteenth century, which cataloged over 250 birds including 39 new species, was later eclipsed by John James Audubon's masterful *Birds of America*, issued in serial form between 1827 and 1838, which identified nearly 500 avian species.[12]

Historical assertions about the bountiful and diverse avian environment had some basis in fact. Modern comprehensive listings of North American birds identify more than one thousand native species. Adding in migratory avian forms, the numbers rise to more than two thousand. This total, however, is dynamic, changing as new species are added or deleted, as natural or introduced arrivals appear, as extinctions occur, and as new findings are discovered in obscure records.[13] Particularly troublesome to categorize are those bird occurrences described as accidental or species identified as vagrants or visitors. Total bird populations are also difficult to pin down. Twentieth-century estimates have ranged from 2.6 billion to 10 billion breeding birds in the United States. With migratory species added, the total rises to nearly 20 billion birds.[14] Seeking historic population numbers produces even greater uncertainty, but the accumulated evidence suggests staggering sums.

In the market hunting era, wildfowl were ubiquitous, despite noticeable localized declines in certain populations and some early extinctions of species such as the great auk and the Labrador duck.[15] Major waterways, even those located near expanding cities, attracted large numbers of migratory wildfowl in the mid-nineteenth century. In Washington, DC, innumerable soras and reedbirds (bobolinks) frequented the surrounding tidal wild rice marshes of the Potomac and Anacostia rivers in early fall, and different varieties of waterfowl arrived later in the season.[16] Wild ducks were so commonplace on the city's rivers "that people used to throw stones at them," though this method was not particularly useful for bagging them.[17] At midcentury, expansive marshes still bordered Chicago, offering prime habitat for curlews, snipe, and other wading birds in fall and spring. Snipe were so plentiful around the city that "you could not walk a rod without flushing one of the long-bills."[18] As one chronicler reported, Chicagoans of the 1870s "could walk a mile from the corner of Dearborn and Washington Streets and shoot prairie chickens and quail," and early settlers were able to shoot "wild ducks on a small pond on the site of the present City Hall."[19] Going a bit farther afield, a hunter might easily bring down "partridge, grouse, prairie chickens, every variety of duck, wood duck, plover and widgeon" in a single day's outing. Passenger pigeons in immense flocks continued to visit midwestern urban centers in great numbers until the 1870s.[20] This avian diversity was directly related to the great variety of environments available. Imposed on North America's physiographic features were astoundingly rich biomes and ecotones that allowed for avian species to be variously distributed across the continent.[21] Wherever one stood in North America, one was sure to find all sorts of birds.

Patterns of avian mobility added to the array of wildfowl available across the continent at any given time and provided market hunters with a rotating smorgasbord of prey. Every bird species exhibits some type of migratory behavior, with at least a portion of the population moving away from breeding territories to forage, for postbreeding dispersal, or in response to environmental changes.[22] Each migrant species possesses its own patterns of movement. Best known are those long-distance seasonal travels between breeding and wintering locations, even though only a minority of bird species worldwide undertakes this kind of extensive travel.[23] Several species of northern breeding shorebirds and upland birds, for example, migrate in circular paths, gathering in the fall along northeastern Atlantic shorelines preparatory to their nonstop flights over to wintering areas in the Caribbean and South America. Some shorebirds that breed in arctic regions follow elliptical migration routes, traveling in a clockwise motion between their far northern breeding grounds in spring and winter homes in South America in the fall.[24] Broad latitudinal highways where multiple individual migration routes converge are called flyways, a concept first conceived in the 1930s by Frederick C. Lincoln of the US Fish and Wildlife Service after extensive waterfowl banding studies allowed researchers to map specific migration routes between breeding and winter grounds onto broad geographic zones of avian movement.[25] The North American waterfowl flyways—Pacific, Central, Mississippi, and Atlantic—funnel birds generally from north to south, serving as migration channels for not only ducks and geese but also many shorebirds and land species.[26] Long-distance migrants benefit from increased space for nesting and enhanced food availability in their summer habitats.[27] Hence, the timing of birds' migratory travels is closely tied to their preferred food sources; insectivorous species tend to migrate earlier in the fall than do seed eaters, for example.[28]

Other, shorter-distance migratory behaviors keep most wildfowl in motion at least part of the time. Some birds find seasonal habitats through altitudinal migration, that is, simply by moving up or down a mountainside.[29] Climate and weather also influence avian movements, prompting some species, such as white-winged doves in the Rio Grande Valley, to vacate breeding areas during heavy summer rainstorms or hurricanes. Adverse weather along the coasts tends to channel migrating birds down shorelines and cluster them on peninsulas and points.[30] Changes in social interactions or food sources can also cause variable migrations away from breeding territories. Cedar waxwings and blue jays are just two types of winged wanderers or "nomads" that travel irregularly, moving on when the food runs out. Some species exhibit partial migration behaviors, with certain individuals departing from their summer resorts in the fall while others remain in place. Even so-called resident or sedentary wildfowl display some form of localized daily or seasonal mobility, either singly or in flocks.[31]

All these configurations of avian distribution and movement shaped the geography and timing of market hunting. Not surprisingly, commonly hunted birds varied by region, and market hunters specialized in the types of game that were locally abundant. Around Washington, DC, and Baltimore, sora rail shooting was perhaps the most celebrated. During fall migrations in the nation's capital, wrote naturalists Elliott Coues and D. Webster Prentiss in the 1880s, a single gunner could "secure from twelve to twenty dozen Carolina [sora] rail and as many reedbirds." Also available nearby were numerous cedar waxwings, pileated woodpeckers, great horned owls, and marsh hawks, all of which made regular appearances in the city's markets and restaurants.[32] Baltimore-area commercial hunters could "make a comfortable living" in the Chesapeake's "far-famed ducking-ground."[33] In Massachusetts, locally common yellowlegs were harvested in large amounts on Cape Cod to be sent to Boston game markets.[34] Market hunters shot plentiful long-tailed ducks on the Niagara River in the 1880s, while they lasted.[35] In the Midwest, commercial hunters exploited still plentiful prairie chickens, while waterfowl were major targets for professional shooters in the South.[36] Plume hunters were prominent in the Gulf states, where they harvested breeding egrets, gulls, and grebes.[37] San Francisco–area market hunters plied the region's estuaries, bays, and the vast wetlands of the Sacramento–San Joaquin river delta for abundant ducks and geese as well as the commercially popular California clapper rail (Ridgway's rail).[38] Gunners took down band-tailed pigeons along the Pacific coast range, and plume hunters and egg collectors invaded Pacific island seabird rookeries.[39]

Although variable avian ranges and species distributions narrowed the availability of prey within a particular region, seasonal migrations temporarily broadened the array of birds to be had. Market hunters were keenly aware of the annual ebb and flow of migratory populations that set the timing and locations of wildfowl harvests. Fall and spring migrations brought hordes of waterfowl, shorebirds, cranes, doves, pigeons, woodcock, and snipe through the major flyways and tributary migration routes crossing the continent. As flocks traveled, they became targets for new rounds of market hunters, especially at traditional stopover locations, and dealers had a near year-round variety of "seasonal food" to sell.[40] In late-nineteenth-century Washington, DC, game markets offered woodcock, snipe, and swans in January; geese, ducks, and pigeons in February and March; plovers in May, June, July, and August; larks, woodcock, reedbirds, and rails in September and October; and woodcock, snipe, and wild ducks in November and December.[41] Other cities had similar patterns of wildfowl availability, which revealed the extent to which the market's seasonal nature was a consequence of biology rather than economics.

Erratic migrants such as passenger pigeons presented professional hunters with a different kind of economic timing. Seemingly random seasonal and daily movements—the birds' biological logic—governed the ways they could be harvested and marketed. Massive pigeon nesting colonies shifted several hundred miles from one year to the next, prompting amateur and professional pigeoneers to descend upon the communal roosts. Although some expert pigeon netters claimed they could "tell with considerable accuracy" where a nesting was likely to occur, others argued that "even the pigeons themselves could not tell you now, where they will spend next winter."[42] When the birds did appear, local and regional markets felt their impact. Simply put, erratic migrations produced erratic markets. The great nestings in Shelby, Michigan, in 1874 and 1876, for example, revived the local economy and overwhelmed markets as far away as Chicago, where the sudden glut dropped prices to fifty cents a barrel.[43] Within a week of the start of another major nesting—this one in Wisconsin—New York market prices plummeted more than five dollars per dozen, slashing the profits pigeon hunters and game dealers expected.[44]

Seasonal and daily avian movements, even among more-predictable migrant species, were as changeable as the weather, quite literally. The "business of shooting for the market was, of course, seasonable," wrote J. Kemp Bartlett of his early market hunting days, but "each season varied with the weather and the flight of ducks."[45] Weather conditions influenced the timing of avian arrivals and departures, the routes taken, the use of stopover points, and daily foraging movements. Experienced waterfowl shooters who knew how to "read the weather" generally agreed that windy days with little rain or sleet were best for ducks, a notion supported by current research in avian ecology.[46]

In the winter, ice hole shooting could be used to kill birds congregating at open water holes on frozen lakes and ponds.[47] "Like an omnibus," wrote one 1897 contributor to *Forest and Stream*, an ice hole "always has room for one more." It was like "a mine which [was] exhaustless for that day."[48] The variability of ocean tides affected some species' availability to hunters. During extremely high tides, sora and clapper rails, for example, depend on access to elevated marsh grass environments for refuge where they are far more vulnerable to hunters.[49] In the nineteenth century, newspapers and magazines published special "rail tides" timetables to draw professional and amateur gunners alike to these events.[50] "The more tide the most Rail, always," averred Charles Hallock of *Forest and Stream*.[51]

Birds' vulnerability to hunters also varied by time of day. Evening and night hunting as birds traveled to and from feeding grounds proved extraordinarily deadly for waterfowl, particularly for usually wary species. At twilight, flying birds were silhouetted against the skies while hunters on the ground or water were concealed in darkness.[52] Resting waterfowl were prime targets for night-hunting commercial shooters who used Big Guns or batteries of guns to rake rafted ducks or geese with a hail of ammunition. Lamps, lanterns, or torches could be used to mesmerize birds at night, freezing them in place to allow the hunter to approach within netting or shooting range. "Shining," "fire-lighting," or "fire-hunting" methods were especially effective for taking waterfowl from punt or sneak boats, for shooting woodcock ashore, and for capturing passenger pigeons from their roosts.[53]

Environmental patterns such as these made a great deal of difference to the birds and, consequently, to the market in them. The severe 1855–56 winter on Chesapeake Bay, for example, pushed ducks farther south than usual, creating a shortage in the local game markets and driving up the prices of "all descriptions of wildfowl."[54] An unseasonably warm spring in 1886 caused "a scarcity of ducks" for Chicago dealers, even though the market gunners were afield in force.[55] Warm weather presented additional challenges for hunters and dealers struggling to keep their stock from deteriorating in higher temperatures before it could be sold. An excess of birds in poor condition worked to depress prices for all wild game in the markets.[56]

Wildfowl behaviors shaped the market as well, since interspecific vulnerability to hunting varies considerably. Differing habits of flight influenced birds' susceptibility to the killing methods employed by hunters and the profits to be made from them. Sporting and professional wildfowlers alike knew that wildfowl were attracted to land-water physiographic configurations such as points and passes—areas that concentrated birds and offered especially productive shooting sites—many of which became legendary in the annals of waterfowling.[57] Open water–dwelling sea ducks such as scoters and eiders that typically flew well offshore could be taken with the cooperative method of "line boating." Hunters in anchored fishing boats arranged in a perpendicular line extending from the mainland could easily shoot passing ducks that had little choice but to fly between them.[58] Waterfowl's propensity for taking off into the wind allowed market hunters to use wind direction to their advantage, sailing upwind of a waterfowl concentration and then turning abruptly downwind toward it. Gunners could often come within range before the birds—especially heavy waterfowl and diving duck species that needed long distances for take-off—became airborne and swung aside.[59]

Gregariousness, another behavioral factor, likewise mattered when it came to which species could be commercially exploited most easily. For market hunters, flocking wildfowl like mallards, pintails, and many geese varieties were far more profitable targets because they could be killed quickly in large numbers. Conversely, solitary whooping cranes, spotted sandpipers, swans, and goldeneye ducks, along with wary and swift-flying species such as the black-bellied plover and Wilson's snipe, were more difficult to hunt in economically efficient ways.[60] Gregarious behaviors also determined the effectiveness of calls and decoys as hunting tools.[61] Pigeon netters deployed "fliers," live birds tied by a cord and thrown into the air, or "stool pigeons," tethered to a platform to attract their prey.[62] Waterfowl hunters staked out cripples and used the wounded birds' calls of distress to lure their companions within shooting range. Dead birds could be propped into lifelike postures to attract others of their kind. Many commercial gunners used artificial calls, homemade and manufactured, to imitate natural avian vocalizations with great felicity.[63] The efficacy of such deceptions, though, varied by species. Sporting writer William Hazelton claimed that bluebills and ringbills were the easiest to decoy, while pintail and canvasback ducks had the reputations of being more difficult to fool than mallards.[64] Among the ducks, the redhead was regarded as being especially vulnerable to market shooting because of its propensity to gather in large, dense flocks and also because of its attraction to decoys.[65]

Another behavioral variable that played a role in the wildfowl market was the extent to which different species were able to adapt to human activities, including hunting. Many birds flee human contact; others seem reasonably tolerant. Expansion of grain farming in the nineteenth century proved to be a boon for some species of waterfowl, most notably the mallards, pintails, and Canada geese that were drawn to farmers' fields. These preferences were well known by experienced hunters who baited areas with corn, wheat, and barley to lure waterfowl. Orin Sabin, an old market hunter, recalled that professional gunners paid farmers to harrow the margins of prairie sloughs so that they could lace the mud flats with grain to draw plovers.[66] Spring-migrating plovers also gathered on burned-over prairie lands to feed on insects, and a few enterprising market hunters took to igniting grass fires near shallow sloughs to further their success.[67] Pigeons and doves could be convinced to land on ground strewn with salt.[68] The human invention of wildfowl refuges at the turn of the century paradoxically encouraged commercial exploitation by attracting substantial numbers of wildfowl and equally substantial numbers of market hunters who waited patiently just outside the sanctuaries' borders for unsuspecting birds.[69] Anecdotal evidence suggests that the wariness of migratory birds sought for commercial and recreational purposes increased in proportion with hunting pressure, a point sportsman-writer Gasper Howland affirmed in 1893. "All kinds of wild fowl," he argued, were "more wary than ever before" because of the incessant shooting by market hunters and sportsmen. As a result, birds experienced in the ways of hunters were "more difficult to circumvent and capture."[70] Twenty-first-century banding studies of game birds tend to bear out these historical observations, showing that older birds are less vulnerable to capture than immature individuals. Although some of the difference can be attributed to lesser physical development, including power of flight, the limited exposure of younger birds to humans makes them less cautious than adults who have survived earlier confrontations with hunters. Vulnerability to hunting, then, is likely higher for younger individuals—a biological factor from which professional shooters of the past profited.[71]

Biology also governed market hunters' choices of prey. Feather hunters sought herons, egrets, and other plumage species in their breeding colonies at the times when their "nuptial" plumage was most colorful. Breeding behaviors that attached nesting birds strongly to their eggs and fledglings left them exposed to capture. Colony-nesting marine birds such as the gannet and tidal water birds such as the clapper rail were particularly vulnerable to egg hunters. Clapper rails laid large clutches of eggs and commonly renested if their first nests were destroyed, characteristics that made them attractive for commercial exploitation. Conversely, nesting colonies perched on inaccessible cliffs or on rugged offshore islands were more likely to escape severe commercial hunting pressure.[72]

Not surprisingly, the growth of market hunting affected bird biology adversely, though its negative effects varied by species and by the type of exploitation each one faced. Eggers caused perhaps the most direct disruption to bird reproduction by removing potential progeny from the population. Disturbances caused by egg hunters caused nest abandonment, destruction of young, and other negative avian health effects. In some cases, colonies were virtually destroyed in but a few days of egg collecting.[73] Plume hunters disrupted birds' breeding behaviors and success, since the peak season for plumes coincided with the breeding, egg laying, incubation, and fledging periods. Plumers routinely killed nesting birds, destroyed nests and eggs, and created immense commotion that upset the reproductive activities of the entire colony. Since plumage colors were generally sex specific, hunters selectively shot males for their brightly colored feathers, destroying in the process the long-lasting pair bonds that sometimes formed between mates. The loss of a mate could thus potentially wipe out years of future offspring.[74]

Game bird populations subjected to commercial harvests were similarly harmed by widespread spring shooting practices that killed large numbers of young birds, upset breeding and migration patterns, or eradicated local avian populations. Intensive hunting during migrations interrupted birds' customary flight paths and disturbed them at stopover points.[75] Market gunners' concerns for efficiency led to wasteful hunting because professionals considered their time too economically valuable to spend it looking for downed and crippled birds.[76] Intense hunting pressure in bird-rich environments diminished or diverted avian populations over time, effectively circumscribing birds' former natural ranges and shifting the geography of commercial wildfowling to new, untapped regions. Favored species disappeared around cities first, especially those with active game markets and high concentrations of sport hunters. Birds inhabiting regions within easy reach by rail or water went next. When Merritt sent his son Clarence to Nebraska in the 1880s, it was to find species that had been hunted to near extinction in Illinois in the years before.[77] At the end of the century, blue-winged teal were almost gone from New England and the mid-Atlantic states, places where they had once been "very common indeed." Once the map of wildfowl haunts had been redrawn, professional

Plume hunter Leigh M. Pearsall poses with a black-crowned night heron on Florida's Santa Fe Lake early in the twentieth century. Commercial plumers in southern Gulf states exploited breeding colonies, harvesting egrets, herons, ibises, and other species when clad in their "nuptial plumage." Plume birds were often shot while sitting on their nests. Courtesy of the State Archives of Florida, Florida Memory (Creative Commons PDM 1.0)

and amateur hunters had to travel to "the inaccessible sloughs, grown with wild rice," in the Midwest or the lagoons of the lower Mississippi Valley to see any flights of wild birds.[78]

For avian entrepreneurs, profiting from birds was as much a biological endeavor as it was an economic one. Birds' natures made their commercial exploitation possible. Abundant, diverse, and broadly distributed wildfowl populations meant that nearly every locale in North America was home to some variety of winged species in numbers that could be harvested for commerce—at least for a time. Breeding, nesting, and feeding behaviors concentrated avian resources in ways that market hunters could easily and efficiently harvest. Above all, biological mobility, particularly seasonal migration habits, brought millions of birds to hunters across the country almost year-round. Provided with an available and vulnerable array of prey, market hunters and game dealers like Merritt were able to transform wildfowling from a short-term diversion into a large-scale, modern capitalist enterprise. Yet even in this industrialized environment, outwardly governed by a capitalist logic, the production of wild birds for the consumer market could not escape the biological logic of birds' lives.

The Avian Economy

Of course, production was only half of the economic equation. Consumption played an equal and complementary role in creating the modern market in birds. Merritt's observation that "fashion . . . set the price of all game birds" was an explicit acknowledgment of the new patterns of consumption shaping the wildfowl market in the late nineteenth century.[79] As avian entrepreneurs harnessed novel technologies to circumvent the biophysical constraints that had previously limited the market's reach, they expanded the distance between wildfowl producers in the field and urban consumers. Frozen game and dealers' storage facilities worked to stretch the biological timing of the seasons so that buyers could forget birds' seasonal mobility. Consumers sought out new avian products, and sellers obliged by advertising and delivering the latest fads in wild merchandise.

Wildfowl migrated from the countryside to cities in these years according to a capitalist logic. Birds traveled along routes mapped by a complex economic network that extracted birds from the biophysical contexts in which they were embedded; alienated them from the hunters, processors, shippers, and sellers who produced them; and stamped them with a price. The late-nineteenth-century market in birds cut avian ecological diversity into neat packages that could be easily bought and sold. Converted into different "species" of economic commodities, wildfowl reappeared to American consumers as meat, eggs, feathers, decoys, targets, or specimens in ever-increasing quantities. In short, the avian economy fundamentally changed the ways in which many Americans encountered wild birds.

Commodified birds were most widely distributed as food for consumers' tables. Eggs were seasonal features of late-nineteenth-century markets, though their commercial sale was geographically limited and never approached the extensiveness of the meat bird market. Fragile and perishable, wild bird eggs sold close to where they were harvested during periods of intensive egg laying. The egg trade in the Farallon Islands, thirty miles offshore of San Francisco, served the gold rush town's hungry populace beginning in the 1850s when domestic food supplies were scarce and expensive.[80] Consumers particularly coveted the large and uniquely colored eggs from the common murre, a penguin-like seabird, during "their peculiar season." Used commonly for baking, murre eggs were "considered to be very nutritious," and some Californians deemed them "a great delicacy." One San Francisco resident averred that fried murre eggs were "a delicious dish," though they needed a little whiskey to "take away the strong taste" that was reminiscent of fish if the eggs were even slightly stale.[81] Others found the sight of the cooked eggs unappetizing, with transparent yolks "of a fiery orange colour—almost red."[82]

By far the wildfowl commodity most familiar to consumers, however, was meat. Long before the height of the market hunting era, merchants throughout the country sold a wide variety of birds including the now-extinct Labrador duck.[83] At first, biological factors largely governed supplies of meat birds, and market hunters worked mainly during the colder parts of the year when migrants were plentiful and meat preservation was simpler. There was a "proper season" for the consumption and marketing of each species, as the *New York Times* noted in 1878. Geese, for example, could be eaten from October until the end of January only. After that, their flesh became "stringy" and they were "not generally so appetizing."[84] Birds eaten "out of season" were considered "unfit, unwholesome, and even poisonous" by some.[85] Consumer demand was conditioned by the seasonality of the supply in those earliest years, with purchasing interest highest during the "social season" and winter holidays from October through March, when local wildfowl was available and at its best. Summers were slack times. As Merritt's midcentury experience in Illinois demonstrated, before the modern game market developed, "nobody seemed to think there was any market for any game anywhere, except in cold weather."[86]

Although the market in birds retained much of its seasonal character in the later nineteenth and early twentieth centuries, refrigeration technologies and faster transportation vastly expanded game availability and extended the seasons of consumption, in the process transforming the consumer market in

"Scottie 'the Egger' South Farallon Islands, July '96." Eggers made a practice of destroying all the eggs on their first visit to ensure the freshness of the eggs collected on subsequent trips. Arthur L. Bolton © California Academy of Sciences, San Francisco

avian meat products. Mechanical considerations—the efficiency of storage facilities and communication, access to transportation—combined with biological dynamics to modernize and industrialize the avian economy.[87] Railroads brought wildfowl from near and far into urban game markets that, in turn, offered customers an expansive array of birds for their tables in numbers that would have been unimaginable but a few decades earlier. Passenger pigeons were arguably the most numerous of commodified species to migrate by rail to city markets in the mid-nineteenth century. At the end of April 1851, seventy-five tons of pigeons were shipped to New York City just from the western counties of Steuben and Allegany.[88] Nesting sites in the Michigan counties of Oceana, Newaygo, and Grand Traverse sent one thousand tons of squabs and nearly two and half million adult pigeons to markets in 1874.[89] During the great 1878 Petoskey, Michigan, pigeon roost, 128 barrels of dead birds and 108 crates of live birds shipped from one railroad station in a single day. These numbers did not include the pigeons sent by steamers from ports on the Great Lakes.[90] This kind of modern intensive marketing was no different, argued Chicago game dealer E. T. Martin in 1879, than the agricultural trade, for passenger pigeons were "as much an article of commerce as wheat, corn, hogs, beeves, or sheep" and it was "no more cruel to kill them for the market by the thousand" than it was "to countenance the killing at the stock yards."[91]

The advent of refrigerated transport and commercial freezers in the 1870s and 1880s only added to the range of available wildfowl to be consumed. Refrigeration preserved the quality of meat birds into the warmer months, though, to some, frozen game held little appeal at first. Much like the domestic livestock trade, frozen wild meats initially generated negative reactions from consumers accustomed to buying freshly killed animals. To the fastidious, the freezing of game birds destroyed their special allure or, even worse, rendered the meat toxic.[92] The "thousands of birds served in Eastern cities," wrote ornithologist Daniel Giraud Elliot in 1897, had "been frozen, tossed about perhaps for weeks like lumps of ice and then thawed out before being cooked." It was little wonder, then, that the "most toothsome morsel in the world, after such treatment, could not be expected to have much more flavor than a dried chip."[93] Yet the novelty of frozen wildfowl that could be consumed in any season soon overcame customers' reluctance to eat meat that was months or even years old. Freezing made "it possible for those people who can afford to pay for the luxury to have their table supplied with anything in the way of game at any time of the year."[94] Restaurants sold frozen wildfowl to people unfamiliar with "its true flavor when eaten in the field." Consumers who ate wildfowl because it was "fashionable to do so" began to expect their favorite game birds to be more readily obtainable in any season, albeit for a price.[95]

Prices for game species quickly became a topic for publication in newspapers, magazines, and government reports.[96] Generally, the unit of sale was based on the species' size. Large birds such as geese and cranes were sold individually, medium-sized birds like most ducks were priced by the pair or "brace," and smaller shorebirds, rails, doves, and others were priced by the dozen. In a few instances, very small birds including reedbirds (bobolinks) or redwing blackbirds were sold by the bushel or bag. Market values for both resident and migratory birds directly competed with the prices for domestic fowl, causing farm-raised poultry sales to slacken during periods of wild-fowl abundance.[97] Prices also varied by species, as Charles Hallock's regular listings of New York game market prices in *Forest and Stream* showed. From 1874 to 1876, Hallock's systematic survey showed that the highest-valued species per pound, adjusted by live weight, was the woodcock at $1.62, fol-lowed by Wilson's snipe at $1.14, the diminutive bobolink or reedbird at $0.90, the bobwhite at $0.76, and the ruffed grouse at $0.47. By weight, these upland game birds cost more than waterfowl such as the canvasback at $0.44 per pound, the pintail at $0.23, or the greater scaup at $0.15. Passenger pigeons sold at price comparable to midrange waterfowl, while prairie chickens brought in from the Midwest were valued a bit higher.[98]

Consumer preferences increasingly drove prices, which varied consider-ably by region.[99] Chicago and New York markets saw higher prices for game than those in New Orleans and Richmond, Virginia. Boston's game birds ran even higher still. San Francisco's buyers preferred game brought in from Sac-ramento, which had "none of that fishy taste," over local birds taken along the coast.[100] Prices in other cities differed considerably, but in general, reported the US Biological Survey early in the twentieth century, "the less important the market, the lower the prices."[101] Regional preferences also governed price dif-ferentials between species. Along the Columbia River in Oregon, for example, most "trash ducks" were hardly marketable, sportsman S. H. Greene told *Forest and Stream* in 1891, while swans shot on the river would hardly "bring more than fifty or seventy-five cents."[102] Canvasback and redhead ducks were always sought in the eastern states, while scaups were less desirable and mergansers were scarcely salable. Chicago's markets could sell canvasbacks for twice the price of other ducks despite the fact that they were "in no wise superior" to pintails, according to ornithologist Robert Ridgeway. Farther west, the can-vasback's appeal declined. Although it was "highly prized by eastern epicures," in California it was "considered a very inferior bird for the table."[103] The same was true in Texas, where the *Galveston News* scoffed in 1897 at New York and Washington elites who paid as much for the "lordly canvasback" as a Texan would shell out for a yearling steer.[104]

Wildfowl abundance in late-nineteenth-century markets meant that some sort of meat bird was within the reach of nearly every household. Even the housekeeper of moderate means could afford to "vary her table" by cooking whatever wild birds were in season and relatively cheap. Most markets carried medium-sized game birds such as partridges and pigeons along with varieties of "other small birds like sand peeps [snipe], reed birds, etc." These smaller species were usually sold with the heads skinned, the eyes removed, "and the sockets filled with some kind of force-meat." They could be baked and then served on buttered toast.[105] New Orleans' markets offered American bittern and mud hens, which were favorites of "the poorer classes" for making gumbo.[106] In the 1880s, robins were "a staple article of food" in Georgia, and "robin pie delight[ed] the heart of the average citizen."[107] Ducks, geese, swans, and a variety of shorebirds from the Chesapeake, wrote sporting author John Mortimer Murphy in 1882, were "shipped by the tens of tons to the northern cities" where they sold at rates so low that even "the humblest of the working classes" could have wild game dinners "at a cost little beyond what they would pay for beef or mutton."[108]

As many desirable species grew scarce toward the end of the century, however, city dwellers of ordinary means found that the better avian commodities were often priced beyond their reach or not available at all. Evolving state-level game codes also constrained consumption by imposing legal seasons for wildfowl harvests and sales that reinforced birds' biological cycles. Urban game dealers advertised what was "in season"—an implicit recognition of both the biological and legal timing shaping the market.[109] Official hunting and commercial sale seasons altered the trajectories of market prices, lowering them in the fall when a surfeit of birds hit the market legitimately, and raising them in the late winter and spring as the game supply dwindled and the legal season ended. Illicit birds could be had during closed seasons, but not cheaply. Over time, as bird populations generally declined and as game laws were more rigidly enforced, prices for wildfowl increased.[110] As the twentieth century approached, many agreed that "for the vast and overwhelming multitude of people of the continent," wild game was "no longer in any sense an essential factor" for ordinary subsistence. Instead, wildfowl had become a luxury item for city dwellers, an extravagance "sold for prices that made it such."[111]

Across the late nineteenth century, many urban dwellers came to count the consumption of wild game as an elevated culinary experience, agreeing with Elisha Lewis's assessment that only the most "ignorant, low-minded fellow" would prefer "a barn-yard duck to a Chesapeake canvass-back."[112] Few of the truly epicurean elite would have quarreled with Merritt's prejudiced claim that for a restaurant to be truly great, it had to serve wild game. Game dinners and banquets became regular features at many prominent big-city hotels, restaurants, and private clubs in this era. In Chicago, the Tremont and Grand Pacific hotels were well known for their annual game dinners, which were instituted by proprietor John B. Drake in 1855 and continued until 1893. These banquets featured imaginative avian delicacies including "Wild Pigeon Compote," "Fillet of Grouse with Truffles," and "Pyramid of Wild Goose Liver in Jelly." An 1886 menu, titled *Procession of Game*, listed twenty-six species of roasted game and seventeen species of broiled game. The grand meals often closed with a cold, ornamental dish such as "Stuffed coon, au naturel."[113]

In Washington, DC, Willard's Hotel specialized in wild game in its early years, focusing on what was plentiful from the nearby Chesapeake Bay and the Potomac River. Willard's offered wild game at banquets and on its daily menu, from which one could order dishes such as "wild pigeon" and "two robins on toast."[114] Denver, San Francisco, Minneapolis, and many other cities boasted hotel restaurant fare with plenty of wildfowl delicacies such as "Becassines au Cresson" [snipe on watercress], "Broiled Plover on Toast," and the mysterious "game pie."[115] Metropolitan Baltimore, according to one nineteenth-century writer, earned its "Land of the Epicure" sobriquet from its access to "the two very choicest delicacies known to the world . . . the canvasback duck and diamond-back terrapin."[116] New York's restaurants were particularly well endowed with wildfowl, a consequence of the city's extensive market. At mid-century, the Astor House menu featured a veritable gastronomical aviary that included "black duck, lake duck, meadow hen, short neck snipe, doe witches, cedar birds, grouse, plover, rail birds, mallard duck, robin snipe, [and] surf snipe."[117] The bill of fare at the Windsor Hotel in New York likewise featured wild birds ranging from wild turkeys and ducks to plovers and quails.[118] Christmas dinner at the Windsor in 1890 included canvasback ducks served with "old fashioned currant jelly" along with different varieties of imported birds. "The game preserves of distant lands," noted the *New York Times*, "had been ransacked for delicacies of the feast" that year.[119]

Chefs at Delmonico's, New York's premier restaurant during the market hunting era, were perhaps the most important contributors to the wild game mystique spreading across the fine-dining landscape in the late nineteenth century. Wild birds done up in the French tradition were always a major feature of Delmonico's menus. Chef Charles Ranhofer's encyclopedic cookbook *The Epicurean* (1894) contained an entire chapter on the preparation of game and an index listing what game was available in each season. Recipes such as "Blackbirds À La Degrange," "Canvasback Ducks Roasted Garnished With Hominy" (which required Havre de Grace, Maryland, ducks), "Wild Pigeon or Squabs Poupeton, Ancient Style," and "California Quails À La Monterey" revealed both a devotion to regional foods and the popular mixture of continental cooking with characteristically American ingredients that was the basis for the "Franco-American" culinary style. Dishes like "Plovers À La Victor Hugo," "Partridges À La Jules Verne," and "Salamis of Teal Duck À La Harrison" demonstrated Ranhofer's penchant for naming dishes after famous personages, especially those who had patronized Delmonico's. Woodcock appeared to be a favorite of Ranhofer's, since he created a dozen different recipes for the species including "Breasts of Woodcocks À La Houston."[120]

If fashion set the price of all game birds, it absolutely dictated the consumption of plume species. Although feathers and down had long been harvested for utilitarian purposes such as quill pens, blankets, comforters, and pillows, the extensive use of plumes and whole bird skins for ornamentation was a turn-of-the-century peculiarity. Trends in women's clothing and household decoration elevated avian feathers and skins to the heights of modern style during the Gilded Age. Plume dealers avidly sought extremely vulnerable wading and water birds, most notably the snowy egrets, pelicans, herons, gulls, terns, and grebes that nested in dense colonies mainly in southern swamps and marshes. Various perching birds also found their way to market, including the colorful Carolina parakeet, the diminutive bobolink, and the even smaller hummingbird—birds whose heads could be set like "precious gems" in earrings and brooches.[121] Customers desired wild birds, often in extravagant nuptial plumage, for hats, muffs, collarettes, dress and blouse decoration, hairpieces, and capes as well as ornamental bric-a-brac for Victorian homes.[122]

This growing demand for wild plume birds energized the millinery trade most of all, as stylish American women donned the latest feather-topped designs from Paris, London, New York, and other centers of fashion.[123] In 1874, *Godey's Lady's Book* reported that consumers' demand for feathers, expertly dyed and designed by craftsmen, "can scarcely be supplied." Bonnets could be decorated with tiny, jewel-like hummingbirds, and birds of blue-gray, tawny brown, or scarlet could be posed to "perch saucily on hats with outspread wings" in order to give "a youthful stylish effect" to women's wear. Although

Godey's editors twinged a bit at what seemed to be a "relic of the savage state," women's increasing demand for dead birds was ultimately just one of those "strange vagaries" enforced by fashion's decree.[124] "The American woman, in nothing, is so whimsical as in her bonnet," *Good Housekeeping* proclaimed in 1888, prompting milliners to keep pace with consumers' desires. Those desires demanded "immense cases of small birds, wings, breasts, montures, and tufts of feathers of all kinds and descriptions." Even "the anathema of humane societies for the protection of birds," fashion advertisers asserted, could not diminish the popularity of feathered trimmings.[125]

The varieties of fashionable design available at the turn of the century were limited only by the plumassier's imagination: "He puts the Dove's head on the Tanager's body, or the wings of the Woodpecker or Trogon are mingled with the plumes of Egrets, or the heads of Snipes and Plovers." Yet women desired natural-looking birds most of all. Owls, pigeons, and gulls could not escape the hat trap when "big birds [were] all the rage." At other times "small birds alone suit the ladies' capricious tastes." The plumassier and milliner were driven to come up with "endless combinations . . . to supply the endless thirst for novelty in the female heart." Consequently, the market in plumes and customers' cravings for wildfowl fluctuated with the whims of fashion.[126]

Artfully arranged in their new habitats, wild birds became a familiar sight to city folk, with fashionable urbanites in their avian accoutrements mingling in great diversity on city streets and sidewalks. A veritable ornithological catalog could be compiled from sightings of ladies' hats—an exercise the National Audubon Society's Frank Chapman undertook in 1886. In a single day's observations in New York, Chapman spied over five hundred hats decorated with some twenty different identifiable species of birds. Over three quarters of the hats he recorded were decorated with some sort of bird feathers or plumes, most from unrecognizable species. A two-day observational census identified forty different species of birds on ladies' hats. Preferred species tended to be small colorful types such as cedar waxwings, flickers, common terns, Carolina parakeets, and bobwhite quail.[127] In many cases, women sported hats with multiple birds or bird parts from a variety species. Another "ornithological friend" counted the birds observed in one Madison Avenue horse car where eleven of the thirteen women in the car wore birds on their hats, including "(1) heads and wings of three European starlings; (2) an entire bird (species unknown), of foreign origin; (3) seven warblers representing four species; (4) a large tern; (5) the heads and wings of three shore-larks; (6) the wings of seven

Bird-hat-wearing women from Seattle, Washington, in the early 1900s. The two women pictured here are sporting hats made with three stuffed Carolina parakeets each. Now extinct, the Carolina parakeet was the only parrot species native to the eastern United States. Its colorful plumage made it particularly attractive to milliners and fashionable consumers. The last Carolina parakeets died in captivity in 1918 in the Cincinnati Zoo, the same zoo that housed the last surviving passenger pigeon. Image from Charles W. Sanders Collection (2014.026.123), courtesy of Renton History Museum, Renton, Washington

shore-larks; (7) one-half of a gallinule [a type of rail]; (8) a small tern; (9) a turtle-dove; (10) a vireo and a yellow breasted chat; and (11) ostrich-plumes."[128] New York's shop windows were, the *New York World* declared with only a hint of irony, "alive with birds" at the start of the twentieth century. Just how far might this "craze for novelty" go? Perhaps trend-setting women would even be willing "to sacrifice the household canary or pet parrot" on fashion's altar.[129]

Birds in far fewer numbers but in even greater diversity were sought by yet another species of modern consumer: the specimen collector. Although natural history collecting had been a widely practiced avocation since the eighteenth century, the modern market transformed the practice into a commercialized enterprise in the latter half of the nineteenth century.[130] Ornithological collectors ranged from scientists to natural history buffs, many of whom took to heart Elliott Coues's 1892 dictum "Birds may be sought anywhere at any time; they should be sought everywhere, at all times."[131] The market in birds opened new fields for collectors to prowl in search of rare and interesting birds to buy for public or personal collections.[132] Audubon's 1830 purchase of 350 passenger pigeons in New York to distribute among English noblemen was an early but by no means unique example.[133] William Brewster, the cofounder of the American Ornithologists' Union, was the proud owner of one of the biggest private collections in North America, with many specimens acquired from markets in and around Boston.[134] *Forest and Stream* editor George Bird Grinnell frequented the markets in New York and reported on the variety of avian species he found, like the rare Labrador ducks that ornithologists snapped up "as soon as a specimen was exposed for sale."[135] In Chicago, the future chief of the Bureau of Biological Survey, Edward Nelson, routinely engaged in a kind of macabre bird watching at the South Water Street game markets, recording his sightings of swans, widgeons, and scoters available in dealers' stalls.[136] Buffalo, New York's newly appointed park commissioner in 1895 was reported to be "an enthusiastic ornithologist" with a collection of "200 sets of eggs" of which "80 came from Europe or remote places in this country." A prize of his collection was "a set of woodcock's eggs that date[d] back to 1864."[137] For professional and amateur collectors alike, the market offered access to a widening inventory of avian specimens they could never hope to amass on their own.

Commodified birds, skins, eggs, nests, and bird mounts, supplied by market hunters or natural history companies, found ready purchasers in this growing community of specimen collectors and taxidermists starting in the mid-nineteenth century. Avian species that were rare, exotic, or skidding toward extinction were particularly coveted. Labrador ducks, for example, appeared in increasing numbers in New York markets between 1840 and 1860 despite declining populations—the consequence of "far greater numbers of scientific collectors" interested in acquiring skins of this species.[138] A good supply of eagles "killed on Long Island and brought to market for setting up as specimens" were available in New York in the 1850s, according to game dealer Amos Robbins.[139] Furthering the trade were organized natural history clubs, journals, and catalogs for hobbyists in which avian entrepreneurs could turn a profit from their finds. In the early 1880s, one collector serendipitously purchased seven hundred Long Island clapper rail eggs in New York markets, priced "at what they are worth for cooking purposes," and resold them to individual collectors. His enormous stock of eggs ultimately depressed the value of rail egg specimens selling in catalogs for two years.[140] Virtually all North American species were taken for specimens, including many that previously had little commercial value. In general, no species was safe from collectors, though the rarest wild birds were the most desired, commanded the highest prices, and therefore were the most exploited.

Although the wildfowl market dealt mainly in dead commodities, not all avian economic migrations occurred postmortem. Live birds made up a slice of the market as well. Some consumers sought live wild birds as pets, and sportsmen bought live ducks and geese to use as wildfowl lures.[141] Advertisers in outdoor magazines offered live decoys, usually pinioned Canada geese or mallards, pitching them to recreational hunters "willing to do something practical" in order to get "more game."[142] By far the largest quantities of live wild birds consumed, though, were passenger pigeons used for recreational and competitive target shooting, a pastime that grew tremendously in the late nineteenth century.[143] Highly ritualized and almost exclusively male affairs, pigeon shoots involved elaborate rules, generated enormous wagers, and drew large numbers of spectators and gamblers.[144] Domestic pigeons were sometimes used for competitions, but gunners considered wild pigeons the more sporting targets. Typical matches allowed 25 to 100 birds per shooter, but larger shoots might dispatch 25,000 or more.[145] One three-day shooting tournament sponsored by the Peoria, Illinois, shooting club in 1876 killed 1,084 pigeons on the first day.[146] That same year, the New York Sportsmen's Association purchased 40,000 to 45,000 pigeons from Michigan for its tournaments. According to one estimate, competitors shot more than a quarter million passenger pigeons at tournaments in the peak-use year of 1878. The sport was widespread and quite popular in numerous cities, so much so that some game dealers, like Chicago's E. T. Martin, made a specialty out of supplying passenger pigeons to shooting clubs.[147] The expanded pigeon trade raised the sport to a level above mere butchery, according to an 1895 wing-shooting treatise, by allowing enthusiasts to "shoot at the best-flying pigeons we can procure."[148]

No matter the commodified form, birds were eminently salable. It was in their natures. For suppliers, birds' availability—their abundance, widespread distribution, vulnerability to various hunting methods, and seasonal mobility—made them attractive targets. Avian physiology added to birds' economic appeal. Their relatively small sizes in comparison to other game animals made them easier and cheaper to process, preserve, and transport efficiently. For consumers, the widening range of avian commodities available—from meat and eggs to feathers and specimens—made wildfowl desirable products. It was this combination of biological and economic factors that allowed for both the development of the modern wildfowl market and the evolution of modern wildfowl consumers.

The Price of Wildness

To be sure, game markets of earlier eras exploited some of the same bioeconomic conjunctures. In fledgling settlements, abundant wild creatures made convenient substitutes for domestic animal products that were often more costly and in short supply. Venison hams or wild snipe were valued in the same way as were pork and barnyard fowl.[149] American-produced deerskins made cost-effective replacements for European domestic cowhides in the eighteenth century. Even the highly commercialized bison trade of the mid-nineteenth century functioned as an alternative source of cheap leather. In those days, wildness was important only to the extent that wildlife was a plentiful and unowned material resource.[150]

The modern market in birds, however, fundamentally changed the value of wildness for Americans. In this newly industrialized capitalist environment, wildfowl worth was derived in no small part from the birds' intrinsic wildness. Although the quail-dining gourmand, the plume-wearing dame, and the pigeon-shooting partisan could be ignorant of the socio-natural relations that brought wildfowl to the market and the ecological consequences of those acts, they could not forget—nor did they want to—that the avian commodities they fancied had once been wild creatures. In that way, birds were never completely commodified. They remained recognizable as birds even as they were standardized and rendered exchangeable in the market. Remaking wildfowl into objects of consumer desire brought nature into the marketplace in a quite different way and assigned it a new kind of symbolic value. In short, the market in birds repackaged wildness for a modern American consumer culture.

Reminders of wildness always accompanied the consumption of birds. Specimen collectors wanted nothing more than to replicate the look of a wild bird, with "a correct and natural attitude" while museum taxidermists arranged wildfowl "in their natural haunts."[151] Women's hats featured birds and bird parts "placed in natural positions." Birds' nests complete with "tiny pearly eggs" nestled in the brims. Hummingbirds frolicked among strings of ribbon representing grass, and bluebirds were "placed low on the back of the bonnet as if flying down."[152] Game birds were commonly sold and sometimes served in their natural states, with heads and feathers intact. Snipe and woodcock, according to one authority, "must always be served with their heads on, as the long bills proclaim them, and are highly prized."[153] Game dinners featured perhaps the most elaborate advertisements of birds' wild origins. Cooked birds were sometimes situated in their native habitats and re-dressed by enclosing them in their former plumage. At such affairs, guests might find entrées such as "Blackbirds at Play," "Boned Quail in Plumage," or "Partridge in a Nest," perhaps overseen by red-winged starlings perched on a tree limb.[154] The Grand Pacific Hotel's 1881 game dinner in Chicago featured an array of ornamental groups of "animals in lifelike attitudes." At the entrance "was seated a large black bear" and a deer "crouching in the herbage," while a variety of "birds were twittering in the branches of a realistic tree—reed birds, grouse and snipe." All were "ready to eat, inside their skins," yet they appeared animated as if they had "been stuffed by skilled taxidermists."[155]

Avian wildness, marketed as meat, acquired great cultural value as quintessentially American food. Lacking the studied sophistication of European cuisine, American fare could nonetheless boast of a "bounteous variety" derived from the natural abundance that had always set the New World apart from the Old. A "good American dinner" was likely to "be a revelation to many" outside the United States.[156] Although Americans might seem to be provincial and "semi-barbarous epicures," wrote Charles Augustus Goodrich at midcentury, their esteem for grouse, canvasbacks, brants, plover, wild turkeys, and the like was quite warranted. "One mouthful" was "sufficient to prove that there [was] a difference between a partridge and a hen."[157] Wildfowl, ably prepared, was evidence of American cultural advancement in the modern age. European disdain for American cuisine could be countered with a bill of fare that included "Canvas-back ducks, the finest game in the world, and our turkeys, so far superior to those in Europe." Indeed, English wildfowl were "uneatable when compared with those of the United States."[158] Other samples of "the intricacies of American cooking" involved "wild duck, squab, grouse, quail, reed-bird, plover, [and] prairie-chicken."[159] Among the first meals that Mark Twain desired after returning from a sojourn abroad were "Roast wild turkey, Woodcock, Canvas-back duck from Baltimore, Prairie hens from Illinois, Missouri partridges broiled."[160]

Twain's geographically oriented menu reflected a growing popular admiration for regional food cultures that aligned with regional wildfowl distributions. Consumers prized dishes made from Illinois prairie chickens, East Coast rail, and California quail for their geographic distinctiveness. Along the gulf coast, the black-bellied tree duck was "considered quite a delicacy."[161] Early culinary innovators, like William Niblo, proprietor of New York's Bank Coffee House, served up "Bald Eagle shot on the Grouse plains of Long Island," wild turkeys "shot in the backwoods of Pennsylvania," hawks and owls from "Turtle Grove, Hoboken," and wild swans from Havre de Grace, Maryland.[162]

No species was more closely associated with a regional food preference than was the Chesapeake Bay canvasback. Its reputation as an excellent eating bird was long-standing. Early-nineteenth-century observers noted the "great quantities" of canvasbacks "shot with long guns on the Potomac" and served up in a variety of ways.[163] The key to the canvasback quality in the region was due, most asserted, to the "water celery" on which the bird fed, though ornithologist Robert Ridgeway argued that this theory for canvasback palatability was "wholly fallacious." Instead, he attributed the duck's regional reputation to "fashion and imagination, or perhaps a superior style of cooking and serving."[164] Nevertheless, many connoisseurs claimed they could distinguish a Chesapeake bird from those taken elsewhere.[165] Ohio senator George E. Pugh tested this conventional wisdom by serving up a mess of Lake Erie waterfowl "to his gentlemen friends as Baltimore ducks," who praised them as the finest of Maryland cookery. Even after Pugh confessed the deception, his friends remained unconvinced.[166]

The market prices of wild birds in the late nineteenth century were clear measures of their cultural currency in a modernizing world where the authentic competed with the artificial for consumers' attentions. Wildfowl's natural authenticity—its associations with the premodern, frontier existence of backwoodsmen—captivated affluent city dwellers largely divorced from everyday contacts with the natural world and preoccupied with concerns about enervating overcivilization.[167] Wildness, albeit purchased most often in a lifeless form, put urbanites in touch with the material reality of the nonhuman sphere. The seasonality of wildfowl available; the diversity of species appearing in game markets, on ladies hats, and in public and private collections; and the geographical specificity of avian commodities all contributed to connecting consumers to the environmental contexts from which their bird purchases originated.[168]

The varieties of commodified avian species and the differences in their qualities gave buyers a personal connection with wild diversity. Commodified in the marketplace, that diversity, however, offered opportunities for sellers' artifice. Dealers often disguised less desirable avian species as more palatable types when they could get away with it. Ruddy ducks could be passed off as teal, as they often were in Philadelphia markets, and unclear distinctions between canvasback and redhead ducks allowed early marketers to sell both at the same price.[169] Washington, DC, restaurants served up bobolinks and reedbirds, which were "luscious morsels when genuine," but it was not unusual for "a great many Blackbirds or English Sparrows [to be] devoured by accomplished gourmands, who nevertheless do not know the difference when the bill of fare is printed correctly, and the charges are sufficiently exorbitant."[170] Consumers thus had to be knowledgeable about avian species and their natures to avoid deception. Luckily, some birds naturally offered clues to help discerning buyers. The prized canvasback duck and the blue-wing teal, for example, changed colors from spring to fall, identifying the better-conditioned fall ducks from less desirable spring ones. Other species did not reveal their ages so obviously. With values increasing for canvasbacks in the late nineteenth century, suppliers took more pains to provide "true and real canvas-back ducks" to "lovers of good eating."[171] Restaurateurs, however, could do more to hide the true identities of their dishes with sauces, herbs, and seasonings. A redhead duck could more easily masquerade as a superior canvasback in this way, especially if its eating were accompanied by ample quantities of champagne.[172]

For city folk, wildness was an exotic commodity and, as wildlife populations receded, an increasingly rare one. Sellers capitalized on this scarcity to convert the trade into a predominately luxury market, and advertisers promoted wildness as a desirable commodity. A "famous epicure in Chicago," recounted a Kansas newspaper in 1886, was so addicted to prairie chicken eggs that he managed each year to obtain "three or four dozen of this delicacy," paying as much as sixty dollars a dozen for them (a believable claim, the reporter noted, because his informant was "not in Chicago politics").[173] Other types of wild bird eggs possessed exceedingly "rare qualities" that, once tasted, would cause one to lose all interest in domestic chicken eggs. The ruffed grouse egg was "simply a morsel fit for the gods."[174] Milliners advertised the "rare feathers" in stock for ladies' hats, harvested from "rarest as well as gayest species," including terns, egrets, herons, bitterns, flamingos, ibises, and all birds of "colorful plumage."[175] Collectors likewise coveted the most unusual specimens, and advertisers made sure to offer them. A "very rare" yellow-throated warbler and a similarly unusual cerulean warbler, shot by a hobbyist in 1883, were reportedly the only ones ever seen on Long Island.[176] For egg-collecting oologists, the eggs of uncommon and especially rare species represented the greatest challenges to acquire and required the greatest investment. A single "California Vulture [condor]" egg advertised in a 1904 catalog was listed for $350, while blue jay eggs went for a mere ten cents each. Even more-common species, though, could command higher prices if they produced particularly attractive eggs that collectors desired in large numbers.[177]

This expanding consumption of avian wildness paradoxically functioned as an expression of nature appreciation, providing genteel nature lovers with a cure for the ills of modern civilization.[178] Although actual encounters with the natural world were best for cultivating these sensibilities, commodified wildness offered reasonable facsimiles that could satisfy both scientific and aesthetic impulses. Ornithologically inclined buyers patronized taxidermy shops and natural history supply companies, searching for mounted wildfowl specimens to add to their collections. Others sought out stuffed birds for household decorations featuring the "wonderful tints and colors . . . furnished by the plumages of the many birds"—hues that could not be reproduced by any "mortal powers."[179] Their purchases fueled the expansion of commercial taxidermy as well as the formation of the Society of American Taxidermists in 1880.[180] In contrast to the staid and conservative museum taxidermy of earlier decades, the modern decorative form of the art was "radical and progressive," according to the Society, and its practitioners were "drawn to it out of a pure love of nature."[181] Egg collecting had its own specialized associations and periodicals aimed at hobbyists.[182] While some magazines had educational content, others were little more than classified advertisements for individual egg sellers and commercial dealers. Two of the most widely circulated magazines were *The Oologist* (founded 1875) and *The Ornithologist and Oologist* (1881). Youngsters could start their egg-collecting pursuits with the advice of *The Young Collector* (1882), and *The Young Oologist* (1884).[183] Several excellent bird egg identification manuals published in the late nineteenth century fast became important reference works for the growing population of specimen buyers and sellers, who numbered in the thousands by the 1880s.[184]

Millinery decorations similarly capitalized on the aesthetic characteristics of a wide range of birds, both unusual and familiar. Common perching birds and even larger starlings and woodpeckers provided a touch of natural, wild beauty to women's bonnets. The contention from some that "our most desirable song birds, such as thrushes, wrens, grenlets, and finches [were] in limited demand on account of their plain colors," was deemed unfounded by other fashion writers.[185] The fact was that even the drabbest wild bird added aesthetic value, a fact wryly confirmed by ornithologist William Dutcher in 1885 upon hearing reports of a woman wearing a dead crow "reposed among the curls and braids of her hair." Dutcher opined that "if the lady in question could have seen the crow during its lifetime perched upon and feeding on the decaying carcass of a horse, she might have objected to the association."[186]

In all, the act of purchasing dead, commodified avian wildness in the modern marketplace functioned culturally as a symbolic interaction with nature. Experiences with wildness, both real and imagined, could be reconstructed and remembered in a game dinner, a wildfowl egg, or a stuffed bird. For sportsmen especially, wildfowl products enhanced their experiences with the wild. Shooters deemed swift-flying passenger pigeons to be the targets more sporting than domestic birds because of their wildness.[187] Live birds, even though costly, made better decoys than artificial ones since their wild behaviors more readily lured waterfowl into gun range.[188] The spoils of a successful hunting trip might be stuffed and placed on a mantelpiece, thus preserved as a priceless reminder of the wild encounter. Or a bird mount sold by a commercial taxidermist could stand in for the one that got away. Menus at waterfowling resorts featured wild game shot by the resort patrons or purchased from local hunters who sold birds to city "sports," who arrived by rail and wagon. The game dinner, a mainstay of elite sportsmen's club culture, put wildness on a plate in the form of roasted canvasbacks or "winter bay-birds" on toast; served with wine, champagne, or "a large pitcher of cocktails," the meal was consumed by "wearied huntsmen" with "a zest they never experienced in the city."[189]

The elite sportsman, the bird hat–wearing woman, the avian collector, and the wildfowl epicure all continued to buy birds far beyond what was economically expedient, because they were willing to pay the price for wildness. With the cultural currency of the wild, consumers in an increasingly industrialized and standardized world could rediscover a nostalgic tie to a simpler American past or a connection to regional distinctiveness. Commodified wildness brought nature's authenticity and aesthetics home to counter empty, modern artificiality. And it served as a reminder of wildness experiences that were missing from urban spaces. The turn-of-the-century market promised much more than food or a fancy hat. It offered deep cultural messages about the value of wildlife in the modern age. The price of wildness was a measure of the difference between wildlife consumption and wildlife consumerism.

In the market, wildness most often came packaged in avian forms. Biology and economy conspired to make wildfowl exceptionally salable, but it was wildness that made birds especially valuable to Americans. "All good things are wild and free," wrote Henry David Thoreau at midcentury.[190] For the modern consumer, though, only half that statement was true. The market in birds proffered the laudable qualities of wild Nature, but they were hardly free.

CHAPTER FOUR

The Sportsman

Go to South Water Street if you want to know where the game is most abundant.

EMERSON HOUGH, SPORTSMAN

Henry Clay Merritt was busier than ever. Though no longer hunting, he was deep in the game business, building a minor commercial empire from his Kewanee, Illinois, home base. Consumer demand drove the growth of his business throughout the 1870s and 1880s as orders from East Coast cities increased. Woodcocks remained a Merritt specialty, and between 1878 and 1885 "very many barrels were forwarded and sold . . . for three hundred dollars per barrel." With few exceptions, everything sold well. Even old, moldy stock from his freezing rooms, simply wiped clean, could bring the highest prices with "no complaint."[1] New requests for game came in from New York, Boston, and elsewhere "every two or three days and sometimes daily." From the 1880s onward, "there were as many buyers as sellers," and birds rarely "had to wait for a market." Merritt's family fortunes grew with the help of his youngest son, Herbie, who worked the freezing rooms, and his eldest son, Clarence, who traveled the western prairies to buy game. The whole "Great West" became the game dealer's own "happy hunting ground."[2]

Some of Merritt's old hunting partners had fled westward as well, seeking birds that had become scarce east of the Mississippi. For his part, Merritt was relatively unconcerned about the dwindling numbers of birds in Illinois. The decline was to be expected—an understandable consequence of settlement, population growth, and urbanization. In his accounting, which he held to be "the universal sentiment," local wildlife losses were insignificant when measured against the technological advancements stimulated by the game business. Indeed, the harm done to game birds by market hunting was "more fanciful than real."[3] Merritt did allow that commercial exploitation was a factor responsible for declining wildfowl populations in Illinois and elsewhere, but in his view, it could not be the only or even the most significant culprit. More important were the myriad environmental alterations that accompanied a growing and urbanizing country. He had seen the changes firsthand. The previously well-watered Henry County was but one example of the larger agricultural and industrial trends. Farmers, eyeing larger profits, had relentlessly reclaimed the county's wetlands, drying them out to the point that the birds refused to come back. The "changed conditions [were] fatal."[4]

Equally destructive was the growing army of recreational wildfowlers. The modern "rage for hunting" encouraged amateurs to pursue birds relentlessly, and new forms of transportation allowed them to do so. Sportsmen's elaborate codes and practices—like wing shooting and warm-weather hunting—were far more objectionable, in Merritt's estimation, than the calculated efforts of the "up and down hunter who hunted for profit." Professionals were unlikely to kill game illegally or to excess, since it was in their interests to maintain the wild supply. Nine times out of ten, it was the insatiable "pasteboard and gingersnap sport" who violated the legal limits, such as they were. Besides, Merritt insisted with a touch of professional conceit, the very idea of hunting for "sport" was unjustifiable. It was wasteful to murder birds simply for the sake of an afternoon's enjoyment. Hunting for the market, supplying a basic human need, and providing a prized commodity to distant consumers, on the other hand, was socially and economically valuable, even honorable. It seemed only fair to Merritt that, so long as there remained enough wildfowl to be harvested, killing for sale should remain at least as legitimate as killing for sport. In this way, "all the rights of citizens in different states would be protected" along with the birds.[5]

Others, too, noticed the rapid and widespread disappearance of wild birds. But they did not always agree on the primary causes or the consequences of this change. Sporting writer William Henry Herbert challenged the notion that agricultural development had any "injurious operation . . . on any kind of winged game." Instead, Herbert blamed overhunting by Americans in general and rural folk especially whose disregard for game laws, sporting codes of conduct, and any kind of restrictions was well known.[6] Writing nearly twenty years after Herbert, George Perkins Marsh similarly conjectured that "all the wild birds which are chased for their flesh or their plumage" would so rapidly decline that "the last of them will soon follow the dodo and the wingless auk."[7]

Herbert traced the fault even further, though, to charge "the wealthier classes" with culpability in hunters' excesses. The urban demand for game birds at all times of the year provided excuse and incentive for enormous destruction. Woodcocks, for example, would soon become "extinct everywhere within a hundred miles of the Atlantic seaboard—and inland, everywhere within a hundred miles of any city large enough to afford a market," he predicted.[8] Politician and sportsman Robert B. Roosevelt likewise condemned the post–Civil War "*nouveaux riches*" who fancied themselves epicures for being willing to "pay for unseasonable game an extravagant price."[9] By the 1880s, sport hunters as far west as Minnesota began to wonder aloud "whether the good old days" were now "gone to never return."[10]

Roosevelt and Herbert were among the forerunners of a new breed of mostly urban, wealthy, and influential hunters who organized and nominated themselves as America's game protectors across the latter half of the nineteenth century. By the century's end, they had become a visible force as the first, and arguably most significant group, to call for national action to conserve wildlife.[11] Though interested in the preservation of larger game animals, these sportsmen-conservationists were especially preoccupied with the protection of birds since, for the urban dweller, big-game hunting was an infrequent recreation, usually involving a protracted, long-distance excursion into the wilds. Game birds, however, could be found much closer to home: in a farmer's stubble field, on a nearby lake or stream, or in a coastal estuary. In their seasons, birds came to the hunters, not the other way around. Hence, the protection of distant creatures such as bison did not have the same immediacy as did efforts aimed to increase the numbers of ducks on the bays of Boston, brants in estuaries around San Francisco, or quail in West Jersey's wooded brakes. The effects of commercial hunting were also far more visible to urban sportsmen who encountered game birds in the market or in restaurants as frequently as they started them in the field.

Alarmed by dwindling wildfowl populations and increasing numbers of wildfowl commodities, these activist sportsmen struggled to find a way to preserve the remnant avian stocks as well as their sport by placing new limits on hunting for amateurs and professionals alike. But their localized efforts fell short. Voluntary codes and local laws had little impact on the far-reaching modern game business. In the last decades of the nineteenth century, these fledgling conservationists turned their attention to controlling the extensive market in birds, eventually adopting a "no sale" position and inaugurating a coordinated, national campaign to outlaw the commercial wildfowl market. A cadre of politically powerful and socially influential sportsmen-conservationists led the way in this campaign, shifting the focus from preserving birds in the field to eliminating avian commodities from the market. Along the way, they grappled with fundamental and long-standing understandings of economic property, rights, and the legal ownership and consumption of wildlife. In the end, activist sport hunters succeeded in redefining what had been a respectable enterprise as an outmoded, mercenary practice of questionable legality. In this new narrative, the "market hunter" and the "game dealer" were cast as villains. The sportsman, in his role as the "game protector," was the hero.

Inventing the Sportsman

Prior to the nineteenth century, the "sportsman" was all but unknown in the United States. Certainly, recreational hunting was an element of American life from the colonial era onward, but few bothered to distinguish between hunting as a vocation and as an avocation since it was often both. Environmental and economic changes in the nineteenth century, however, shifted the role of hunting in American society. For city dwellers, hunting was not only unnecessary but also expensive and inconvenient. The differences between those who pursued wild creatures for subsistence or profit and those who shot solely for recreation became more distinct in the nineteenth century, largely due to the efforts of an emerging group of outdoor writers. Among the earliest was the transplanted Englishman Herbert, whose sporting articles and books written under the pseudonym Frank Forester introduced American readers to English-style sportsmanship in the 1840s and 1850s. He volunteered himself as "the champion of American Sport and Sportsmanship," advocating a code of hunting ethics that emphasized restraint.[12]

Herbert's efforts to refine the image of the sportsman were reinforced by numerous sportsmen's organizations and publications appearing at midcentury. From the start, these vehicles for sportsmen's self-identification intertwined sport hunting advocacy with efforts to protect and propagate game. New York sportsmen led the way in many of these developments, and the New York Sporting Association, founded in 1844, was the earliest of these organizations.[13] Renamed the New York Association for the Protection of Game (NYAPG) in 1873, the unincorporated group was based in and drew its elite membership primarily from New York City.[14] Hundreds of other sportsman's clubs and game protective associations emulated the NYAPG's pioneering model in the following decades.[15] At the same time, many new, nationally circulated sporting periodicals, including the influential outdoor weekly *Forest and Stream*, emerged as important forums for crafting and advertising both an image of ethical sportsmanship and a movement for wildlife conservation.[16]

An explosion of mass-produced hunting accessories, books, and magazines in the late nineteenth century meant that anyone with an interest and money to spend could aspire to the sportsmen's ranks.[17] Once admitted to that club, the sport hunter was supposed to use his moral authority to encourage others—"the great mass of those who shoot"—to adhere to the sportsman's largely voluntary code of behavior.[18] As part of the larger community of hunters, activist sportsmen had "great missionary work" to do "among the unlettered," exhorted *Forest and Stream* editor Charles Hallock in 1875. The "small farmers, bushrangers, and frontiersmen (to say nothing of the negroes of the South, who all use guns)," Hallock explained, lacked "the instincts of sportsmen."[19] The true sportsman was different, and distinguishing his identity from the many other gunners, whose activities appeared quite similar on the surface, quickly became a major goal of outdoor periodicals and sportsmen's associations.

The market hunter presented a particularly salient counterpoint to sportsmen's project of self-definition. The "real sportsmen, who shoot wholly for the health, pleasure, and excitement the pastime affords," observed the *Chicago Tribune* in 1872, set themselves apart from the professional gunners they regarded as "pot hunters." Market hunters utilized whatever means were available to kill or capture game, since the faster they could obtain birds, the more profitable their business. Restraint was not part of their practice. The sportsman, however, faithfully followed the rules of fair chase and took no more game than needed (though defining what constituted enough remained elusive). *Forest and Stream* drew even clearer categories, defining a market hunter as "one who hunts for the purpose of selling his game," and a game hog as "one who kills an unreasonable amount of game." The "pot hunter" epithet could be applied to both.[20] "It is a rare exception," one of the magazine's correspondents wrote of market hunters, "to find one of these men who is known as a thrifty, honest citizen."[21] Texas was "cursed with the abominable market hunter and the netter of quail," another contributor complained.[22] Average market hunters were "industrious despoilers" and "agents of wholesale game depletion," certainly not persons of any "idyllic character."[23]

Despite such criticisms, the sportsman-crafted popular image of market hunters was initially neither unitary nor wholly negative. Sportsmen conceded that professional gunners were hard workers engaged in a business that required physical and mental toughness.[24] Commercial hunting did not promise the "life of elegant ease and indolence" that those "unacquainted with the average market hunter's routine" might assume, explained *Forest and Stream*. It required instead "ability and pluck, which if properly directed, would insure success in a more honorable pursuit."[25] Professionals' outdoor skills were also unquestioned, leading not a few sportsmen to include market hunters among their wildfowling friends. Emerson Hough, whose column Chicago and the West was a regular *Forest and Stream* feature, bragged about his trips with commercial hunters such as Billy Griggs of Texas and "Sam Gibault, a French market-hunter and 'guide of Momence.'" Well-connected sportsmen used insider knowledge about waterfowl migrations in southern Wisconsin acquired from the market-hunting Kleinman brothers, or about plover-shooting gleaned from Chicago's "Italian Joe," to improve their chances in the field.[26] Other sportsmen challenged the pot hunter characterizations appearing in outdoor periodicals, arguing that such terms were based on highly subjective and often-fluid ethical standards.[27] *Forest and Stream* contributors pointed to the inconsistencies in sportsmen's rhetoric that allowed them to castigate the "fortunate market hunter" for "killing fifty birds a day to support his wife and babies" while lauding the sportsman who "kill[ed] a hundred to give to his friends or rot in the garbage barrel."[28] Throughout the 1870s, the game market itself received at least tacit approval from *Forest and Stream*'s editorial leadership in that the magazine provided regular updates on current game market prices.[29] Readers routinely requested advice from *Forest and Stream* about becoming commercial hunters themselves, clear proof that the publication presented an inconsistent position on the profession in its earliest years.[30]

Despite offering these somewhat mixed messages, newly popular sporting publications and organizations sought to distinguish between sportsmen and market hunters, even as disagreements arose over the exact qualities of each. "A market-hunter may possibly be a good fellow," as one speaker at an 1884 sportsmen's convention opined, "but a sportsman cannot be a market-hunter."[31] The crucial difference most often came down to the purity of each group's reasons for hunting. Sportsmen, whose hunting was not motivated by economic interests, possessed the moral authority to speak for wildlife protection. The "moral force of the entire army of conservative sportsmen," *Forest and Stream* insisted, was necessary to infuse market hunters and game hogs with the "elevating and ennobling influence of the gospel."[32] No other group had the ethical influence or social power necessary to alter American wildlife use.

Activist sportsmen took to the field in the late nineteenth century to investigate and advertise market hunting excesses, such as the accelerating slaughter of passenger pigeons. Two members of the Michigan Sportsman's Association, William Mershon and Henry B. Roney, took on the task of pigeon protection, convinced that the state game laws related to those birds were "a bungling piece of business" designed solely to benefit game dealers. Mershon, a wealthy lumberman who had his own railcar to take on his frequent hunting trips, and Roney, a somewhat less wealthy music teacher from Saginaw, coauthored an *American Field* article to publicize the destruction wrought by the army of netters and buyers working the massive 1878 Petoskey nesting.[33] Writing about his undercover experiences among the pigeoneers, Roney immodestly described his mission as "a Herculean one" undertaken "in behalf of justice and humanity." But the call to action was "backed up by such true sportsmen as well as by the sentiment of every humane citizen of the State," and therefore he and Mershon "could do no other." For Roney, the protection of Michigan's wildfowl was the sportsman's obligation, and he would never "shrink from the consequences in the discharge of that duty."[34]

High-minded arguments about wildlife protection, especially for passenger pigeons, rang hollow when sportsmen stepped onto the trapshooting range. To many outside the sporting community, live pigeon–shooting contests smacked of immorality, and the popular competitions provoked one of the earliest anti-cruelty campaigns launched by the American Society for the Prevention of Cruelty to Animals (ASPCA). In 1869, Henry Bergh, the organization's founder, began interrupting shooting contests in New York and New Jersey, ordering the well-heeled bankers, brokers, and other competitors to desist from the "brutal, blackguard business."[35] Bergh's anti-cruelty crusade took on a national character when it spread beyond New York in 1875 to ensnare the famed wing shooter Captain Adam Bogardus in an animal cruelty charge during a St. Louis shooting match.[36] Sportsmen's clubs argued that attacks on these "contests of skill" interfered with their abilities to raise funds for game protection and impinged on the "chartered rights" of a group of the "most excellent citizens, whose characters place them above the suspicion of any cruelty to animals."[37]

Nevertheless, sportsmen-sponsored public trapshooting exhibitions, reported *Forest and Stream*, provided traction for the anti-cruelty cause and weakened sportsmen's developing image as "the true protectors of game."[38] The NYAPG's sizable 1881 Coney Island trap shoot, in which some twenty thousand pigeons were killed, for example, gave "the little hellbender" Bergh valuable ammunition. Had the tournament been confined to private shooting clubs—"an afternoon's shoot with a few pigeons"—no criticism would have arisen, the journal opined. But given that the contest was supported by "a society . . . organized ostensibly for the protection of game," negative public sentiment about sportsmen was bound to result.[39] Consequently, many sportsmen's associations gradually adopted policies to limit and even eliminate pigeon-shooting competitions from their activities, and famed shooters like Bogardus redeemed their reputations by designing glass ball and clay targets to replace wild pigeons at their matches.[40] The national scandal that the ASPCA provoked, activist sport hunters realized, damaged the "popularity and efficiency" of their game protective associations and tarnished sportsmen's claims to the moral high ground in wildlife matters.[41]

Moral authority alone, however, was not enough to reverse wildfowl population declines, and activist sportsmen soon sought the legal backing necessary to enforce their versions of proper wildlife use. Wildlife lawmaking had previously been inconsistent and mostly local in application, with game laws varying by county and custom.[42] In the mid-nineteenth century, however, many states—often at the behest of politically connected sport hunters— began reorganizing and consolidating disconnected statutes into uniform state game codes, as well as establishing new state commissions on fish and game—also led by activist sportsmen—to direct future wildlife regulation.[43] These newly devised game codes, though, lacked effective enforcement. As it

"The great pigeon shooting match at Greenville, New Jersey, March 6th and 7th, between Mr. John Taylor and Mr. William Seeds, for \$1,000 a side." This print, from *Frank Leslie's Illustrated Newspaper* in 1866, depicts one of the many popular sportsmen's club contests responsible for consuming enormous numbers of passenger pigeons. Courtesy of the Library of Congress Prints and Photographs Division, Washington, DC

had been since the colonial era, game law execution continued to be the responsibility of individual citizens, local justices of the peace, or in some cases, overseers of the poor.[44] "The game-laws of most of our States," complained one sporting writer in 1851, were "a mere *bagatelle*" or "dead letter," since no one was specifically charged with the duty to implement them.[45] Even as game warden systems began to appear after 1850 (such as the "ducking police" force created for Maryland's Cecil and Hartford counties in 1872), observance of the law remained patchy and mostly voluntary.[46]

Activist sportsmen stepped into this void, taking on a policing function that became crucial to their developing identities as game protectors. Sportsmen's clubs and game associations, often composed of politically and socially influential elites, operated as unofficial (and, in some cases, officially sanctioned) enforcers. The NYAPG was one of the most active. The organization counted among its early members not only Herbert of "Frank Forester" fame but also some of New York's most prominent lawyers, judges, financiers, and politicians. During its most energetic phase, in the 1860s and 1870s, the association was headed by Royal Phelps, a leading businessman and former state legislator.[47] With Phelps at the helm, the NYAPG embarked upon an innovative program of legal action to combat wildfowl depletion while building a network of activist sportsmen, linking clubs together in the shared project of game law enforcement.

The challenges facing this project were many. As an organization, the NYAPG had no formal authority to enforce compliance with the law. In contrast, the West Jersey Game Protective Association (WJGPA), incorporated in 1873, had extensive authority, at least on paper. The legislature granted the club powers to execute the game laws in six New Jersey counties. A similar association for the central portion of the state was established five years later.[48] Conferred on all the members of these associations were "the powers of constables" and the authority "to arrest without process violators of the game laws."[49] The WJGPA instituted the first nonresident hunting license in the United States, requiring hunters from out of state to purchase a five-dollar club membership to legally hunt in the western counties.[50] But within one year of incorporation, the association was already struggling to carry out its charge. In 1874, John W. Hamer complained to NYAPG president Phelps that the game law was practically "a dead letter." Scores of Philadelphia residents came into New Jersey by "every afternoon train" to kill "pheasants, quail, woodcock, etc. without having paid to this society the legal fee," he wrote. Hamer had dutifully paid his fee, which was *to be expended for the protection of game,*" and he therefore wanted "all sportsmen [to] be treated alike."[51] For Pennsylvania resident B. A. Hoopes, however, the association's nonresident fee rankled. In his opinion, the law incorporating the WJGPA, or as he called it in a letter to Hallock, the "W. J. Pot Hunters & Trappers Society," needed to be repealed. The organization's powers were "certainly in conflict with the Constitution of the United States, and [would] probably be soon tested by appeals." Hoopes and other members of his Philadelphia sportsmen's club would not continue to pay subscriptions and still "run the risk of having their members and their friends fined" by overzealous association members.[52]

Although violations by amateur hunters were of concern to the larger community of activist sportsmen, the unregulated actions of market hunters loomed even larger. The Mad River Sportsman's Club in Oneida County, New York, proposed a state law against partridge snaring to Phelps in 1875 and argued that the possession of birds out of season ought to be presumptive evidence of guilt.[53] The Rochester, New York, sportsmen's association, reported state fish commissioner and NYAPG member Seth Green, had been enforcing the game laws, and consequently "there [were] now some parties in the

Penitentiary paying their penalty for breaking the game law." But game-protecting sportsmen wanted to go even further by taking the law out of the hands of local politicians who courted "the poacher vote."[54] Farther west, asserted a St. Paul, Minnesota, attorney in 1875, the issue became more complicated. "It is easy to enforce the laws when there is but little game," he explained to Phelps, but in Minnesota, where hunting was widely pursued, game law enforcement "arrests the sports and injures the profits for many." Consequently, the "courts, people, and press" all stood in opposition.[55] Phelps responded that the same conditions had pertained in New York a decade or so earlier. "If Western people can look with indifference on this sacrifice" of game to eastern markets, he retorted, "on them let the shame rest." It was necessary to rile public sentiment against the wastage. "We here can stop the retailing in the market," the NYAPG president continued, "but it is doubtful whether we could legally stop this kind of trade between different states."[56]

Indeed, policing retail markets was exactly what the NYAPG was doing. New York's midcentury game laws, like early game laws in many other states, allowed individuals to sue game law violators and recover fines.[57] This the association did as early as 1844, when it brought its first case against a Fulton marketman.[58] Phelps, however, took urban enforcement still further, convinced that "the peculiar circumstances" of the club, located as it was in the preeminent game market city, positioned it to "do more good than any other" sportsmen's organization in the country. Zealous prosecution of the city's game dealers, restaurateurs, and hotel keepers for the sale of out-of-season game had potentially far-reaching effects. If illegal game could not be sold in New York markets, market hunters across the country would have little incentive to violate closed-season laws in their states. In the 1870s, NYAPG's elite members began to aggressively pursue game law transgressions in the market, shaking down merchants, hotel keepers, and restaurateurs for fines and court costs related to violations of closed-season possession laws.[59] By 1873, *Forest and Stream* reported that the NYAPG had prevailed in twenty-four out of the twenty-seven suits it brought against the city's game dealers. This winning record led Phelps to castigate other sportsmen's groups for allowing "the most flagrant cases of killing and marketing game to occur" without repercussions.[60] What kinds of repercussions could contain the far-flung market in birds, though, Phelps did not say.

The Legalities of Consumption

At a meeting of the recently organized American Fish Culturalists Association in New York City in 1874, *Forest and Stream* editor Charles Hallock proposed that fish and game protective associations ought to bring the haphazard tangle of existing game laws into some sort of rational order. The problem, according to Hallock, was at once political, biological, and economic. Decisions about wildlife protection were a product of "the accidental selection by the legislatures" of laws in other states or territories. Politicians paid little heed to the biological and climatological factors governing the migration, breeding, and incubation periods of different kinds of game. Adjoining states that shared the same latitude, climate, and fauna often set very different open seasons. These variations in protections had especially negative effects on migratory bird populations since adequate protections in one state were nullified by inadequate protections in the one adjoining. The solution, Hallock suggested, was cooperation between states to devise game laws "based on scientific principles" and responsive to the biological needs of wildlife.[61]

Hallock was an early proponent of national game law uniformity, and he assumed that organized sportsman, in their role as game protectors, would spearhead the process. With a committee drawn from the NYAPG that included Royal Phelps, Robert Roosevelt, and Seth Green, Hallock sought to create a national body capable of designing cooperative game laws between the states that were grounded in biological principles. Numerous sportsmen's associations jumped on the idea, proposing a national convention with the goal of "the procurement of intelligent and efficient legislation for the Protection of Game Birds and Fish." Although he left no room for game dealers in his organization, Hallock acknowledged "the great market value" of birds such as quail as a justification for greater protections. The most problematic matters in the law were the state-level inconsistencies that made it possible for market hunters and game merchants to evade restrictions. Avian entrepreneurs had economic incentives to claim that any out-of-season game they sold was killed legally in another state, making prosecution an uncertain and risky prospect for accusers. State game laws designed to protect wildlife within the state were poor mechanisms for policing the far-reaching economic activities of the game market. Legal uniformity would thus help to close loopholes that market hunters and game dealers routinely exploited.[62]

Hallock was not speaking in hypotheticals. In 1874, the NYAPG (which included Hallock as an active member) was pursuing a case of illegal possession in which out-of-state game birds were at issue. Joseph H. Racey, a New York City merchant, admitted to the charges that he had in his possession, and exposed for sale at the city's Central Market, quail and pinnated grouse weeks after the close of the legal season. But unlike the accused dealers in some of the NYAPG's earlier suits, Racey was not willing to simply pay his fines to the organization after he lost his case in a lower court. He carried his defense to the New York Court of Appeals, maintaining that the birds he had stored in his "patent[ed] preserving apparatus" had been killed legally in Minnesota and Illinois the previous year.[63] Hallock's 1874 call for uniform game laws were no doubt provoked, at least in part, by this contentious case working its way through the New York court system at the same time.

When the *Phelps v. Racey* case was finally concluded in 1875, with Racey on the losing end, the conflict between the sportsmen's association and the New York marketmen had broken new legal ground. The district and appellate justices concurred that the recently amended New York game law of 1871 had specifically banned the sale of out-of-state quail and pinnated grouse during the closed season.[64] As to the novelty of the birds' preservation by an innovative cold storage method, Chief Justice Charles P. Daly of the Court of Common Pleas wrote that he was "wholly at a loss to see upon what ground it can be said that the possession which existed in this case was not the kind of possession which the statute meant." While he conceded that lawmakers might not have foreseen the possibility that "game killed in the autumn of one year could be preserved (as in this case) so as to be sold a year afterward," he could not say definitively "as a matter of law" that the statute was not intended to apply in those instances.[65]

The principle first expressed in *Phelps v. Racey*—that possession during the closed season was illegal no matter the time or place of capture—was revolutionary. In refuting Racey's defense, the New York court maintained that the birds' origins did not matter; it was within the state's police power to regulate all game, dead or alive. Key in this decision was the court's acknowledgment that game dealers' possession and sale during the closed season provided market hunters economic incentives to violate or evade game laws in the field—incentives that sportsmen presumably did not pursue. The particular birds in question might not have been killed illegally, but their presence in

Racey's market opened the possibility that other birds, potentially New York wildfowl, might be.[66]

Six years later, a similar case of closed-season wildfowl possession and sale worked its way up to the Illinois Supreme Court. In *Magner v. People*, the justices heard a Chicago game retailer's appeal of his conviction on numerous violations of the 1879 game law related to possession and sale of wildfowl during the closed season. James Magner had purchased quail from Kansas and subsequently sold them to consumers in Chicago and New York after the end of the legal season. Magner's defense primarily rested on two main assertions: the first was that the state's game protection laws applied only to the state's wildlife; hence, it would be "absurd to hold that the inhibition against the purchase and sale of game imported from the State of New York or Kansas" protected wildfowl in Illinois. Second, the state had overreached its authority by interfering with interstate commerce, a regulatory function constitutionally reserved for the federal government. In the first objection, as with *Racey*, the Illinois high court determined that it made no difference when and where the birds were killed. While the court conceded that allowing quail to be killed in Kansas had no direct effect on Illinois birds, the justices recognized the broader consequences of consumer market forces. It was "obvious that the prohibition of *all* possession and sales of such wild fowls or birds during the prohibited seasons would tend to their protection, in excluding the opportunity for the evasion of such law by clandestinely taking them, when secretly killed or captured here . . . or by other subterfuges and evasions."[67]

Rejecting Magner's second claim, Chief Justice Schofield asserted, in no uncertain terms, the principle that the state, as a representative of the people, owned wild game within its borders and thus had wide latitude in regulating access to wildlife in the public interest. The hunting, killing, and consumption of wildlife was a privilege, not an individual right, and hence there were no individual property rights to be considered. As to the interstate commerce issue, Schofield opined that wild game possessed in violation of the law was neither legal property nor a legitimate article of commerce.[68] Borrowing language from the 1847 *Licenses Cases*, which prohibited sales of intoxicating liquors even when such prohibitions interfered with interstate commerce, Schofield wrote that "'the State is not bound to furnish a market' for game, and, by parity of reasoning, is not bound to furnish game for a market." The court concurred with the prosecution's position that "the States can better control this question than Congress."[69] Activist sportsmen in Chicago celebrated the *Magner* decision. Nicholas Rowe, president of the Illinois State Sportsman's Association and editor of the *Chicago Field*, called it "the most important game case ever tried in America and perhaps in the world," because "every condition of the sale and traffic of game was considered." Rowe was confident that the judgment would bring lawbreakers into line.[70]

It was no coincidence that these foundational wildlife cases involved game dealers and birds in New York and Chicago—the eastern and western hubs of the commercial wildfowl business. The ideas articulated in both *Racey* and *Magner* were legal innovations with effects that swept outward from these urban centers of the modern game market at the end of the nineteenth century. In *Racey*, the state's police power to regulate all wildlife, alive or dead, no matter its origins, broke new legal ground. In *Magner*, the Illinois Supreme Court indicated that it was simply following the *Racey* precedent, but its decision went further than that in *Racey* not only to assert the state's police power to regulate game, but also to articulate for the first time the state's absolute ownership of wild creatures. These cases laid the groundwork for future challenges to commercial hunting across the country, and the principles they asserted would be contested repeatedly over the following decades.[71] Even more important, *Racey* and *Magner* signaled that the key wildfowl protection battles would be fought in the urban market rather than in the rural field.

In these and other disputes, organized sportsmen positioned themselves as competitors with market consumers for the birds that remained. A growing "prejudice against marketing game in America," opined *Forest and Stream*, was a direct consequence of the limited supplies; there simply was "not enough game to go around among sportsmen," let alone enough to satisfy the consuming desires of the entire country.[72] Commercialization could be blamed for the inefficacy of state game laws that attempted to limit wildfowl harvests. "The game dealers try to make it appear that they are obliged to sit supinely down on their own stalls and take all the game outside parties send to them," *Forest and Stream* suggested in 1885, yet it was "quite within their power" to restrict their game purchases to only that which they could sell during the open season.[73] Instead, the market's volume and reach had only grown over time. On Nebraska's prairies there were at least five "permanent shipping establishments, with refrigerative annexes and shipping departments" along the Burlington and Missouri River Railroad, that shipped prairie chickens to eastern markets year-round in contravention of the game laws.[74] Prairie chickens in Iowa were killed off by "a class of men" who supplied the game freezers springing up in almost every small town. Sporting clubs, wrote one correspondent, needed "to spend more time in prosecuting the market hunter and less attention to the city sportsman."[75]

The plume industry presented sportsmen with perhaps the clearest-cut example of the environmental costs of unlimited market exploitation. Unlike the consumption of game birds, which recreational hunters found awkward to condemn entirely, the anti-plumage cause was one they could wholeheartedly endorse. *Forest and Stream* led the way in this campaign when, in 1886, George Bird Grinnell, who had taken over the editorship of the magazine from Hallock in 1880, founded the Audubon Society to organize protection for nongame plume species and launched the *Audubon Magazine* to educate readers about plume and songbird conservation.[76] Grinnell's intention was to make "the Audubon movement a national one" that would be "dedicated to halt the decline of American birds by arousing public sentiment." In particular, he singled out fashionable women's consumption of bird skins and feathers as the impetus for the creation of the society.[77] Society-sponsored public presentations, such as the one titled "Woman as a Bird Enemy," challenged female consumers' humanitarian sensibilities, and sportsmen wondered aloud if "the ladies themselves really care for the birds."[78] Without changing public attitudes and stimulating public interest—especially among bird-buying women—conservation laws were unlikely to make a difference. "Ah! if only the women would think a little," *Forest and Stream* complained, and "realized the incalculable harm they are doing." A simple alteration in women's habits of consumption, however, would quickly "bring to bankruptcy those who make their living out of this traffic in bird skins."[79]

As activist sportsmen and bird protection organizations focused on songbirds and "other non-edible birds of plume," some milliners and their customers shifted to game birds for hat feathers, provoking competitive concern among recreational wildfowlers. An 1886 *Forest and Stream* article noted that the new "bird wearing craze" was now for snipe and other game bird species of interest to recreational gunners. This harvest might be "more difficult to discourage," the magazine acknowledged, since it could be argued "that it is as legitimate to destroy them for their feathers as for their flesh."[80] Sportsmen's moral authority on this kind of consumption was less clear, and the law was on the milliners' and market gunners' sides.

The wildlife laws that reform-minded sportsmen advocated, they insisted, were regulations that limited everyone, not just market hunters and game

"The cruelties of fashion—'fine feathers make fine birds.'" Sportsmen found common cause with bird lovers in the antiplumage movement. Cruelty was a charge routinely leveled against bird hat–wearing women, as this drawing from *Frank Leslie's Illustrated Newspaper* demonstrates. The pictures trace the progression of a bird from hunter's prey to milliner's creation and finally to woman's decoration. Courtesy of the Library of Congress Prints and Photographs Division, Washington, DC dealers. They denied that new game codes targeted any group or that the restrictions amounted to class legislation. "No one section or class of people is responsible for the condition that the wild animal and bird life is in to-day," insisted a *Forest and Stream* correspondent. The solution for everyone, sportsman and market shooter alike, was "to give our guns a rest, and give more attention to the preservation and welfare of the wild life around us." While market hunters certainly deserved the lion's portion of the blame for the depletion of small game, wrote C. M. Stark, the man who killed for sport was a close second.[81] Proposed laws limiting shooting to three days a week during the open season met with approval from *Forest and Stream* readers who argued that such laws would "reach just the ones who are killing off the game, that is, the market-hunter, who plies his trade every day of the week, and the man of leisure who can hunt every day."[82] Limits to restrain market hunters and sportsmen might make it possible "to save our game from extermination"[83] Legislation that had "the same consideration for high and low, rich and poor, wise and foolish, with justice for all, and discrimination for none" was necessary, explained W. W. McCain. "I do not champion market-hunting," he continued; "neither do I champion the club man, nor the yacht owner, who kill by 'the hundred in a morning's shoot.'" Only laws that furthered "honest, consistent game protection" and applied equally to all would be effective.[84]

Even sport hunters were prone to overhunting, and too many needed that "elevating and ennobling influence" of self-restraint.[85] "No doubt market and pot-hunters are entitled to a share of the blame for the rapid thinning out of game" in many places, maintained some observers in the 1880s, but it was sportsmen's clubs and out-of-state "sports" who were the most culpable for wildfowl destruction.[86] These elite sportsmen of "high professional standing" wielded their influence among legislators to demand alterations in game laws and hunting seasons "to meet their own wishes."[87] All the "'cussing' of the 'market hunters' in the columns of our sporting publications," complained "Hoodoo" to *Forest and Stream*, was so much "twaddle." Commercial shooters amassed large bags because of their skill, not from unfair or illegal methods.[88] While crusading game protectors publicized the "champion outrage of market hunting," they all too often remained silent about sportsmen's excessive consumption.[89] There was a great difference between the sportsman who killed vast quantities of birds, wrote "Nom de Plume," and the market hunter who killed a similar number to support "his starving family."[90] In the end all "the interests of the rational sportsman," some insisted, were not injured by the professional shooter, but rather by the amateur hunter "who kills for score and to brag."[91]

Insisting that recreational and commercial hunting could coexist, some sportsmen argued that confrontations with market hunters and game dealers were bound to be unproductive. It was better to pursue protective policies based on the common ground that commercial and recreational hunters shared. The Long Island Sportsman's Association maintained "that the true theory of protection is by the mutual conservation of the several interests" in an effort "to find that plane upon which we all can stand." The Long Island

"Hunting Car 'City of Saginaw,' Return from Dawson, ND, 1889." William B. Mershon—prominent Michigan lumberman, one-time mayor of Saginaw, and avid recreational hunter—was active in numerous sportsmen's associations, advocating for strict conservation measures to protect wildfowl from commercial exploitation. Like many sportsmen, however, his own hunting could appear less than restrained. Mershon's railcar, which he used for hunting parties, is shown here festooned with scores of birds taken on one trip to the northern plains. Image from the William B. Mershon Papers, courtesy of the University of Michigan Bentley Historical Library, Ann Arbor club—which counted market hunters and dealers among its members—refused to carry out what it deemed the "'star chamber' work" advocated by anti-market activists.[92] In St. Louis, a major game market city, "the dealers [had] been so thoroughly loyal to the sportsmen" in the 1870s and into the 1880s that demand for out-of-season game had all but ceased. Through their joint efforts, legislation had been crafted that was "most satisfactory all around," reported the St. Louis *Missouri Republican*. Game protection would be best served if sportsmen considered the dealer's "side of the question before forcing him into an opposition from which he will probably come out second best, but into which he should never have been driven."[93]

Besides, it seemed an impossible task to legislate the market out of existence. Actions against dealers like those pursued by the NYAPG were, according to critics, attacks on private property, and many doubted the constitutionality of such measures. Why, asked a *Forest and Stream* reader, was it "not entirely proper for a man to sell the game he shoots if he chooses to do so?"[94] As an 1881 *New York Times* editorial pointed out, possession conferred ownership in wild game. Wildfowl in the wild state were not property; they "belonged to their captor" once caught. This had been the legal principle governing property rights in wild creatures since the earliest days of the republic.[95] It was inconsistent, argued "Nor'east" in *Forest and Stream*, to pass laws that allowed sportsmen to transport their wild property across state lines while denying the market hunter the right "to market his spoils to almost any extent."[96]

"Honest, consistent game protection," however, was difficult to achieve with game codes that operated on the state level. Market hunters and dealers, insisted reform-minded sportsmen, were much more likely to exploit legal inconsistencies and evade legal limits than were recreational hunters.[97] *Forest and Stream* reported in 1884 that laws prohibiting the export of game from Maine, New Hampshire, and Connecticut created "a tendency to greater slaughter in Massachusetts" by market hunters supplying Boston's markets.[98] Stringent protections in one state, sportsmen argued, would do little to save migratory wildfowl whose populations were diminished in others. The Monmouth Shooting and Fishing Club pointed out that the ducks passing through Illinois were most threatened by market hunters in southern swamps and egg hunters in northern nesting grounds. "The market has never been glutted with birds killed in this state," the club claimed. Laws limiting hunting in Illinois would thus only harm the sportsman while market hunters continued their slaughter elsewhere. Better to work first, the club members argued, on restricting commercial shooters "at the ends" rather than in "the middle" in states like Illinois and Missouri.[99] Even when state game laws explicitly outlawed out-of-season possession, market men could claim the wildfowl they had on hand during the closed season came from other states. "The willingness of the game dealer . . . to deal in contraband goods," *Forest and Stream* observed in 1885, "is only a fresh illustration of the don't-give-a-continental-about-the-game-law spirit of most game dealers and shows how hollow are their pretensions when they pose as friends of game protection."[100] Legal sport hunting "would never exterminate" wildfowl, insisted the magazine, but the market hunter's "consummate skill, born of his needs and opportunities," necessarily created the conditions for a "merciless war of extermination."[101] Speaking for many activist sportsmen, the Utica (New York) Fish and Game Protective Association identified "the unprincipled market hunter" as the key culprit responsible for "diminish[ing] the domain of the sportsman."[102]

Game protections advocated by sportsmen and their publications from the 1870s onward increasingly targeted commercial interstate wildfowling, even as activists insisted that the measures they supported amounted to uniform limits on all wildlife users. Discrimination against nonresidents became a common feature of state game codes, especially in the South and West. Laws revoking nonresident hunting privileges or requiring nonresident hunters to purchase licenses clearly aimed to restrict commercial hunting. Arkansas established the first nonresident market hunting license in 1875. Other southern states followed suit in the late 1870s and early 1880s.[103] When several states implemented nonexport laws, designed to prevent killed birds from crossing state boundaries, *Forest and Stream* contributors held them up as fair limitations on all classes of hunters, recreational and commercial alike.[104] Politically active sportsmen also pressed for better enforcement of existing game laws, which was partially accomplished during the 1880s when many states created game warden systems that shifted responsibility for wildlife protection from individuals to the government.[105] Sport hunting organizations and publications took an active interest in the selection of game wardens and supervision of their work. The Illinois State Sportsman's Association and *Forest and Stream*, for example, touted their efforts in 1891 to get Chicago's "notoriously inefficient and incompetent" game warden H. P. Brusewitz dismissed for his unwillingness to take on the South Water Street game dealers for possession violations.[106] Calling on the governor to find "a warden that will make a national name for himself" in the style of the ASPCA's Henry Bergh or Connecticut's Abbott C. Collins, *Forest and Stream* argued that a "one-man power" like Collins would start a string of "old offenders," including market hunters, bird trappers, game dealers, railroad workers, and steamboat hands, "finding their way into court."[107]

Yet the sportsmen's crusade for consistent limits on wildfowl use raised opposition from others, who complained that the laws' effects weighed most heavily on amateur hunters. Market shooters, who were quite willing to hunt illegally, were left untouched by game codes enforced in the field. State regulations against spring waterfowl shooting, for instance, were more likely to benefit commercial hunters who could go elsewhere for game, wrote one Illinois shooting club in 1887, while the sportsman of moderate means could not.[108] Nonexport laws intended to cut off market hunters raised criticisms from hotel keepers who feared recreational hunters would not take hunting trips out of state if they were barred from sending their catch back home.[109] Bag limits that put the same restrictions on both professional and sport hunters, maintained "Cohannat" in an 1889 letter to *Forest and Stream*, would be better replaced by laws to outlaw hunting methods used by market hunters and to tax guns for the revenues to pay wardens. Such measures would "equalize the sport more" for consumers.[110] Selective policing by game wardens also rankled sportsmen. *Forest and Stream* columnist Hough complained that the Illinois state game warden, Maurice R. Bortree, "passed by a whole street full of offenders" in 1890 to pick on Henry Sloan, a Chicago game dealer but, more important, a member of the Mak-saw-ba Club and friend of Hough. He was "the very man they ought to have left alone," in Hough's opinion, because Sloan was "more sportsman than he [was] anything else."[111] And a sportsman, by definition, was a game protector, not a game exploiter.

By the 1880s, sportsmen had become an organized force working to reform game laws, particularly those related to wildfowl. They pressed for uniform legislation to govern hunting and possession seasons and bag limits while insisting upon more effective enforcement mechanisms to restrain wildlife consumption. Many activists initially insisted that their remedies were fair, designed to prevent overhunting in all its forms. Nevertheless, the actions of commercial exploiters more often took center stage in game protection debates. Although the commercial gunner could be "as manly and honorable as a sportsman," the commercial market was "threatening the annihilation of the game that [was] left." Killing for profit had thus become the equivalent of "bad sportsmanship."[112] Laws aimed directly at curtailing the workings of the wildfowl market found increasing support from sport hunters. But not every sportsman agreed that a hard line on commercial sale was required or was effective. Debates over the legality, efficacy, and morality of anti–market hunting laws swirled in sportsmen's associations, conventions, and publications. Time and again, however, the argument rolled back to the market's far-reaching effects on wild birds and the limited reach of state laws in protecting them.

What was becoming clearer was that application of game laws in the field had little chance of success so long as the modern wildfowl market continued to operate as it had. The "great trouble in trying to enforce the game laws," wrote "Brant" in 1882, was "that we have all the while been trying to get at the market hunters instead of the venders of the game after it is killed." The editors of *Forest and Stream* had done much to educate the public on the evils of commercialization, yet the market persisted and even grew. Indeed, the "nefarious business" was carried on "within a stone's throw" of the magazine's offices.[113] The legal actions and the public relations campaigns carried out by activist sportsmen, however, were having an effect on urban game merchants, according to Hough. Despite the dealers' "professed scorn of sportsmen and their efforts," they were "a little bit uneasy on account of the recent outlook." The crusading columnist told his fellow sportsmen that with increased effort "we can make them more trouble."[114]

Outlawing the Market

The kinds of trouble Hough had in mind included laws targeted to restrict the possession and commercial sale of game in the markets. Although reforms in most states during the 1870s had transformed scattered and poorly enforced hunting laws into consolidated state game codes, the rules regarding commercial possession and sale remained confusing and inconsistently applied. Many places with nonexport laws on the books still allowed shipping and sale of out-of-state game under certain conditions. The 1885 revised Illinois game laws, for example, made it illegal to offer for sale a number of species of native wildfowl, but the act permitted the state's inhabitants to "receive and ship game from other States, and expose and sell the same on the markets" between October and February.[115] Cold storage had further muddied the clarity of possession laws that were based on the assumption that out-of-season possession was "evidence of killing the game out of season."[116] Game law enforcement in most states was limited, and game wardens were often tasked with policing violations both in the hunting fields and in urban marketplaces. Convinced that wildlife commodification was the central cause of wildfowl destruction, activist sportsmen began to target the market more directly in the 1880s.

Consequently, sportsmen's enforcement campaigns focused on urban game dealers and retailers rather than individual market hunters. Wildlife protection advocates, taking a cue from the NYAPG's early efforts to regulate New York City's wildfowl trade, began more aggressively to pursue game dealers, transport companies, restaurants, and hotels for closed-season possession and sale of wildfowl in the 1880s and 1890s. At the same time, reformers agitated for legal changes to prevent commercial exploitation of wildfowl, convinced that the "powerful backing of the large game dealers in the cities" prevented "any really effectual laws" from being passed.[117] Sportsmen's magazines publicized urban examples of game law violations such as the 1890 "Delmonico Woodcock Case," which *Forest and Stream* exposed as an example of New York's refusal to enforce sanctions on elite restaurants for the illegal sale of wildfowl out of season.[118] In Chicago, the state sportsmen's association led the way in a string of "restaurant cases" that went to court in 1891. Activist sportsmen not only boasted of an unbroken winning streak in those cases but also claimed, according to Hough, that they "won victory without a struggle."[119] These successful raids on city retailers raised notice in circles beyond sportsmen's publications. In 1894, the *Chicago Tribune* reported on the efforts of the state game warden to prevent the illegal sale of quail at the Richelieu and Saratoga hotels, despite owners' objections that the wildfowl they possessed were legally killed and kept in cold storage.[120] In California, the *Sacramento Daily Union* warned consumers that shipments of illegally trapped game birds had found their way into the market and that buyers should "be careful to note whether or not" the birds they purchased were "free from shot-marks," which was a clear sign they had been trapped in contravention of the law.[121]

Faced by a rising tide of well-organized political and legal activism from sportsmen, game dealers fought back, creating their own associations and advancing game legislation favorable to the market. The formation of the American Association for the Protection of Game, Game Dealers, and Consumers in 1885 was just one piece of evidence that the "unwritten treaty" that had once existed between sportsmen and marketmen was broken.[122] The laws advocated by sportsmen, protested the rising class of avian entrepreneurs, constituted a restraint of trade and discriminated against marketmen. Dealers particularly resisted sportsmen-led efforts to restrict the sale of out-of-state game that had been killed legally and kept in cold storage.[123] Dealers also complained that sportsmen's game protection societies were "erratic" in their "prosecution of the offenders of the laws."[124] Game wardens, argued the trade publication *Ice and Refrigeration* in the early 1890s, should start trying to "earn their salaries by actually protecting the *live game* (the only kind worth protecting)." Otherwise, "the virtuous activities by the game wardens" would begin to look "suspiciously like horse-play and deliberate posing."[125] Marketmen were unfairly demonized in all the "game and fish literature," which cast everyone as "a scapegoat . . . except the 'sportsman.'"[126]

Working to blunt the most restrictive game legislation advocated by influential sportsmen were politically well-connected game merchants. In some cases, marketmen crossed into the game protection ranks, serving on state fish and game commissions and in sportsmen's associations.[127] Others actively lobbied for legislation favorable to their commercial interests. One of those was the Chicago commission merchant Colonel E. S. Bond of Bond & Whitcomb, who told *Forest and Stream* in 1891 that the dealers "owned" the Illinois legislature.[128] The South Water Street merchant further declared that "sportsmen could pass no laws in Illinois except as the game dealers allowed them to do so, as the game dealers controlled all legislation on that head." When Hough interviewed Bond for his Chicago and the West column, he found the dealer intransigent in his disregard for game protection. "You fellows aren't doing any good with your restaurant cases," Bond claimed. Crackdowns in Chicago simply sent the birds to Boston and other eastern cities, or sometimes to Europe, where market enforcement was less rigorous. Hough doubted dealers' public denials of game law violations. "Go to South Water Street if you want to know where the game is most abundant," he wrote. If you wanted to find prairie chickens, you could simply scan the letters from the many western towns in Bond & Whitcomb's files. Most appalling to Hough was Bond's breezy indifference to the disappearance of wild game. "If I had the last pair of game birds on earth," the game dealer declared, "I'd ring their necks."[129]

By the last decade of the nineteenth century, the market's damaging effect on wildfowl populations was a well-rehearsed discussion in activist sportsmen's circles and especially in outdoor periodicals. Writing about New York's "fish and game interests" in 1890, *Forest and Stream* asserted that while market hunters were "the active, combative forces that threaten the extinction of game," the market was "the complement of these destructive agencies" and the repository of ill-gotten spoils.[130] A few years later, the magazine made its opposition to wildlife commercialization official and absolute by adopting a "no sale" plank. It was a "general consensus of opinion," *Forest and Stream* asserted in 1894, "that the markets are answerable for a larger proportion of game destruction than any other agency or all other agencies combined." The effects of sportsmen, settlement, or agriculture were "trivial in comparison" to market hunting. It was time for sportsmen to organize in an effort to outlaw the sale of game in all seasons.[131] The magazine's uncompromising proposal made clear there was no common ground left on which both sportsmen and marketmen could stand.

Forest and Stream readers quickly jumped on the bandwagon. Although sportsmen still occasionally voiced support for market hunters through the latter half of the 1890s, a significant plurality of the journal's audience approved of the anti-market measure.[132] "I have long wanted to drive a spike in that platform of yours," wrote "Spike Horn," adding that complete abolition of commercial game sales was the only way to "put an end to the wholesale slaughter."[133] The campaign quickly grew beyond *Forest and Stream* as other outdoor recreation periodicals and game protection organizations joined the movement, advocating for uniform laws "prohibiting the sale of all kinds of game."[134] Edwin W. Sandys observed in *Outing* that the idea of outlawing the market was one he had broached years earlier, but he had thought that "to keep game from the market entirely would be too high-handed a measure to find favor in this country." Times had changed, though, and now "every sporting journal [was] clamoring for laws to prohibit the sale of game."[135]

Vocal sportsmen insisted that the linchpin in wildfowl protection was the abolition of the consumer market, not the eradication of wildlife consumerism as sport. The commercial shooter would be "compelled to turn his attention to some other purpose" if his market were eliminated. Plume species, proclaimed the revitalized, female-led Audubon movement in the mid-1890s, could be protected only by discouraging those who bought and wore "the feathers of any wild bird."[136] Massachusetts's deputy fish and game commissioner advised consumers to learn to shoot if they wanted game birds on their tables; otherwise they could eat pork.[137] One New York sportsmen's association member went even further to declare that "people can live without game" to buy but "the coming generation" needed a supply of birds in order to "have a little sport."[138] Wildlife conservation efforts would come to naught until the market in birds was closed. Dismayed by the quantities of wildfowl being shipped into cities, activist sportsmen clamored for laws to prevent "the handling of game in Chicago, St. Louis, New York, and a few other of the large centers of population."[139] If these principal wildfowl markets were eliminated, smaller centers of commercial sale would necessarily follow.[140] Game dealers, retailers, and shippers—in short, all the apparatus of the interstate market—had to be regulated out of existence before the market hunter would disappear.[141]

Game protectors found the legal backing for their anti-market cause in cases like *Magner*: early articulations of state ownership doctrines that extended state control over dead wild commodities as well as living wild creatures. Game codes in many states began to explicitly declare the "full control of game and fish . . . to be the property" of the state, even after capture. Such statutes, including Minnesota's 1891 laws, "embodied certain radical principles, never before incorporated in a game law," according to *Forest and Stream*. These represented the best in "practical game protection." With the complete right to and title in game, the state's authority to place "exceedingly stringent regulations respecting possession, transportation and sale" was practically unlimited.[142]

The state's ownership of dead marketed wildfowl was the central issue in a variety of court cases during the 1890s. In *American Express Company v. People*, a transport company insisted that its delivery of quail to a Chicago commission merchant could not be considered a violation of the 1889 Illinois game law, because the birds in question were the property of the killer. Preventing sale thus infringed upon the company's due process rights. The Illinois Supreme Court disagreed, arguing that "the legislature has never conferred an absolute property in quail upon the person who might kill the same." The state retained proprietorship even after the game was killed, leaving their captors with only a "qualified property in the birds after they are killed."[143] *Ex Parte Maier*, another significant state ownership case, came before the California Supreme Court in 1894. Taking even more direct aim at the interstate game trade, the justices found that "game lawfully killed elsewhere and brought here as private property" to sell was still subject to the state's control and regulation. Although the court conceded that, normally, "the power to regulate is not the power to destroy" property, the state's power to regulate wildlife was so complete that it carried with it the right to restrict the use of game even to the extent that it, in effect, destroyed such private property.[144]

Legislative and judicial actions that reached into merchants' storerooms to uphold state ownership of wildlife commodities struck particularly hard in

Chicago. State game warden Maurice R. Bortree, who had replaced the much-maligned "Stonewall" Brusewitz in Cook County at the behest of the Illinois State Sportsmen's Association, argued in his first annual report that the "extermination of our game is largely, if not entirely, to be charged to the market hunters and the great market centres of the United States."[145] After some aggressive policing, Bortree touted his success in gaining compliance with the law early in 1893.[146] Dealers' refusal to sell out-of-season game, the warden boasted, had "close[d] the greatest game market in America to violators," and their actions had done "more to preserve game not in Illinois only, but throughout the entire West, than all the official agencies that could be set in motion."[147] But less than a year later, Bortree's successor, Charles H. Blow, found that game dealers continued to circumvent restrictions on out-of-season game. Bringing them to heel was a daunting task. His first actions as game warden, including several seizures of game from Chicago marketmen, initially met with approval from *Forest and Stream* and columnist Hough.[148] Blow, however, quickly grew discouraged with the judicial outcomes of his hard work. Local courts and sympathetic juries, the warden complained to the *Tribune*, made the game laws ineffective. Despite his best efforts to prevent the illegal sale of game in the city's hotels and restaurants, he was not confident that any successful prosecutions would be forthcoming. "Justices of the Peace were in the habit of dismissing cases for violation of the game laws no matter how conclusive the evidence," Blow explained.[149]

It did not help that the new warden had a tough act to follow. Bortree was a game protection association insider, and he had the support of the city's activist sportsmen, especially the hard-to-please Hough. Bortree had made a name for himself among the South Water Street dealers as the "most cordially hated" man in Chicago.[150] Blow, in contrast, was an immigrant, having lived in Illinois less than ten years. He had been working as an undertaker before he was handed the game warden position through the intercession of his well-connected father-in-law, who happened to be a friend of the Democratic governor, John Altgeld.[151] Blow's new job, though, was nothing but an exercise in frustration. Governor Altgeld had filled only one of the three open state game warden positions, leaving Blow with "the protection of the game of the whole State on his hands."[152]

Without effective enforcement, the flow of commodified wildfowl through Chicago ran practically unabated. The city's markets remained well stocked with relatively cheap game, so much so that partridges and mallards brought in from points westward, often illegally, "rival[ed] plain beef in cheapness."[153] Fines collected from game law violators paid the warden's salary, but unsuccessful prosecutions meant no fines and no wages for Blow. The new warden's failures to get convictions against influential Chicago game merchants convinced him that it was a waste of time and effort to pursue game law enforcement in the market. There was "no use shouting after the game is killed," Blow insisted; "it is then beyond resuscitation."[154] The place to protect wildlife was in the field.[155]

Hoping to effect a rapprochement between activist sportsmen and Chicago game merchants, Blow ventured into legislative waters early in 1895, suggesting revisions to the Illinois game code that included expanded arrest powers for game wardens, increased numbers of wardens, and higher salaries for them that did not depend on funds derived from fines. Excluded from his proposal were protections for nongame bird species, prohibitions on the sale of Illinois game, and periods when sale of out-of-state game would be illegal. *Forest and Stream*, the *Chicago Tribune*, the local Audubon Society, and the Illinois State Sportsmen's Association exploded in opposition, accusing Blow of being in league with the city's game dealers and their allies. Critics argued the warden's proposal amounted to opening the wildfowl market to year-round sales. As proof of the alleged collusion between the warden and the city's game dealers, the *Chicago Tribune* reported that money furnished to print and circulate Blow's bill had been furnished by the South Water Street marketmen. The merchants were "in favour of protecting in every way possible the game of this State" so long as they could sell out-of-state game freely. "The lines are now sharply drawn," the newspaper opined, and it had become "a question of protection of game or extermination."[156]

Forest and Stream offered an even more negative depiction of the game merchants' efforts to sell the "Blow Bill" to sportsmen. Hough turned hyperbolic as his Chicago and the West column recounted the details of the "lively" meeting between the Illinois State Sportsmen's Association and the smug South Water Street merchants who had come to support the game law revisions (or force them down the throats of sportsmen, as Hough described it). He reviled the *Chicago Herald* and the Chicago sporting publication *American Field* (which Hough renamed *The Game Dealers' Friend*) for advocating Blow's bill. This was an "utterly vicious" proposal, "fatal to sportsmen's interests," Hough declared. "The animus of South Water street was shown up completely," he asserted, "and it was brought out unmistakably that there could be no union of interest between the sportsmen and South Water street."[157] In a slew of articles published on the Blow Bill, *Forest and Stream* repeatedly warned readers that "the game laws of this State were never in so critical a condition." The threats to wildfowl were coming from the very man charged with their protection.[158]

The embattled game warden broadcast his defense in the *Chicago Herald*. Admitting that he had made some unintentional mistakes in his original proposal, Blow undertook revisions and added provisions such as those protecting insectivorous species, plumage birds, and songbirds to satisfy the local Audubon Society. At the same time, he excoriated the naysayers. He had given sportsmen many chances to comment on his bill, he had lawyers vet the wording, and he had been open to changes, but his corrections met with equal disdain from activist sportsmen who were working to eliminate the wildfowl market. Blow testily responded by arguing that the way to stop commercial exploitation was to stop the game from being killed in the first place. Prosecuting dealers for birds held in cold storage for years was a waste of time. "You can't legislate a dead bird back to life," he argued. Honest merchants, commission men, restaurateurs, and hotel keepers were being "persecuted by the sportsman's associations and their wardens," who had nominated themselves as wildlife protectors and had forced through anti-market legislation. "How is such a law to protect Illinois game?" Blow asked. For his part, he was willing to work hard—he already had. In his short time as warden, he had "traveled all over the State" and had made 483 arrests for game law violations. He had spent his own money and worked for the cause of bird protection "for the pure love of it." He had "risked his life often for the birds," enforcing the laws against men who were members of the same sportsmen's association that was now treating him so badly. If he were given the powers he wanted, Blow promised, he would "wipe illegal killing and snaring out of existence," with neither "politics nor 'pull'" to sway his actions.[159]

The effects of the proposed "Blow Bill" rippled outward to other states where it found just as many supporters and detractors as it had in Illinois. Minnesota sportsmen took issue with Blow's parochial vision of game protection, which failed to recognize "that the more sparsely settled States must of necessity be the source of supply from which game must come to restock" other states' markets. While local markets might drain the game supply, major urban centers, especially Chicago with its massive wholesale and retail markets, posed the greater threat to western wildfowl.[160] In Nebraska, a proposal to institute nonresident licenses to keep market hunters out of the state was prompted, according to one *Forest and Stream* reader, by "the infamous Blow bill of Illinois."[161] Similar measures appearing in other states during 1895 were directly connected to the pending Illinois legislation that promised more-open commercial sales of game.[162] The NYAPG condemned an addition to the state's game law, dubbed "New York's 'Blow Bill,'" that permitted the year-round sale of game and put the state "in the position of a fence" receiving game stolen from other places.[163] The president of the North Dakota State Sportsman's Association made the connection between state game laws, the market, and wildfowl protection explicit. The legal situation in Illinois was "of vital interest to all law-abiding people in the North-west." If the warden's proposed bill should pass, C. E. Robbins wrote to *Forest and Stream*, "game protection in North Dakota would receive a fatal blow." It was surprising to Robbins that a game warden could sponsor a law so obviously "in the interests of game destroyers."[164]

The criticisms leveled by Illinois sportsmen and *Forest and Stream* followed the Blow Bill into the legislature, where it died an ignominious death.[165] A chastened Blow renewed his efforts to police the game markets, but he had spent his goodwill with the reformers. "The renegade warden," reported Hough doubtfully, "was saying now that he is the sportsman's friend."[166] Blow's actions, however, proved his scorn for recreational hunters and his sympathies for Chicago game dealers. "Some one else should be entrusted the duty of enforcing the law in South Water street," *Forest and Stream* insisted.[167] The market was an unstoppable force leading to the "utter annihilation" of wildlife, and sportsmen could broker no compromises with the business in birds. It had become clear to them that "market hunting and game-protection were perfectly incompatible."[168]

❧

In little more than two decades, much had changed. As wildlife populations visibly diminished, politically active elite recreational hunters embarked upon two complementary campaigns. The first was a project of self-identification: the invention of the sportsman-conservationist whose purportedly noneconomic interests in wildlife, especially wildfowl, uniquely qualified him as a game protector. Between the mid-1870s and the mid-1890s, newly formed associations and mass-produced publications helped to establish sportsmen's influence and authority on wildlife issues. The successes these activists realized were in large part due to the broader scales of organization and communication they developed during those years. Policy challenges and innovations in one state were quickly disseminated through interlocking networks of sportsmen's clubs, state and national game protective associations, nationally circulated publications, and state fish and game commissions. Elite, often urban sportsmen took the lead in advocating game law reform and more effective law enforcement to preserve wild birds. The organizational and public relations work of activist sportsmen in this period transformed wildlife conservation from a local concern into a national issue.

Sportsmen's conserving vision initially focused on limiting all forms of wildfowl consumption, recreational and commercial alike. They quickly realized, though, that modern, capitalist markets in wildfowl—often located in the same major urban centers as elite sportsmen and their organizations—exerted an independent force on birds that could not be contained by local hunting and possession laws. Game dealers, commission men, game shippers, and restaurant owners were not bound by geography or seasons; they operated across ecological zones, state lines, and jurisdictional boundaries. Activist sportsmen's engagement with the game bird and plume markets refocused their attention outward, inexorably pushing them toward national, or even international, perspectives on wildlife consumerism while convincing them that wildfowl commercialization was the crux of the conservation problem. Sportsmen's second major campaign was thus an anti-market one: a movement for legal uniformity to preserve migratory birds for recreational consumption alone.

The legal patchwork of state legislation, however, remained in place. Hallock renewed his plea for uniformity in 1897 when, at a meeting of the National Game, Bird, and Fish Protective Association, he called for "a code of cooperative legislation." Hallock's plan involved three geographically defined "concessions" (northern, southern, and Pacific) in which game regulations would be as uniform as possible. At *Forest and Stream*, Hough endorsed the former editor's "scheme for a system of game laws to cover the entire country" that was "stringent enough and wise enough to please any sportsman." But the "one concession" left out, Hough observed, was the one securing wildlife for the sole use of sportsmen in the future. This was "the concession which our friends the game dealers will not be likely to make."[169]

As Hough's comments suggested, the power of organized sportsmen to eliminate the market in birds proved insufficient. Despite some legal reforms and increasing restrictions, game merchants continued to ply their businesses, and overall wildfowl populations continued to decline. The passenger pigeon, whose numbers plummeted from millions of birds in the 1870s to scattered flocks of a few dozen by the mid-1890s, represented only the most visible evidence of the avian destruction. Game dealers had organized themselves to oppose limitations on the market and had begun to wield influence on the legal apparatus of state wildlife protection. For his part, Hallock had all but given up on sportsmen's organizations to effect any wider changes in Americans' consumption. He characterized efforts to organize a new National Sportsman's Association in 1892 as an "ancient scheme" and "a delusion" in regard to "any practical game or fish or forest protection." While Hallock lost confidence in organized sportsmen to provide solutions beyond those motivated by their own consumerism, he expressed his faith in governmental action and "the united efforts of groups of contiguous States to secure uniformity of laws and a certain union and co-operation in game and fish protection."[170]

Where activist sportsmen had succeeded, however, was in changing the conversation about the market itself. By the end of the century, debates about the commercial exploitation of wildlife, once confined mostly to sportsmen's circles, spilled out into a general public discussion about the ecological effects, economic ethics, and environmental legalities of wildlife consumerism. American audiences became far more willing to believe the worst about commercial hunters' depredations, as the Great Alaska Duck Egg Fake of 1894–95 proved. Reports appearing in numerous publications from Seattle to New York alleged that a mass destruction of millions of waterfowl eggs, harvested for their albumen to supply the photographic, leather-making, and confectionary trades, was being carried out in Alaska annually. This exploitation threatened the entire western population of ducks and geese, according to game protectors. "The greed of gain," trumpeted the *Chicago Tribune*, would consign all game birds to the passenger pigeon's fate. In response, senators from Oregon and Massachusetts proposed sweeping legislation to prohibit the importation of game bird eggs. In 1895, however, *Forest and Stream* launched an investigation into the story and found that no evidence from customs or transportation records showed any massive shipments of eggs coming from the Arctic. The fabulous story, editor Grinnell asserted, was simply an invention designed to explain the decline in waterfowl populations that were caused largely by excessive market shooting within the United States.[171]

Although the egg story turned out to be a hoax, the idea of containing or even abolishing the market in dead birds and other game gained political traction in the 1890s. State-level wildfowl sale and export prohibitions, upheld by key court decisions, revealed that the bright line of legitimacy between recreational and commercial hunting, which was not quite so obvious in the field, became much clearer when it was drawn directly through the market. When the US Congress discussed new game law measures for Washington, DC, in 1898, the issues of possession and commercial sale during closed seasons raised a particularly contentious dispute as politicians argued about circumscribing the free market. A law designed to constrain the city's businesses by prohibiting the sale of game killed legally in other states, argued Georgia's Representative William Fleming, was "such an obnoxious provision" that it constituted "an outrage" not only to hotel keepers and restaurant owners but also to "every man who has to eat in the District." Others disagreed. Iowa representative George Curtis insisted that Washington's relatively open market had made the city "the storehouse for 'pot-hunters' throughout the country." The lack of uniformity in game laws between states was part of the problem, but the ultimate cause for fewer wildfowl was the market. "The only way to enforce the game laws" elsewhere, declared Missouri's William Cowherd, was "to stop the man who hunts for the market, and the only way to destroy the market-hunters [was] to close the market."[172]

Watching these developments from Kewanee, Merritt was dismayed by the governmental actions which ensured that what little legal wild game remained would "fall into the hand of the city sport." Those elite sportsmen could always "outclass" and "outbuy" the ambitious young man who sought to make a name for himself "in the field with dog and gun" as Merritt had done. The situation was so obviously unfair, unethical, and undemocratic. More galling was the fact that the formerly immense flocks of birds would continue to decrease as the wild grass was plowed under and the watercourses were drained. The citified sportsman-consumer, protected by his powerful friends, would kill off the final stragglers.[173] Caught between ecological change and sportsman-dominated politics, the future of Merritt's game business looked increasingly uncertain.

CHAPTER FIVE

The Criminal

You can't legislate a dead bird back to life.
    CHARLES H. BLOW, ILLINOIS STATE GAME WARDEN

From his perspective in rural Kewanee, Illinois, Henry Clay Merritt saw all sorts of collusion and corruption in the game law apparatus, but it had been safely confined to Chicago. Only major urban dealers had time to lobby for legislation favorable to their interests, and bribes paid to game wardens like Blow kept the South Water Street marketmen in business.[1] Merritt's operations, far from the city, had escaped official notice—that is, until 1895. That summer, the game laws came to Henry County along with Warden Blow, who was desperate to revive his game protection credibility. The incident began in July with a decoy letter, sent by Blow under the signature of a prominent Chicago sportsman, asking Merritt to supply out-of-season birds for a game dinner. Merritt unwittingly filled the order and offered even more from his freezers if needed. Blow seized the shipment, hustled down to Kewanee, and confronted the game dealer with the evidence of his violations of the 1889 game code. According to Merritt, Blow offered to make the case go away for the cost of a two-hundred-dollar bribe. Merritt refused. In his opinion, there was no crime.[2] The next day, after Merritt left for Chicago on a business trip, Blow and his deputies broke into the dealer's freezer, where they discovered barrels of quail, prairie chickens, and other birds "finely preserved" with "some of them having been there for ten years."[3]

News of the Kewanee affair spread quickly. Headlines in the *Chicago Tribune* touted Warden Blow's "Big Haul" at "a Game House" and the "Illegal Kewanee Business" he had shut down. Depending on which of the several conflicting *Tribune* accounts one read, the game warden had seized anywhere from eight hundred to nearly one million birds, and the game dealer faced fines from $7,500 to $15,000,000. The *Tribune* also variously reported that Merritt had resisted the warden's attempts to enter his storerooms, or that Merritt had fled Kewanee prior to the raid, or that he been arrested at the train station.[4] The *Kewanee Courier* called the contradictory accounts highly exaggerated. "The reports of Mr. Merritt's arrest, his attempt to run away like some condemned criminal, and the fine detective work of the game warden and other officials" made sensational items for readers' consumption, but they were hardly accurate depictions of the simple out-of-season possession and sale violations allegedly committed.[5]

Merritt secured legal counsel in Chicago, while his son Clarence did the same in Kewanee. On the day before the trial, Merritt's attorneys, William Keep and F. M. Smith, arrived from Chicago and "proceeded to make themselves merry with benzene" while telling all who would listen "that they had an important case" to try.[6] In the meantime, Warden Blow went to the Illinois State Sportsman's Association to ask for legal help in prosecuting what was sure to be a significant case. "People were growing rich," he told members, "carrying on a business that was no better than safe-blowing."[7] The association complied and put one of their own lawyers, F. S. Baird, on the case.[8] Although the *Chicago Tribune* claimed that Merritt's neighbors were "anxious to have him prosecuted," local papers were divided in their assessments of Merritt's guilt.[9] The *Kewanee Courier* opined that rigid protective laws were necessary to preserve "legitimate sport" in Illinois. For that reason, "many sportsmen (not professional hunters)" were going to take a great interest in the outcome of the trial.[10] *The Independent* took a different view, arguing that it seemed "Merritt has a perfect right to buy game in season and hold it as long as he may wish—selling doesn't affect the point of the law." Either way, this was sure to be made a test case by Merritt and other game dealers in the state.[11]

On the day of Merritt's trial, which commenced less than two weeks after Blow's freezer raid, a standing-room crowd of spectators made up of out-of-town sportsmen and locals gathered in the hall of Kewanee's public library to watch the high-profile proceedings. The game dealer confessed that he had thousands, or even tens of thousands, of birds in cold storage, but he insisted that the wildfowl in his possession had come legally from outside the state. The prosecution's cross-examination of Merritt was accompanied by a "running fire of objections, repartee, and side talk" from the defense attorneys who argued that "property once legally obtained cannot be outlawed by the lapse of a few months." The prosecution countered with the revised Illinois game code and the 1881 *Magner* precedent.[12] When Judge Isaac Pyle, a longtime friend of Merritt's, announced the decision, a surprised buzz swept across the room: guilty.[13]

Merritt's defense team immediately moved for appeal to the circuit court. Attorney Keep insisted that the guilty verdict was "the best thing for Merritt and all gamekeepers" since the constitutionality of anti-market game laws would be settled once and for all. This was "what Merritt and game keepers want," the lawyers argued.[14] The 27,000 birds held in Merritt's freezers were left in legal limbo while his appeal worked its way through the system. The widely publicized massive fines were likewise held in abeyance, waiting for the conclusion of the case. News of Merritt's conviction spread quickly, nevertheless. The *New York Times* and a number of other papers across the country ran stories about the massive amount of "contraband game" seized and the potentially enormous penalties Merritt faced.[15] In early August, *Forest and Stream* claimed this case was Illinois's best chance to settle the question of whether there could be property rights in wildfowl and to rein in the cold storage houses threatening to "despoil the entire Western country of game."[16]

In the meantime, Warden Blow redoubled his efforts, seeking out other violators among Chicago's marketmen. His actions proved to the *Chicago Tribune* that the game possession laws were finally being enforced in the city. The newspaper also reported that Merritt was "suing for peace," but there was "no settlement in sight."[17] Hough and *Forest and Stream*, however, were not quite so sanguine about Blow's sudden change of heart regarding the market. Hough grudgingly allowed that the game warden deserved credit for breaking Merritt's far-flung wildfowl business, but the columnist could not forget that just a few months earlier, Blow had been "howling against the wrong of persecuting cold storage men who had legal property in game."[18] Hough was convinced that Blow's interest in the Merritt case was nothing more than a bid for notoriety.

A second warrant to search Merritt's freezers, sworn out by State's Attorney Emery Graves, resulted in another suit against the dealer for the illegal possession of ducks and quail. Merritt found himself before the county criminal court once again in mid-August.[19] The game dealer fired his blustering attorneys Keep and Smith and retained new counsel. State's Attorney Graves filed more than three thousand pages of information on the case—making "one of the most peculiar documents ever filed in an American court."[20] The criminal court jury returned with a guilty verdict, but only for the 161 birds Merritt shipped, not the 27,000 birds left in his freezer. Assessed the minimum fine, Merritt paid a little over eight hundred dollars, not the hundreds of thousands of dollars that had been breathlessly predicted by many newspapers. After being denied a request for a new trial, the defense quickly moved their case to the appellate level.[21] Though unhappy with the modest fine, *Forest and Stream* was pleased to see that the state's attorney had taken up the cause. Finally, Hough claimed, "proper, legal, statutory action" would be taken and "the actual enormity of Merritt's sin" would become clear. United in their opposition to the market, sportsmen would see to it that henceforth "the life of the game dealer will not be a happy one."[22]

During the final years of the nineteenth century, *Forest and Stream* hardened its no-sale stance, convinced that the market was antithetical to wildfowl conservation. The modern game business was far too extensive to be controlled by local actions, its commodities far too mobile to be contained within existing legal frameworks. Although the no-sale campaign seemed to activist sportsmen as the obvious solution to a growing environmental problem, it represented a major shift in American attitudes about wildlife consumption. Singling out and criminalizing a particular mode of wildfowl use—market exchange—was a significant departure from previous wildlife laws that had stressed open access to living wild creatures and private possession of dead ones. Not surprisingly, resistance to such a dramatic change came from a variety of places. Charges of criminality and corruption flew between different wildfowl-interested constituencies, each accusing the other of seeking to monopolize wildlife for its own gain. As the twentieth century dawned, the debate over the market in birds sharpened as hunters, dealers, shippers, customers, game protectors, and lawmakers wrangled over what happened to dead birds' bodies.

The Poacher

The market hunter was the most obvious target for sportsmen's anti-market campaigns, but threading the needle of legitimate hunting was indeed tricky. If restrictions on killing were necessary to preserve the wild stock, both commercial and recreational hunters would be subject to limits.[23] No one had the inherent right to continue the exploitation "in a way inconsistent with maintaining the normal stock."[24] Game law advocates claimed not to "champion the club man" over market hunters but instead sought "honest, consistent game protection" for the market and for sport.[25] A few writers went so far as to insist, sometimes a bit disingenuously, that commercial hunting could be tolerated "so long as the system [was] recognized and legalized."[26] The repeated message from sportsmen's periodicals at the end of the nineteenth century was that hunting limits were required across the board and that they should be enforced equally on all wildlife users.[27]

After numerous state-level court decisions like *Racey* and *Magner* established the rights of states to regulate and control common property in wild game, legislatures imposed new restrictive hunting laws. Nonresident hunting licenses appeared in some states in the 1870s and became much more widespread after 1895. Nominal-fee, general resident hunting licenses were adopted in some places by the end of the century. States progressively shortened open seasons on wildfowl and set bag limits on certain game species.[28] Hunting methods such as baiting fields, shooting from sinkboxes or batteries, deploying live decoys, and using automatic weapons all became subject to critical scrutiny. Repeating shotguns, insisted *Recreation* editor George Shields in an open letter to the Browning Brothers Arms Company, were unnecessary and "more destructive" innovations that should be outlawed.[29] Such restrictions, activist sportsmen argued, constrained everyone. The high cost of nonresident licenses hit visiting sportsmen and smacked of unfairness—they could even seem un-American.[30] Bag limits and shortened seasons reduced their hunting opportunities.[31] Sportsmen might find that the increasingly restrictive laws would "pinch now and then," admitted *Forest and Stream*, yet it was sportsmen's duty to respect them. To do otherwise would be to become "worse than criminals."[32]

Although sportsmen felt the effects of new regulations, there could be little doubt that much of the growing corpus of late-nineteenth-century state game legislation targeted the market, and many game law supporters said so as they castigated market hunters for being "game hogs" and poachers.[33] Bag limits, shortened seasons, and costly nonresident and special market hunting licenses were enacted to make market hunting less profitable, and state-appointed game wardens circumscribed the actions of commercial hunters disproportionally. The "army of expert market-shooters," and not the sportsman's club member, wrote William Henry Atherton in an 1897 letter to *Forest and Stream*, was responsible for "destroying the game of the nation."[34] Professional hunters pursued their prey across the country, "emigrat[ing] with the birds" from north to south each year. "One market hunter will destroy more game in a season than twelve men who shoot for pleasure," claimed T. S. Ford, since it was the professional's business to hunt day in and day out.[35] With profit as his motive, the only limit to his killing was his supply of ammunition and his stamina. "Do away with the market hunter who hunts for the money there is in it," and the return of small game to the sportsman-consumer would undoubtedly be assured.[36]

Sport hunting game protectors disparaged professional gunners as criminals, claiming that they refused to accept reasonable limits. Commercial hunters had to be restricted, a California state sportsmen's convention insisted in 1896, "to a certain time in which to follow a rather lazy and indolent occupation, which is slaughtering ducks and quails." New York's game protectors argued that market hunters needed to be educated on the game laws, for if "left alone, these men become outlaws."[37] In 1908, the Virginia State Sportsmen's Association noted that the game laws they advocated were met "with the approval of every one except the market hunter."[38] Old-time market hunters, such as Jesse Poplar in Havre de Grace, Maryland, routinely found themselves at odds with the new regulations, caught up in the "usual number of arrests made" by the state's "ducking police" at the start of the season on the Susquehanna flats.[39]

In California, the contest of legitimacy between sportsmen and market hunters, which had previously been mostly confined to special-interest publications, became a public debate at the end of the 1890s. Controversy raged between the local newspapers in the Sacramento valley over game laws that favored sportsmen's bird killing. While the *Sacramento Bee* gave voice to commercial hunters trying to earn their livings, the *Sacramento Daily Union* was far less sympathetic to the plight of the "poor devil of a market hunter" and the effects of anti-market game protection.[40] "The market hunting 'industry,'" the newspaper opined in 1894, was "one in which a few men make a little money at the expense of everybody else." The business had to be "restricted and rigidly regulated" to ensure that common workingmen had access to living wildlife and recreational hunting.[41] Sport hunting critics possessed "the same disregard for property rights exhibited by the communist," the *Daily Union* scoffed. The truth was that "the men who engage in market hunting kill more than double the number of birds each season than all the club members and other sportsmen combined."[42]

By the first decade of the twentieth century, many market hunters saw the handwriting on the wall. Commercial wildfowling was fast becoming an illegal activity, and some professional hunters shifted to allied professions or began new endeavors. Quite a few reformed market hunters became game wardens or game protectors for private shooting preserves. Elmer E. Pedlar, an erstwhile California market gunner, learned to view his former life with shame after he got a job policing the game laws in San Francisco in the early twentieth century.[43] Many of the Chesapeake's commercial wildfowlers and big gunners were drafted into the area's special "ducking police," and in Iowa, game wardens Fred Hinshaw and B. V. Palmer saw little incongruity between their earlier experiences as market hunters and their new enforcement roles.[44] The fact that many market hunters "turned toward legal and proper pursuits" such as working as sportsmen's guides proved "the majesty of the [game] law" and the good effects of enforcement, supporters argued.[45] Many one-time market hunters agreed that the anti–market hunting laws had ameliorative consequences. Iowa's Robert Miller described himself as "the strictest crank on observing the game laws," though when he was younger, he "couldn't appreciate the law." Echoing Merritt's self-description, Miller explained: "You see, I was born with a gun in my hand and I couldn't believe that there should be any game laws passed . . . game was here to be shot . . . and we didn't think it was wrong to shoot all the time."[46]

A significant number of other professionals, however, continued wildfowling for profit, often working across state lines and beyond states' jurisdictional boundaries. One outlaw recalled that all the laws prohibiting waterfowl market hunting merely raised the price and made the hunting more fun. Evading law enforcement in the field and concealing illegal wildfowl during shipment provided new challenges to hunters and a sense of accomplishment when they were able to outwit game wardens successfully. Commercial shooters worked out elaborate signals with railroad workers and bridge tenders to warn them of law enforcement operations. Hunters tossed their illegal big guns, tied to a float and a block of rock salt, overboard to escape detection. When the salt dissolved, the gun rose to the surface to be recovered.[47] Even when caught violating the game laws, professional gunners could hope that sympathetic judges and juries would not convict them, especially in smaller jurisdictions where justices of the peace or county judges had authority.[48]

As the customary practices of market hunters were progressively banned, contentious conflicts between them and game law enforcers erupted, sometimes with violent consequences. Game wardens grew accustomed to being physically threatened by professional shooters. In Illinois, Warden Harry W. Loveday reported that a hunter violating the new nonresident license requirement in 1900 promised to shoot the official trying to enforce the law.[49] California fish and game officer Pedlar likewise recorded the "threats to kill certain people" leveled by a poacher.[50] The dangers that game wardens faced went beyond verbal threats in some extreme cases. South Florida game warden Guy Bradley was killed in 1905 while investigating illegal plume hunters operating at the Lake Cuthbert heronry. Even though there was no question about who shot Bradley, his killer was not indicted. The anti–market hunting laws Bradley was enforcing had little support in the Florida everglades, and the grand jury took the plume hunter's word that he shot in self-defense.[51] Bradley himself had predicted that vengeful professional plumers would "do him in the end."[52] Another Florida warden, Columbus G. McLeod, disappeared in 1909. His sunken, bloodstained patrol boat was found but his body was not. Once again, no one was charged with the murder. Killings like these proved it was "dangerous business to interfere with the honourable industry of plume-hunting in Florida."[53] California's market hunters were known to perpetrate similar acts of violence when cornered. A pair of poachers killed one California game warden near Los Banos in 1914 after being "halted to submit to a search to see if they had more ducks than the law allowed." Gunfire erupted between the parties, killing Warden Rodolph and fatally wounding Earl Farnsworth, the market hunter.[54] These violent confrontations were the clearest proof that the market hunter—practiced in evasion and illegality—was more criminal than not.

Market hunters' public image was transformed at the turn of the twentieth century. What had been a legitimate—though at times questionable—profession, had become illicit. The change was rapid. In the early 1890s, the public could still believe that an affinity existed between recreational and professional hunters. "The shooting of a game-keeper at Suisun by a market

"Game Wardens with Confiscated Armada." Market hunter weapons, such as big guns or battery guns (sometimes called armadas), were progressively outlawed at the turn of the twentieth century. Game wardens, posing with their cannon-like trophies, used photos like these to publicize their efforts to police illicit commercial hunting. Image from the Dr. Harry Walsh Collection (accession no. 213-0006), courtesy of Chesapeake Bay Maritime Museum, St. Michaels, Maryland hunter," reported the *San Francisco Call* in 1891, for example, caused not an outcry but rather a "feeling of regret among sportsmen" who believed in "the society of good fellowship which exists among lovers of the gun and dog."[55] By the early twentieth century, however, the line between legal sport and unlawful commercial hunting was much more starkly drawn. Sportsmen and bird lovers led the crusade against market hunters, advocating any kind of legal restriction on "the game slaughterer," while newspapers and magazines highlighted the differences between "sportsmen, market hunters and butchers" for the modern public.[56] The game butcher killed for the sake of killing, and the market hunter was "out for business," Edgar F. Randolph explained in a 1906 issue of *Forest and Stream*. Both were immoral and illegitimate forms of wildlife use. But all too often "the sins of these two classes" were "indiscriminately laid on the shoulders of the sportsman by people who have a misty idea about real sport," Randolph complained. The nonhunting public needed to be educated about the correct ways to kill and consume wild creatures.[57]

Criminalizing the Market

While license laws, bag limits, and shortened hunting seasons applied to sportsmen as well as market hunters—a point often repeated by amateur hunters to counter professionals' claims about unfair treatment—other late-nineteenth-century legal innovations more directly targeted the vast marketing system responsible for remaking birds into commodities. A smattering of state-level nonexport laws, which banned the transportation of all or some game species across state's borders, sought to prevent the commodity mobility that kept urban game dealers in business. Those laws were challenged before the US Supreme Court in the 1896 case of *Geer v. Connecticut*. Game dealer Edward M. Geer disputed his conviction for transporting legally killed and purchased Connecticut ruffed grouse, woodcocks, and quail to New York, contending that the state's prohibition on the export of game was an unconstitutional interference in interstate trade and commerce.[58] The court concluded that no such interference existed. The key question, wrote Justice Edward White, was "the nature of the property of game." Since wild creatures were the common property of the people of a state, the state had "undoubted authority to the taking and use of that which belongs to no one in particular, but was common to all."[59] It was an open question whether the killing and sale of such wild property constituted commerce at all.[60] With the High Court's landmark ruling, the state ownership doctrine gained a new level of legal legitimacy, and many sportsmen rejoiced in the wide latitude these decisions appeared to give states in wildlife protection.[61]

After the 1896 *Geer* decision established their constitutionality, more states added nonexport laws to their game codes.[62] *Forest and Stream* included nonexport for market sale in its official anti-market platform, convinced that the court's action provided "clear sailing" for the imposition of even stricter penalties on game shippers.[63] In the late 1890s, however, these laws were still novelties. They were hardly universal, and they were particularly difficult to enforce. In many places, dealers could still "purchase and sell all the game they wish[ed]" provided it came from other states.[64] Determining the origins of wildfowl transported and sold, however, was no easy task. Commercial hunters and dealers employed a variety of stratagems to supply the consumer wildfowl market, using fictitious names to cover their excessive killing after daily bag limits were imposed and disguising game bird shipments as other products, often domestic poultry. Railroads and shipping companies claimed ignorance of illegally transported game, arguing that they lacked inspection personnel or enforcement authority to police wildfowl shipments, but they had little incentive to stop the trade.[65] Inadequately funded and staffed state conservation agencies proved to be incapable of checking every shipping record plus the actual contents of the variety of barrels and boxes being moved, making it almost impossible to determine who was liable for unlawfully transported game.

Despite the lawmaking to protect wildlife within states, dealers and merchants in some of the largest urban game markets found that they could easily acquire illegal wildfowl "if they went about it right."[66] In the 1890s, Illinois and New York prohibited the sale of certain types of game killed within the state but permitted the marketing of the same kinds of game imported from other states.[67] Section 249 of New York's 1895 game code expressly exempted the transportation, sale, and possession of out-of-state game, so long as it came from at least three hundred miles away, thus opening the New York City game markets to year-round sales.[68] In Chicago, commission merchants and game dealers insisted that they did not want to sell Illinois game, but they jealously protected any encroachment on an open market in which to sell imported birds. So long as markets existed for imported game, especially those in large commercial centers like Chicago and New York, the temptations to violation were too great. Since many states had nonexport laws on the books, much of the game arriving in Chicago's markets was "stolen property," the *Chicago Tribune* insisted in 1899.[69] New York's liberal imported-game statute ensured "the steady progress of game from the country districts into the city markets."[70]

Transportation of such out-of-state game posed thorny legal issues and complicated game law enforcement structures designed to work within state jurisdictional boundaries.[71] The birds that found their way into markets in Chicago, New York, and other cities in the last decade of the nineteenth century exposed the limitations of nonexport laws. If game shippers could manage to remove wildfowl or other game from a state, they could market them to consumers in another. Both intrastate and interstate laws were inadequate or ineffectual when market hunters took or shipped game from one state to another. Game law enforcement stopped at state lines. Once in the market, a bird killed and transported illegally in another state was indistinguishable from legal game. The existence of game markets anywhere also proved to be an incentive for poaching within a state. Dealers could always claim an out-of-state provenance for the birds they sold. These realities of the market hindered the work of game wardens since they had to ascertain the true origins and destinations of game. The onus was on them to prove the illegality of transported wildlife.

In 1900, Congress waded into this confusing welter of state-level game laws and, in the process, took an important first step in reshaping the state ownership doctrine that *Geer v. Connecticut* had firmly established only four years earlier. Representative John Lacey of Iowa proposed a federal law banning the interstate trafficking of game killed in contravention of state laws. The measure, designed to supplement the states' wildlife conservation efforts, would extend the reach of states' legal control over their wild animal property. The impetus for Lacey's insistence on federal action was clear: the market in birds. It was "the facility of commerce in these days of rapid transit" that enabled "the violator of the State law to market the product of his crime at a distance, and thus defy the laws of his own Commonwealth." The revolutionary federal action to protect insectivorous birds, plumage species, and migratory wildfowl, he argued, was not motivated by sentimentalism. He considered himself a "true sportsman" who was the best friend of birds. Indeed, "the person who does not love birds ought to be classed with the person who has no love for music," opined Lacey, "fit 'only for treason, stratagem, and spoils.'"[72]

Opponents of the proposal raised the issue of constitutionality, arguing that such a law came "very close to producing a conflict between the States over their jurisdiction and the Federal Government." Not so, Lacey insisted; he advocated no "national game laws." Instead, the act was limited "to the prohibition of interstate commerce in those things which the State prohibits."[73] Although Lacey argued that the federal legislation he advocated was supplemental to state laws, its premise for constitutionality contradicted many of the legal precedents concerning the nature of the state's property rights in wildlife. Since the 1881 *Magner* opinion, courts had maintained that states' game codes were based on the understanding that individuals could not acquire property rights in game, even once killed, and therefore wild game could never become legitimate articles of commerce. In this way, state nonexport laws did amount to unconstitutional intrusions into the federal government's jurisdiction over interstate commerce. Lacey, however, reversed the argument in order to justify federal responsibility for wildlife protection. It was because birds and other game were transported between states as articles of commerce that Congress had the authority to regulate the trade.[74]

When the Lacey Act was signed into law on May 25, 1900, activist sportsmen celebrated its passage, convinced that "a growing public sentiment in favor of more stringent game laws and their more rigid enforcement" was sweeping the land.[75] The new federal law was "one of the best agencies ever put in force" to establish legal uniformity between states, insisted state fish and game officials. The effects of the federal legislation had yet to be fully realized, but they were certain to be "far-reaching." Federal action promised to eliminate the illegal marketing of wildlife. *Forest and Stream* anticipated that the market hunter and game dealer would become extinct within the following decade.[76]

Those predictions were premature. The passage of the Lacey Act did not immediately eliminate the marketing of wildfowl, and commercial game interests struck back at anti-market measures. Utah's state game commissioner, John C. Sharp, warned the state's legislature in 1901 that commercial shooters and game dealers would petition to have the state's nonexport laws repealed. These market men were persistent violators of the game laws, Sharp contended, who shipped "wild ducks and perhaps other game out of the State under false pretenses, packing them in boxes and packages falsely marked as butter, poultry, fruit, or something else." Altering the current laws would concede the state's ownership of wildfowl to the "avaricious and unreasonable requirements of the markets and game dealers."[77] Game shippers and transportation companies continued to operate during the first decades of the twentieth century despite sportsmen's insistence that "the shipping as well as the killing is unlawful."[78] Increasing restrictions left Chicago's marketmen undaunted. The lack of stock in the 1900 season, reported one South Water Street merchant, could not "be attributed to State Laws or to the Lacey Act." There were simply too many shooters after too few birds. In his view, supply in the field, not law in the marketplace, controlled the availability of wildfowl for sale.[79]

Shippers and dealers did more than simply evade the game laws; they mounted numerous, and usually unsuccessful, legal challenges to state and federal anti-market measures. The 1906 New York case of *People ex rel. Hill v. Hesterberg* expanded the Lacey Act's reach beyond US borders, with the justices concluding that the federal law was "sufficiently comprehensive to include all game brought into the state," even imported, foreign wildfowl.[80] In the same year, the Arkansas Supreme Court upheld the state's nonexport law and determined that the Lacey Act extended the state's jurisdiction over all living and dead game brought into the state, not just the wildlife living within the state's borders.[81] Charges against shippers were filed in courts in Oklahoma, Georgia, Illinois, and other states where justices held that prohibitions against the transportation of wildfowl across state boundaries were clearly constitutional.[82] "Certainly, since the passage of the so-called Lacey act," went the opinion in the 1912 Georgia case of *Allen v. State*, "there can be no question as to the validity of the game statutes in the different states, so far as interstate commerce is concerned."[83]

Increasingly, commercial game wholesalers, commission agents, shippers, and retailers were charged as criminals who knowingly sought to evade state nonexport laws. Courts largely resisted defendants' charges of unconstitutionality, maintaining, as the California Supreme Court did in the 1902 case of *Ex parte Kenneke*, that state game laws gave "all persons the same right" to kill birds within the same limitations.[84] The game dealers' dodge of claiming they sold birds imported from other states quickly lost its effectiveness as a defense. In the 1905 Minnesota case of *State v. Shattuck*, the justices found that ruffed grouse killed in Wisconsin and sold in Minnesota were "a part of the mass property of the state and subject to its police powers" once the dead birds were brought across state lines. "If there was ever any doubt about this proposition," they continued, it was "set at rest" by the Lacey Act.[85]

Marketmen had few illusions about the targets of federal and state wildlife protection laws. Nonexport laws and restrictive possession limits impacted the market directly, and game dealers resisted what they saw as naked class legislation. In 1896, the *San Francisco Call* reported that game merchants had "instructed the hunters to continue shipping game" out of season and "to pay no attention to the law."[86] According to *Forest and Stream*, St. Louis commission merchants, restaurant owners, "and others interested in the preserving of game by the cold storage process" during closed sale seasons were organizing to revoke some of Missouri's restrictive possession laws in 1898.[87] New York City game dealers, relayed one trade journal in 1899, were "more or less exercised of late owing to the activity of the State game authorities," and they, too, were "ready to organize in an effort to have some of the Game Commissioner's rulings overruled."[88] In the same year, some fifty butchers and poultrymen in Brooklyn formed the Retail Butchers' Protective Association "for the purpose of taking some united action in the various game suits threatened by the State authorities against butchers and dealers" who were caught with wild game after the close of the legal season.[89]

Several high-profile "freezer cases" in the early twentieth century tested the limits of newly enacted possession laws and exposed the complicated interactions between disparate state game codes, interstate wildfowl trafficking, and the modern consumer market in birds. New York's freezer cases, like Merritt's earlier run-in with the state of Illinois, publicized the expansive scale of urban avian commercialization. Staggering amounts of game held in dealers' freezers exposed "the astonishing extent of [the market hunters'] ravages." The 48,000 birds seized from the Arctic Freezer Company during the closed season in 1901 represented just "a small fraction" of all the birds held at "the great cold storage companies" in New York, complained *Forest and Stream*, that allowed the wildfowl business to operate year-round.[90] Game protectors advanced the novel position that possession in the closed seasons—even of birds that had been lawfully purchased and transported during the open season—was a crime. These provisions could "be enforced alike against hunter and dealer, agent and employer."[91]

The New York state supreme court, however, saw things differently when ultimately deciding the case against the Arctic Freezer Company in 1904. The state's game law of 1900, read the opinion in *People v. Bootman*, could not be interpreted as applying to game acquired and imported legally during the open season. "Wherever these birds became the property of the defendants and lawfully possessed within this State," opined the court, "such ownership and possession were as much protected by the Constitution as was any other kind of property similarly situated." The subsequent passage of the Lacey Act could have no effect on the interpretation of the state's previously enacted game law. The justices insisted that the intent of the current game code was limited to the protection of New York's bird life. The state's authority did not extend beyond that.[92]

At the same time, though, the court opened the constitutional door to more-restrictive laws by arguing that the state could revise its game codes to make any possession during the closed season illegal. Laws to "close the game market throughout the state during the period of prohibition, in order to remove temptation from poachers and pot hunters," were perfectly valid. Legislators could go even further to "prohibit the possession of game . . . during the close season, even if it was taken in another state and brought here [legally] during the open season," since there was no way "birds killed in Michigan can be distinguished from those killed in New York." The Lacey Act had "taken away all questions of interstate commerce, so that the state can act with entire freedom."[93] That the state of New York did, as Chief Game Protector J. Warren Pond noted in his 1902–3 report. Lawmakers repaired "the defective points" so that the possession statute "conform[ed] to what the courts held it should have been." Consequently, "the masses of people interested in the protection of song and insectivorous birds" had "a law to stand on" despite the New York Supreme Court's ruling in *Bootman*.[94]

Urban marketmen found it inconceivable that game laws could abolish their rights to hold legally acquired wild property in storage, and, in the first decade of the twentieth century, game sellers initiated a variety of closed-season and imported-game possession legal challenges. In *People v. Martin* (1908), grouse, quail, and plover found in a New York café by game protectors after the open season's end were determined to be held in violation of the state's game laws, and their seizure was not "unconstitutional as a deprivation of property."[95] In the same year, the possession case of *Silz v. Hesterberg* went to the US Supreme Court. New York game dealer August Silz had imported golden plover from England and blackcock from Russia to sell during the closed season. He argued that foreign birds were exempt from the provisions of the game law and the Lacey Act. The justices disagreed, holding that the right to prohibit possession was a justifiable use of the state's police power "independent of any authorization thereof by the Lacey Act." Although the law's purpose was "not to affect the legality of the taking of game in other States" or countries, the government had a duty "to protect the local game in the interest of the food supply of the people of the State."[96] Implicit in the

tangled decision was the recognition that wildfowl commercialization complicated the issues of consumption and preservation, obscured the distinctions
between local and distant, and confused simple definitions of property and
commerce.

Sportsmen and conservationists predictably cheered the High Court's action, claiming that the "celebrated Silz case of New York . . . settle[d] forever the
constitutionality of the question of the power of the Legislature in the matter
of bird and game laws." Although the market in birds continued into the twentieth century, the business was more often operating clandestinely. Game birds
were "no longer exposed openly in the markets" in most northern states,
claimed sporting writer Dwight Williams Huntington in 1903, even though
under-the-counter sales persisted.[97] Despite confiscations from cold-storage
warehouses and large fines levied against dealers, marketmen proved willing
to continue their businesses illegally.[98] Game sellers' actions placed them
in the same categories as their compatriots the game shippers and market
hunters—as criminals in the conservation crusade and as villains in the public
eye. Wildfowl merchants found their enterprises confined within narrowing
legal limits and crippled by vigilant game wardens and self-appointed game
protectors. "A new era has dawned," declared *Forest and Stream* in 1900, one
that would require marketmen as well as market hunters to obey the law.[99]

It had become evident to activist sportsmen and their allies that the market
itself was the corrupting influence. Conservationists could abide no agreement with market shooters and game dealers once they determined that commercial sale itself was a crime against the state's wildlife. Utah's state game
commissioner argued in 1901 that the only course of action was to immediately "prohibit by law the marketing and sale of all kinds of game animals
and game birds," before commercialization destroyed what remained. To do
otherwise would be "to legalize a great wrong."[100] *Forest and Stream* adopted
a firm no-sale stance, leading the call for "thinking sportsmen" to "unite on the
doctrine of 'Stop the sale of game.'"[101] Many states established new, no-sale provisions in the first years of the twentieth century, and Hough reported that
Texas was "edging toward the [no-sale] plank" in 1903.[102] Other sportsmen advocated passage of similar laws in their home states.[103] Only an absolute prohibition against commercial sale would "rid the land of the plumage gatherer,
the brutal destroyer who tears off the wings and heads of all species of birds,"
and "put down the market shooter" who decimated the once plentiful stocks
of wild birds.[104]

But state-level bans on wildfowl sales evolved slowly and in piecemeal fashion at the start of the century. "Legislation restricting the sale of game," wrote T. S. Palmer of the US Biological Survey, was "passing through a transition stage" as states variously outlawed the sale of all or certain kinds of game: some only during closed seasons, others during open seasons, and still others year-round. Exceptions and exemptions in state game codes abounded, resulting in a confusing patchwork of regulations.[105] Marketers' challenges to no-sale laws were sometimes successful and exposed legal deficiencies in early attempts to criminalize the game market.[106] No-sale campaigns aimed at the consumer market in plumes likewise met with spotty success. State and federal laws prohibiting the commercial use of plumage or skins of wild birds met serious resistance from milliners.[107] Trade publications called on "every man or woman connected with the millinery trade" to "lend his or her aid" in "crushing such a measure."[108] When lawmaking failed, the resurrected New York Audubon Society pursued a voluntary compact with the Millinery Merchants Protective Association of New York in 1903 that assured sellers of no "illegal interference on the part of game wardens" in their businesses. In return, dealers promised not to import feathers or skins of hummingbirds, terns, gulls, grebes, or songbirds and to publish a list of prohibited species in the association's trade journal.[109]

Activist sportsmen and conservationists, however, found this continued "vulgar taste in millinery" to be evidence that voluntary measures were insufficient to rein in the market's depredations. Cooperation with dealers and emotional appeals to consumers were of little use. The only real protection for birds, *Forest and Stream* asserted, would come after commercialization was outlawed. To do so, however, required a significant alteration in the public's understanding of wildfowl's worth. Statutory changes promised such a transformation. Laws could criminalize the market in birds and convince people that wild animals were "possessed of certain rights."[110]

From Merritt's perspective, it was not those dealing in birds who were the criminals. Market hunters, express companies, commission agents, cold storage warehouse owners, and game dealers were simply carrying out a business—heretofore legal—that served consumers' desires. Of course, there had been game laws, but in earlier times they had been of little consequence. Revised game codes in the 1870s and 1880s, however, were increasingly crafted with the big-city game dealers' interests in mind. Suddenly, the nonexport and possession laws that had been on the books for years were being applied in ways that raised the old game dealer's suspicions. Merritt was convinced that the Chicago merchants on South Water Street had "concocted a plan to consolidate the trade and shut out others" in the mid-1890s. With the extra profits gained, those urban dealers bribed game wardens to look the other way while they went on with their businesses. The corrupt bargain between the city dealers and game law enforcers, Merritt thought, was the source of all his legal troubles.[111]

The unresolved outcome of Merritt's case left Blow equally perturbed. Since penalties had yet to be assessed and the confiscated game remained in cold storage until the question of ownership was resolved, Blow was left without any compensation for the arrest. When Illinois state's attorney Emery Graves took out the second search warrant on Merritt's freezers, he cut the game warden out of the second trial and left him with no claim on any future fines levied against Merritt.[112] In response, Blow attacked the prosecutor in the press, claiming that Merritt must have struck some sort of deal with Graves. "The whole thing from beginning to end bears the stamp of fraud, deception, and collusion," he complained. It was clear from the start that Merritt was to be let off with the minimum penalty. "Why do you suppose the state's attorney prepared all that information and took all his pains, not only without consulting me, but with an effort to conceal it from me?" Blow asked. "It savors too strong of bribery and I mean to prove it."[113] Attorney Graves refused to stand quietly in the face of the game warden's public accusations. He charged Blow with libel and secured a fifty-thousand-dollar judgment against him in 1896.[114]

Blow found no sympathy in the pages of *Forest and Stream* either. Hough's Chicago and the West column raked the game warden over the coals for his handling of the Merritt investigation. Hough was certain that Blow was far more focused on confiscating Merritt's stock and selling it at public auction for his own profit than on actually pursuing punitive fines against the dealer.[115] As Merritt's conviction was being upheld by the Second District Court of Appeals in 1896, the *Forest and Stream* columnist mused that it was "curious to reflect that Warden Blow never made any attempt to have the man Merritt fined at all" but instead "vigorously went after the illegal game." Blow had the support of "all the machinery of the commission business which openly sends out circulars to the wardens and offers to buy their game."[116] But in the Merritt case, Blow might have found himself "hoist on his own petard," Hough wrote, since the state's attorney had taken over the case.[117]

By the time *Merritt v. People* was decided at the Illinois Supreme Court early in 1897—with the state prevailing—Charles Blow was out of a job. *Forest and Stream* reported that he was off in Wisconsin somewhere, working in "the fish business." According to Hough, a warrant had been sworn out for his arrest in Illinois for impersonating a state fish commissioner and bilking an Illinois sportsmen's club out of seventy-five dollars.[118] Meanwhile, the new state game warden, Harry W. Loveday, went after Merritt's business again, seizing three barrels of prairie chickens at a Chicago railroad depot a few days before Christmas in 1897. Loveday maintained that because the birds confiscated were consigned for New York, the shipment constituted a violation of Illinois's nonexport and closed-season sale laws.[119] Merritt contended that the game warden's actions were literally warrantless. "The birds were not taken under a warrant, as so far as we can learn the Warden sold them for his own account and held the profits," Merritt later recalled.[120]

Merritt conceded that the game law, if strictly construed, prohibited the export of wildfowl from the state. But there was a loophole. The Illinois game code included a provision that allowed dealers to receive game from other states from October to February each year so long as it was intended for sale "in the villages and cities of the State."[121] To outwit the game warden, Merritt conceived of a plan by which he would ship quail and partridges to a poultry dealer who "by marking the barrels poultry . . . would escape the vexations of seizures and trouble and expense afterward."[122] His plan worked once, but when subsequent shipments in 1898 were seized and sold at auction, Merritt proclaimed his innocence. He sold his birds in the state—albeit to a shipper whose intent was to export the game—and therefore the nearly thirteen thousand possession and sale violations charged against him were unjustified. Both the warden's seizure of the birds and his pocketing of the profits from their sale were illegal, Merritt believed.

Although the trial court dismissed Merritt's suit against Loveday for recovery of his confiscated stock, the Kewanee game dealer came out of this new confrontation at least partially victorious. Much to the surprise of Loveday, the state's attorney, and game protection forces, the jury returned a verdict of not guilty, agreeing with Merritt and presiding judge Ramsay that selling prairie chickens within the state was not a crime. The state appealed the lower court's decision, arguing that it was clear Merritt's intent was to ship the birds to New York, but the appellate court upheld the original ruling in 1900.[123] Activist sportsmen criticized the decision. When coupled with the outcome of the earlier cases, which resulted in only a minimal fine and left Merritt in possession of his frozen game, it appeared that, once more, the Kewanee game dealer would go "unwhipped of justice."[124]

The Corrupt Game Protector

This last indictment, even though it ended in his favor, left Merritt more disgusted with the uneven workings of the game law and the capricious way game wardens enforced it. He found it hard to believe that game legally shipped into the state could not legally be shipped out of the state, but that was what Illinois's nonexport law decreed. He also found it hard to believe that the business he had carried on openly one year was deemed to be illegal the next. The laws had not changed in the interim. And he found especially hard to swallow that, once again, the large Chicago dealers found ways not only to circumvent the law but also to use it "to demolish the trade of outside dealers." That they did so with the apparent assistance of so-called game protectors made it even more galling.[125]

On this point, Hough concurred with Merritt. The *Forest and Stream* columnist had long been convinced that Blow was the creature of Chicago's marketmen. Blow's replacement, Loveday, likewise seemed a bit too solicitous toward the city's commercial game interests for Hough's tastes. The new game warden "believed in allowing a restricted sale at certain seasons," and he was convinced that a sweeping no-sale measure would never succeed in Illinois. Loveday also accepted the legality of game held in cold storage, arguing that game protectors "must not interfere with the markets too much."[126] Like Hough, Merritt was sure that South Water Street merchants colluded with the game wardens who kept close watch on his dealings (and lied about his operations) while ignoring surreptitious wildfowl shipments from the city.[127] Both Hough and Merritt agreed that the wardens applied the laws selectively in ways that benefited favorites at the expense of others. As Merritt struggled to continue his business under intensified scrutiny, he wondered if all the confiscations and legal wars waged against him would have any real effect on the birds or if they had been simply a ploy "primarily to fill the pockets of the Game Warden."[128]

While Merritt was out winning his latest case against the game law, his old nemesis Charles Blow resurfaced briefly in the press late in 1900. Authorities were searching for Blow, reported the *Chicago Tribune*, on a charge of attempted murder. Blow's wife alleged that her husband had turned up the gas jets during the night and blocked the doors of their home with chloroform-soaked rugs while she and her three children slept. She awakened before succumbing to the fumes and confronted her husband, who promptly packed his things and fled the family home.[129] Blow, who had been working as a billiards hall proprietor, was reportedly in "financial trouble." He left home with little to no money in his pocket. Although he was believed to still be in the city, police failed to locate the ex–game warden. After a brief flurry of newspaper stories about Blow's alleged misdeeds, he faded from public view.[130] His family seemed uninterested in pressing the case, as Blow's disappearance was "agreeable to his wife."[131]

That Blow was a fugitive from justice would not have surprised his critics in Chicago. Indeed, Hough would likely have enjoyed the spectacle—though it seems to have escaped his notice at the time. But Hough's campaign against the "ugly loophole" in Illinois game law that granted wardens like Blow the power to confiscate and sell game for their own financial gain provided ample proof that the market tainted all who touched it—even those entrusted with the law's enforcement.[132] When game wardens turned game suppliers, they became part of the problem with wildfowl protection, not its solution. They allowed the market in birds to continue—not clandestinely, but openly.[133] Illinois's "risky" confiscation and sale clause, which formed the basis of the initial suit in Merritt's freezer case—"probably the worst bungled prosecution that ever happened under any game laws," according to Hough—gave game wardens a financial interest in the continuation of the wildfowl trade.[134] Market hunters and dealers also acknowledged the weaknesses in game law enforcement that allowed them to go on with their commercial operations. In Iowa, R. C. Darr boasted that he had "bought the State Game Warden and the County Attorney and the County Game Warden off" so that he could continue shooting for a Chicago commission house. In fact, he had "corrupted the whole bunch."[135]

Accusations of corruption within the expanding ranks of official game protection grew ever more common at the turn of the twentieth century. Stories like the one Buck Lufkin read about a Louisiana game commissioner caught with 565 geese and 180 ducks were proof to many that wildfowl conservation laws were not intended to protect wildlife. They were instead simply a ruse to allow certain groups to monopolize the remaining birds.[136] State game wardens as well as game and fish commissioners came under fire for actions observers characterized as self-interested, politically motivated, and elitist. The Lacey

Former game warden Charles Blow's picture in the *Chicago Tribune* after his alleged attempt on the lives of his wife and three children by chloroforming, December 4, 1900. Image courtesy of Newspapers.com

Law added another level of surveillance and another outlet for corruption. Under the new federal law, special US deputy marshals responsible for policing interstate shipping wielded "powers which were clearly outside of those of any one state."[137] When each new administration brought revisions or retractions of the game laws, charges of capriciousness and corruption followed.

Chief among the critics of game law enforcement were the marketmen whose economic futures were jeopardized by stricter anti-market laws. But their opposition campaign now took a different tack than it had in the nineteenth century. Rather than making arguments about individual property rights, restraint of trade, and interference with interstate commerce, game dealers and market hunters sought to make theirs a common cause. Game laws and their enforcement, they now claimed, were blatant class legislation. The game protection system was designed to eliminate American consumers' free access to wildfowl and other game. Game law authorities used their powers irresponsibly, and sometimes illegally, to deprive the people of their traditional, common rights to wildlife.

On this point, marketmen and some recreational hunters agreed. Stricter laws that established licensing fees, shortened hunting seasons, and reduced bag limits seemed designed to suppress lower- and working-class hunters while leaving elite sportsmen untouched. As the following three accounts of state no-sale legislation controversies demonstrate, game protectors' efforts to eradicate the commercial market in birds exposed the political limitations of conservation policies that legitimized one form of wildfowl consumerism and criminalized others.

The St. Louis World's Fair was a remarkable event for many reasons, but one of its lesser-known features was its effects on the illegal game trade. During the 1904 fair, reported Missouri's fish and game warden, nearly four million pounds of wild game was sold in a place already well known as a major game market city. Most of what amounted to "the equivalent of 125 freight carloads of game" had been killed and sold in violation of the law, though no prosecutions were reported. Between 1903 and 1904, about a half million of Missouri's ducks were killed for the market.[138] The visible toll on the state's wildfowl populations galvanized lawmakers, and in 1905, Missouri's legislature passed the Walmsley Act, which prohibited "all commercial sale, trade, and transport of wildlife in the state." The new law also set open seasons for game species, protected nongame birds, and established new licensing rules.[139] Although public support for the progressive law seemed strong, the new restrictions on recreational hunters quickly proved unpopular. The target of hunters' ire was the state game warden, J. H. Rodes. Chief among their complaints was Rodes's expansive interpretation of the new, nonresident license law. The law required that hunters purchase a license to hunt anywhere outside their county of residence. The game warden, however, applied the law universally, demanding licenses from those hunting in their home counties. Rodes's legal logic was supremely flawed "in a syllogistical way," the *Iron County Register* maintained. The game warden's actions impaired the hunter's "old-time right to hunt, unlicensed in the county in which he lives."[140] That the game warden was "over anxious for 'fees,'" complained the Marshall, Missouri, *Republican*, was "no reason that people should be parted from their money."[141] On this matter the courts agreed, overruling Rodes's interpretation of the license law in 1905.[142]

Wildfowl merchants in St. Louis were equally outraged by game law authorities' overreach. Early in 1906, two cold storage companies sued Rodes and his deputy, Tell Grether, for more than one thousand dollars in damages, alleging that the pair seized birds held in their freezers in contravention of the law.[143] Another, more serious challenge occurred later the same year when Fred Heger, the secretary and treasurer of the Missouri Country Produce Dealers' League, proposed an organization of the state's game dealers to oppose sportsmen's club–sponsored anti-market laws. Chief Warden Rodes warned that any compromise with dealers would undercut game protection in Missouri. In states where game merchants had won concessions for limited sale periods, he reported, they used the open season "to fill larders and cold storage houses with an ample supply of game" that retailers could sell illegally "at great profit" during the closed season.[144] Heger, as the game dealers' representative, pushed a test case of the Walmsley Law's constitutionality by selling wildfowl taken in Missouri and Oklahoma. The Missouri Supreme Court upheld his conviction in 1906, however, finding no merit in the constitutional challenges Heger's attorneys posed.[145]

Undeterred, game dealers and market hunters "set to work to devise ways and means to restore former conditions which prevailed prior to the enactment of the Walmsley Law."[146] They scored a legislative success in 1907 when the state's general assembly and a sympathetic new governor gutted the statute, thereby legalizing the sale of local game "in the county of capture" and the sale of imported game during the open season.[147] The legislature also all but eliminated the state's game warden system and turned collection of hunting license fees over to county road commissions. The game dealers' influence on Missouri's political machinery represented "a disgrace to the age we live in," wrote one *Forest and Stream* contributor, and was evidence of the market's corrupting influence. "A Tarbellian study of the springs of action having to do with the passage of such a law," George Kennedy wrote to the magazine, "might afford valuable instruction in the abstruse science of legislation."[148]

Missouri's liberalized sale provisions, however, lasted only until 1909, when the state reversed itself again, repealing "that 1907 game law farce" and reenacting the major features of the former game protection law, including abolition of sale of all game protected by the state.[149] Dealers responded with attacks on both the legal restrictions and the state game warden, Jesse Tolerton. In 1911, Heger and the Missouri Country Produce Dealers' League threatened to "get busy" and "'go down the line' on a fight to repeal the law." He enlisted support from some half-dozen professional associations "to unite in a common fight against the game law" unless Tolerton recommended legislative changes that would open a two-month legal-sale season. The chief warden refused and dared Heger to "get busy." Almost immediately, a new, dealer-sponsored game bill appeared in the General Assembly. At the same time, Senator Carter Buford and Representative Frank Farris launched an investigation into the affairs of Tolerton's office based on charges of misappropriation of funds and political partisanship on the part of the warden and his deputies.[150]

For Tolerton, the events seemed all too reminiscent of those endured by his predecessor Rodes. As in Rodes's case, criticisms of the warden's department were the direct result of the "unpopularity of the law." Rodes had endured an investigation into his conduct as well, the result of which had been complete exoneration, according to Tolerton. The current warden's disgust over the "star chamber proceedings" to which he was subjected saturated the pages of his annual report for 1911. He was "the victim of a deliberate plot" orchestrated by people "who had been employed by large wholesale dealers." Adding insult to injury, Missouri's official state printing commission refused to publish Tolerton's excessively illustrated annual report, saying that it was submitted too late and was too costly to produce. Commission chairman Cornelius Roach did not regard "a score or more full-page pictures of yourself and the Governor 'Afloat on White River' as necessary information to be printed." Neither was, Tolerton waspishly retorted to Roach, "the picture of your family on page six of Missouri's Official Manual issued by you."[151]

Tolerton's self-published report laid out the new tactic disingenuously pursued by game dealers seeking to overturn Missouri's no-sale law. "Certain attorneys," he wrote in 1911, "wax eloquent during legislative game law discussions and paint the little towhead boy on the farm as being opposed by the game and fish laws which prevent him from shipping a few quails to market."

Wildfowl marketing, however, ultimately profited only commercial dealers who would "send out their expert shooters to kill every bird in the little towhead boy's neighborhood."[152] *Forest and Stream* agreed, noting with approval that the "farcical investigation" of Tolerton's office had proven unfounded and that the warden was now able to "put a quietus on the energies of the market hunter."[153] Yet the new attacks on protection of game from commercial interests, cloaked in the language of democratic access and cast against a backdrop of game warden corruption, gained public traction. Missouri's legislature cut funds for conservation, and the state's wildfowl laws continued to shift in the political winds between protection and exploitation.[154]

In New York, the market in birds and other game was still active in the first decade of the twentieth century. The state's game laws prohibited the sale of some native upland game birds but permitted seasonal possession and sale of a number of waterfowl species as well as imported woodcocks, quail, and grouse.[155] The restrictions on sale, however, irritated game merchants, who chafed at the "unjust and 'grafting' methods of the deputy game wardens in New York City."[156] Trade publications and merchant organizations agitated for investigations into game protectors and demanded the return of fees and fines. In 1907, the New York Game Dealers' Association announced its intention to protect "all dealers in and purchasers of game" from "the legalized blackmailing operations of certain employees of the State Fish and Game Commission." The game laws were hopelessly confused, the *National Provisioner* claimed, and "every winter the game faddists and sportsmen get the Assembly to add a few more." Sending representatives to Albany, the marketmen sought to influence bills proposing to eliminate fees from game wardens' compensation, thus removing wardens' "temptation to graft."[157]

Conservationists and activist sportsmen saw the game merchants' organizations quite differently, convinced that their opposition was an attempt to remove any market restraints. New York's millinery interests organized to oppose recently enacted state laws prohibiting the sale of wild bird plumage in the state, while the "market gunners of Long Island formed three strong organizations" with the avowed purpose of rescinding "all the laws for the protection of feathered game." This was a call to war, according to the prominent conservationist William T. Hornaday. Nothing but abolition of the game market would answer the call.[158]

New York's 1911 Bayne-Blauvelt bill was the conservationists' rejoinder to marketmen. The bill prohibited the sale and importation of all wild game, both native and imported, except for rabbits and hares. A last-minute conces-sion to game dealers allowed for the sale, under strict licensing conditions, of some privately reared and foreign game species.[159] The legislature's ap-proval of the sweeping law, with only one dissenting vote, was a signal victory for conservationists. The battles in New York, Hornaday asserted after the law's passage, represented "the Gettysburg of the war between the Army of Destruction and the Army of Defense."[160] This was only the beginning. Evok-ing a martial tone, Hornaday predicted that no-sale statutes like New York's would march across the country and "the tide of slaughter of wild life" would "steadily be rolled back."[161] Indeed, the Bayne-Blauvelt law attracted attention from game protection forces throughout North America. Missouri game war-den Tolerton included Hornaday's "Gettysburg" proclamation in his second annual report, and the *Rod and Gun in Canada* reported that New York's game protection measures were a model to be emulated across the continent.[162] Already by 1911, forty-four states and territories, as well as most of the Cana-dian provinces, prohibited sale of at least certain kinds of game.[163]

Reverberations of the Bayne-Blauvelt bill were heard in California, where field biologist and *Condor* editor Joseph Grinnell organized with other, like-minded conservationists in 1913 to introduce no-sale legislation like New York's.[164] The resulting Flint-Carey bill provoked backlash from commercial hunters and game dealers, the outlines of which were a template for game law opponents elsewhere. Pro-market forces responded to the legal attacks on the sale of wildfowl and other game by advancing a two-pronged argument. The first prong was an attack on no-sale laws as "class legislation" designed to re-strict democratic access to wildlife in favor of a small elite of club-member sportsmen. The second prong stabbed at the heart of game law enforcement, characterizing the game warden system and the state's Fish and Game Com-mission as rife with corruption, political favoritism, and criminality. Accord-ing to marketmen, excessive powers given to enforcement authorities were being wielded unfairly and sometimes illegally. The controversy over the Flint-Carey law revealed that much of the public did not necessarily consider the consumer market in birds to be an unalloyed evil.

As in other states, the continuing activity of game markets in California's major cities, especially San Francisco, proved a challenge to game law enforce-ment. Although California prohibited the sale of upland wildfowl, shore-birds, and big game, waterfowl were still heavily exploited for commercial purposes.[165] During the 1906–7 season, a single San Francisco game transfer company received more than 117,000 ducks from market hunters, not a little of it illegally obtained.[166] State game wardens such as Elmer Pedlar took pride in their abilities to ferret out offenses and bring criminals to justice. Pedlar reported having seized two sacks of ducks from a market hunter and predicted he would "stop all of these illegal shipments" sent by the "horde of outlaw hunters and rounders from Los Banos."[167] A few years later, Pedlar told Fish and Game Commission board members that he had spent several nights inspecting sacks of ducks bound for market to determine the legality of their killing. "Beyond question," he wrote, "most of the ducks coming to the market are killed unlawfully." Pedlar identified the main culprits, "known as 'Bullhunters,'" responsible for the vast majority of duck marketing in San Francisco. If the board would act on his information, the warden confidently asserted, they would dramatically reduce the number of ducks coming to market.[168]

Despite all of his hard work, Pedlar felt unrecognized by his superiors. In 1910, he wrote to the commission to renew his repeated requests for a raise. The fines he had made in that year amounted to only two thousand dollars, while much of his work assisting in other cases or filling in as a license clerk provided no opportunity to add to his income. He attended court dates and inspected game arriving in and leaving the city. Although he had worked as a warden since 1894, in all that time he had not been able to accumulate even enough money "to buy a lot in a cemetery."[169] The public, too, had little appreciation for the efforts of California's official game protectors. In a 1909 letter, Chief Deputy Commissioner Charles A. Vogelsang complained about the "misleading statements" circulating in the popular press charging the Fish and Game Commission with "doing politics." A controversy involving a board member dismissed "for drunkenness and insubordination" had been publicized by *Forest and Stream* as well as other eastern publications as examples of the board's mismanagement—this even though more arrests and fines for game law violations had been secured than ever before.[170]

Under a cloud of critical scrutiny, Vogelsang resigned his post late in 1910 and was replaced by John Babcock who fared little better than his predecessor. He complained to newly appointed fish and game commissioner Frank M. Newbert that certain influential people in the Sacramento area were "doing their level best to create the impression that we are doing nothing in their vicinity."[171] Battling public perceptions of favoritism and corruption, the new chief deputy aimed to reform the commission's workings. Babcock cleaned house, investigating charges of misconduct within the state's game law enforcement system. Pedlar, along with other game wardens operating in San Francisco, was caught up in the investigations and was fired early in 1911 for extorting protection money from Chinese fishermen.[172] In a screed written to the board after his dismissal, Pedlar complained about the commission and its chief deputy. For the previous six years, the commission had been, according to the outgoing warden, a "farce comedy." The chief deputy's qualifications for his job were limited to "histrionic experience, and sufficient hypnotic power, to control the desires of his superiors when opposed to his own." Advancement came only with "sycophantic subserviency to a man as ignorant of the real object of this Commission's existence as Noah was of maritime law." Pedlar took his dismissal as a chance to be "a martyr in the cause" for true game protection.[173]

Charges of corruption and mismanagement continued to plague Babcock. In a move reminiscent of Missouri's inquiries into the state game warden's office, the California legislature initiated an investigation of the Fish and Game Commission's enforcement arm. Market hunters testified against state game wardens, charging them with favoritism toward sportsmen's clubs and capricious application of the game laws. One Orange County game warden was accused of using the state's patrol boat to scare up ducks on Newport Bay and drive them into nearby gun club preserves. It did not help that one of the club members was also a Fish and Game Commission official. Commercial hunters convinced legislators that game regulations were routinely being violated "in the interests of the rich gun clubs."[174] Attacks like these encouraged Babcock to resign his office less than a year after his appointment, and the state subsequently abolished the chief deputy position.[175]

Amid the controversies over corruption, game protectors moved forward with new legislation to stop all commercial sale of wildlife in the state. The Flint-Carey bill, introduced in 1913, was a measure necessary, according to *Forest and Stream*, to meet the "present day needs in game conservation." Repeating a familiar refrain, the magazine argued that market hunters were primarily responsible for the destruction of wildfowl in the state. The consumer market in meat and plumes had driven some species, including white herons, quail, trumpeter swans, and California condors, to the point or nearly to the point of extinction.[176] The *Western Wildlife Call*, a publication of California's associated conservation societies, devoted itself to the Flint-Carey bill, publishing articles authored by leading wildlife advocates, biologists, and ornithologists, including Joseph Grinnell, T. S. Palmer, Edward Forbush, and

William T. Hornaday. Following the articles were lists of those California legislators, as well as Fish and Game Committee members in the California assembly and senate, responsible for the bill's success.[177]

Advocates of the no-sale legislation would accept no compromise with commercial interests, arguing that an unequivocal statutory response was required to prevent further destruction. Various concessions—ranging from a complete cessation of hunting for two years and subsequent reopening of the market, to a proposal requiring sportsmen to sell part of their wildfowl spoils—were all hastily dismissed by conservationists. Commercialism, in any form, was the problem.[178] Wildlife activists scoffed at the opposition's "class legislation" arguments, pointing out that these criticisms were raised by commercial special-interest groups such as the Hotel Men's Lobby that had a financial stake in the game trade.[179] A public debate in the newspapers over the no-sale law raged. When the *San Francisco Examiner,* which had taken a pro-market position, published a piece favoring open sale of wildlife on the grounds that it would provide "cheap food for the people," the anti–market hunting *Fresno Republican* was quick to denounce its "absurd editorial." The *Examiner*'s arguments about democratic access for consumers were illogical. "Which is the more democratic," asked the *Republican,* "to open access to the game to every one who will personally go and take it for himself, or to open it only to the man who can pay a price so high as to be prohibitive to most?"[180] Game protectors highlighted the stark choice facing legislators: the market or wildlife. California could not have both. Lawmakers listened and returned conservationists a victory with the passage of the Flint-Carey bill in the summer of 1913.

Marketmen, unwilling to give up without a fight, organized themselves as the "People's Fish and Game Protective Association" and renewed their public charges of unnecessary and unfair class legislation. They abandoned capitalistic arguments about the sanctity of property and free trade, as well as claims about "the poor market hunter" just trying to make a living—arguments that had been more compelling in the nineteenth century—to recast their position as concern for a common right to wildfowl consumption. If wildfowl and other game belonged to the people, then it was "farcical for the masses," opined the *Marin County Tocsin,* "to support laws against its sale, so that a favored few alone may enjoy it."[181] Other California newspapers posted similar pieces, calling for a public referendum on the Flint-Carey law in order to combat the self-interested, elite shooting club consumers who were bent on monopolizing the state's birds for themselves. The People's Association's first proposal was a revision to the law that would allow shipping of wild ducks and pigeons in the open season—a provision which would reopen the market in those species. A second, more sweeping measure called for an almost complete repudiation of the no-sale law that would "give everybody a chance to eat game."[182] While the more draconian, open-sale proposition failed to garner enough support to be put on the November ballot, the more limited revision on duck and pigeon sales did.[183]

The pending referendum on the Flint-Carey law agitated game law advocates into a frenzy of activity leading up to the 1914 general election. Articles in newspapers, outdoor magazines, and other conservationist publications transformed the state initiative into a national concern, with advocacy pieces coming from all quarters.[184] Conservation supporters proffered several lines of logic leading toward the game trade's abolition. One pointed toward a scientifically based, legal uniformity in wildlife policies. California had a responsibility to neighboring states that had outlawed commercial sale. The birds that Californians exploited migrated in their seasons to Oregon, Washington, Nevada, and other states where no-sale laws were already in effect and market consumers could not purchase wild birds.[185] California, went another argument, was behind the times in its conservation efforts. Eighteen states (or seventeen or nineteen or twenty-one, depending on the source) had almost-complete no-sale provisions on the books, and more than thirty states had prohibited the sale of wild ducks.[186] Scientific experts supported the ban on the sale of migratory birds as well.[187] Clearly, a yes vote on the referendum would "bring California up to date."[188]

Wildlife activists also unmasked the persons behind the "mercenary, revolutionary, and wholly unjustifiable state-wide movement" against the "essential and righteous game laws" so as to counter pro-market claims that the proposed law was undemocratic or un-American.[189] In a nativist tone, Flint-Carey supporters listed some of the founders of the People's Fish and Game Association: game dealers John Corriea and John Campodonico and market hunter L. A. Sischo.[190] Under the title "To Whom Will You Entrust the Making of Our Game Laws?" the *Western Wild Life Call* paired the names of sympathetic, safely Anglo-American state legislators with a list of mostly Italian commercial dealers and market shooters who formed the core of the Flint-Carey opposition.[191] Hornaday similarly disparaged the "alien association," directed by "Corriea, Sischo, Giannini, and Campodonico," as a "game-destroying gang that cares only for 'da mon.'"[192] These "notorious" game law violators' appeal to the people under "the slogan of personal liberty" was simply "a lie and a fraud on the face of it," he maintained.[193]

Most importantly, Flint-Carey supporters directly attacked the opposition's central argument, denying that no-sale laws were class legislation antithetical to non-elite interests. The People's Fish and Game Protective Association, wildlife activists pointed out, was merely a front for commercial dealers, their attorneys, and well-heeled city folks who could afford to eat out at fancy restaurants serving expensive wildfowl.[194] Their motivations were monetary, especially since the upcoming 1915 Pacific Exposition offered a particularly juicy opportunity for profits in bird sales.[195] Ducks brought upward of $2.50 each at San Francisco's "highest class restaurants and cafes." At those prices, waterfowl was no cheap food supply for the common urban consumer.[196]

Frank Newbert, now president of the California State Fish and Game Commission, provided a particularly compelling rejoinder to the People's Association in a pamphlet titled *Swat the Market Hunter and Give the Boy a Chance*. Turning the class legislation argument on its head, Newbert claimed that it was market hunters who had taken away the people's access to wild game. Ducks, quail, prairie chickens, grouse, pheasants, and songbirds had been ruthlessly slaughtered by the unfair methods commercial hunters used. The result was that the adorable, gun-toting, dog-loving, oversized-hunting-coat-wearing lad featured on the cover (and as a coda on each page of the pamphlet) was denied the wholesome "sport and recreation" that was "an inheritance handed down to us by our early ancestors."[197] The Fish and Game Commission's goal was for everyone in the state to have a copy of the pamphlet, which helpfully demonstrated, on its last page, how to mark the referendum ballot correctly.[198]

For wildlife activists, the arguments they advanced seemed air-tight. It was inconceivable that an association of game dealers, retailers, gunners, and gourmands could disguise their motives in an appeal to the common man. But only days before the vote, a new public relations snafu for the state's Fish and Game Commission threatened to reinvigorate the class legislation arguments and the corruption charges against game protectors. The *San Francisco Examiner*—one of the key players in the anti–game law movement—published a picture of Newbert and other sportsmen posing at a private gun club with nearly two hundred dead ducks, a number well more than the legal daily bag limit. The "game hog" picture of the president of the Fish and Game

Commission succinctly confirmed for many that official game protectors were in reality elite-sportsmen protectors. Laws and limits were for others. The *Examiner* dismissed Newbert's anti-market propaganda pamphlet, calling it "the most gorgeous and flamboyant flapdoodle recently emanating from any press anywhere." Apparently "the boy" was not "invited to sneak into one of the Newbert blinds," the article continued. Youngsters, however, were free to gaze upon the photo of the sportsmen's excesses to "see how Newbert does his conserving and preserving."[199]

Although Newbert denied he had killed more birds than legally allowed (the dead ducks were bagged by several hunters not pictured), the damage had been done. The non-sale referendum was narrowly defeated by a margin of about 8,100 votes.[200] Newbert admitted that the unfortunate timing of the *Examiner* photo probably cost Flint-Carey a favorable vote.[201] To conservationists it seemed that Californians had "voted to support the market hunter," but the results also suggested to them that a majority failed to grasp the real issue. Leaving even a sliver of the wildfowl market open, wildlife activists maintained, would inevitably lead to utter extermination. Sacks and barrels of ducks still landed in California markets, but the numbers declined each year. In the 1915–16 hunting season, an estimated 125,000 ducks were sold statewide—little more than the number handled by a single transfer company back in 1906.[202] The dwindling market supply was a harbinger of things to come. "Beware the wild pigeon spectre ye Californians," *Forest and Stream* declared.[203]

The controversies over no-sale initiatives left much of the nonhunting public unwilling to believe that game law officials and activist sportsmen were the altruistic wildfowl protectors they claimed to be. As California's voters showed, many were not yet ready to accept the sportsmen's idea that killing for the market was more destructive than killing for sport. The wardens and commissioners who were given considerable amounts of power over wildfowl resources had all too often wielded that power irresponsibly and sometimes illegally. The highly politicized struggle over California's market in birds seemed to prove that game laws were written and enforced to privilege some kinds of wildlife use over others and to favor some wildfowl consumers over others.

❧

The market in birds looked quite different at the end of the first decade or so of the twentieth century. "The game market of to-day," wrote biologist Henry

Many Californians blamed this photograph from the *San Francisco Examiner* for the defeat of the Flint-Carey no-sale legislation by referendum in 1914. State Fish and Game Commission president Frank Newbert (*fourth from left*), pictured here with ducks killed at a private hunting club in excess of the state's bag limits, was held up by the law's opponents as an example of so-called game protectors' self-serving motivations. Image courtesy of Newspapers.com

Oldys in 1910, was fraught with "inconsistencies and irregularities." Rapidly dwindling wildfowl populations had incited equally rapid changes in wildlife legislation, and laws prohibiting export and sale of wildfowl were the most novel of the new game policies. Criminalizing the consumer market represented a major shift in American wildlife use, which had previously emphasized the value of unowned wildlife as a common resource. The early-twentieth-century legal innovations that aimed to strangle the game market proffered a different vision for wildfowl use in the modern age. In the process, the new laws "excited the antagonism of the vested interests affected" and provoked a "constant contest between officials charged with enforcing the new laws and market hunters and dealers."[204] For sportsmen and conservationists, the logical progression of the game laws pointed toward the abolition of the game market.[205] The "market hunters' days are numbered," declared Connecticut's state ornithologist in 1912, as more and more states were "taking up the important work of retiring the market hunter."[206]

But the work of retiring the market hunter was not as clear-cut as many activists made it out to be. This was no simple contest between destruction and preservation, between the law abiding and the law violating, between self-interest and self-restraint. It was instead a clash between different visions for wildlife consumerism in a modern age. Conflicts over the market in birds exposed a far more complicated story, and the self-proclaimed game protectors were not always the heroes of that tale. Merritt had made that point before. "If anyone has killed game illegally," he insisted in 1904, "it has not been an up and down hunter who hunted for profit."[207]

CHAPTER SIX

The Conservationist

For the crusading conservationist William T. Hornaday, state-level no-sale laws passed in the early twentieth century represented "great victories" for wildlife, victories that would open the door to even greater federal protections.[1] But the struggle proved too divisive for the alliances that had once drawn activists, sportsmen, bird lovers, and scientists together in pursuit of a common cause. Hornaday's estimation of sportsmen underwent a dramatic change in little more than five years. In 1907, the prominent zoologist and director of the New York Zoological Society defended sportsmen as "game savers" in the pages of the *New York Times*.[2] The battles over the Bayne-Blauvelt law in New York and the Flint-Carey bill in California, however, soured Hornaday on sport hunters' altruism, and he came to rank "sportsmen and so-called sportsmen" first on his list of causes of declining bird life, above market, plume, and pot hunters.[3] "The sportsmen of America, and of the world at large," he wrote in 1913, "preserve game to-day in order to kill it tomorrow, a motive essentially selfish and merciless."[4] In the case of California's anti-market statute, it was clear to him by 1912 "that if wholly left to themselves the 'sportsmen' and game-butchers of California never would better the situation in any marked degree."[5] When the law was overturned by referendum, Hornaday claimed that this outcome was what organized sportsmen had desired all along: legalized sales to provide a market for the surplus ducks and geese killed at their private ducking clubs. Any credit for curtailing wildfowl commercialization properly belonged to the real conservationists: primarily scientists such as ornithologists Joseph Grinnell, Edward Forbush, and, especially, Hornaday himself.[6]

The fight for no-sale legislation further proved that wildfowl policy needed to be lifted out of the realm of state politics and away from the influence of sportsmen's associations. The fundamental needs of wild birds, according to Hornaday, were, first, federal protection for migratory species and, second, "the total suppression of the sale of native wild game."[7] Others agreed. In a 1911 letter to John Burnham, the head of the American Association for the Protection of Game, Theodore Roosevelt argued that while national protective legislation for migratory game birds was necessary, relying upon sport hunting consumers to advocate for federal oversight was unfeasible. Recent history had demonstrated that "too often both sportsmen and manufacturers of guns and ammunition, not to speak of dealers in game," displayed an almost inexplicable "short-sightedness as to their own interests"[8] If protective laws had been enacted earlier, Roosevelt suggested, avian populations would have been maintained at levels that could have supported both sport and market hunting well into the future. Over time, however, commercial and recreational overkilling had made more drastic legislative action necessary.[9]

Henry Clay Merritt had already figured out where the game laws and the game business were heading. As the twentieth century dawned, he was finding it harder than ever to carry on as a game dealer. Game codes that had been mostly ignored in western states such as Kansas, Missouri, and Nebraska, from where he had been drawing most of his supplies, were now being upheld more often. Express companies, hemmed in by nonexport laws, began to refuse to guarantee safe shipment, and it seemed that it was only a matter of time before all shipping of wild game would be outlawed. "Game has largely been withdrawn from the tables of the wealthy," Merritt observed, "and hotels and restaurants do not call for it in the cities where it was once so plenty." It was inevitable. A growing human population coupled with a declining avian one made it unlikely that wild birds could continue much longer as "a standard article of diet." The snipe and woodcock that had been so numerous in Henry County a short time before were now hard to find, and waterfowl numbers were decreasing year by year.[10]

Harassment by game officials was the last straw for Merritt. In the midst of his legal troubles, he sold off his frozen stock and turned his attentions toward his business properties and other ventures.[11] Now in his early seventies, and nearly thirty years after he put down his gun, Merritt picked up the pen. He began a writing project: not an autobiography, he insisted, but a history of his profession and, more important, a disquisition on the gun. The reminiscences, Merritt promised, would be brief.[12]

The Kewanee game dealer was convinced that the market in birds was nearing its end, and he was writing its postmortem. Wildfowl numbers were too low to support the business much longer. There would be no market, Merritt predicted, when his volume, *The Shadow of a Gun,* was published in 1904. "Where there is nothing to sell there will be no one to buy."[13] Demographic and environmental changes would drive avian populations still lower until only the scattered remnants of once-vast flocks would be found in the uncultivated fringes of farmers' fields. "Some help might come from the laws," he supposed, "but the men who hunt for sport, under some pretext, manage largely to ignore them."[14]

Conservationists held out much more hope than did Merritt for the power of legislation to reverse the decline and establish new patterns of human-wildlife interaction for the modern age. They also were less willing than Merritt to allow the game business to die a natural death. Wildlife advocates called for federal regulation to put an end to commercial wildfowl exploitation once and for all. National protections were necessary, many argued, because birds' biological migrations when alive and economic mobility when commodified had proved impossible to contain within the jurisdictional control of individual states. Game law uniformity, which had been advocated by activist sportsmen for decades, seemed achievable only under the umbrella of federal authority.

These concerns about declining avian populations and the need for federal action drew a wider, national audience in the early twentieth century, one that went beyond sporting circles. In the process, sportsmen found their authority to speak as wildfowl protectors appropriated by the scientific community, game law officials, and the general public. The conservation coalition of the late nineteenth century fractured as some constituencies advocated for restrictive federal laws that went beyond commercial wildfowlers to limit recreational hunters' consumption in ways they had not known before. Although market hunters continued to operate well into the new century, their remaining, increasingly clandestine activities were no longer central to the conversation about wildfowl conservation. Sport hunting game hogs and poachers replaced the market hunter as villains in the conservation storyline as the fate of wildlife in the twentieth century was debated and legislated on a national scale. Sportsmen still tried to assume the conservationist role, but they did not appear to be the monolithic force in game protection as they had been before. A new coalition, increasingly critical of sportsmen's conservation claims, arose to strip wildlife consumerism of its consumptive character.

The Costs of Consumption

Avian conservation causes broadened their appeals in the early twentieth century, with new national organizations and conventions convened to bring the modern plight of wild birds more squarely into public focus. The resurrected, twentieth-century Audubon Society exemplified this nationalizing of wildfowl issues. George Bird Grinnell's original organization had operated at the local and state levels in the 1890s, popularizing the subject of avifauna protection while the American Ornithologists' Union (AOU) kept its influence contained within scientific circles. In 1905, however, the National Association of Audubon Societies was formed with ornithologists Frank Chapman and William Dutcher at the helm of the larger, cooperative group.[15] Other, national-level popular wildlife conservation organizations followed. During his presidency, Theodore Roosevelt convened several nationwide conservation conferences, and he created the National Conservation Commission in 1908. As a parting gesture when he left the presidency, Roosevelt also planned for the first North American Conference on Conservation of Natural Resources, to take place early in 1909 shortly after William Howard Taft's inauguration.[16] These and other associations elevated wildlife conservation as a matter of national, and even international, interest.

Widely publicized examples of ongoing commercial wildfowl exploitation in the early twentieth century also served to increase public awareness and emphasize the costs of wildfowl consumption. The 1909 Laysan Island "Bird Tragedy" offered a particularly cogent warning about the market's destructive force.[17] Laysan Island, a tiny atoll located some eight hundred miles northwest of Hawaii, was a heavily used rookery, serving as a nesting site for Laysan albatrosses, black-footed albatrosses, numerous tern species, shearwaters, boobies, and other birds. The island was "quite alive with birds," reported one late-nineteenth-century observer, so tame they could be picked off their nests and handled.[18] The massive nesting site attracted not only birds but also the North Pacific Phosphate and Fertilizer Company, which set up a guano-mining business there in the 1890s, superintended by Max Schlemmer. At first, Schlemmer took seriously his role as the island's protector. Walter Fisher's report of his zoological expedition to Laysan in 1902 was prefaced with thanks to Schlemmer and the assertion that the company manager took "the best care of the birds on the island" and had "a genuine interest in their welfare."[19] But in 1909, after Schlemmer orchestrated a contract with a Japanese firm allowing it to remove any natural products from Laysan, a team of workers stripped the island of its birds, removing feathers and wings for sale in international millinery markets.[20]

By the time a US revenue cutter arrived at the island to investigate rumors of poaching, the island's bird life had been decimated. Guano sheds were filled to overflowing with feathers and wings. A 1911 report of the island from zoologists Homer R. Dill and William A. Bryan described the island as barren and strewn with sun-bleached bones. Birds were kept in a dry cistern to starve so their skins would be free of grease. In other buildings, workers ripped wings from still-living birds, leaving them to bleed to death. Smaller birds were captured alive and sent in cages abroad. The "wholesale killing . . . had an appalling effect on the colony," reported another scientist. It was nearly impossible to imagine numbers of birds "willfully sacrificed on Laysan to the whim of fashion and the lust for gain."[21] The notorious event prompted President Roosevelt to create the Hawaiian Islands Reservation for Birds in 1909, which included Laysan and twelve other atolls. "By this act," declared Hornaday, the islands were made "secure from further attacks by the bloody-handed agents of the vain women who still insist upon wearing the wings and feathers of wild birds."[22]

Laysan Island added new fuel to the anti-plumage movement as other examples of feather hunters' destruction garnered growing public attention. California's trumpeter and whistling swans, for instance, were brought "near or quite near extermination," according to Joseph Grinnell, due to the value of their feathers and down.[23] A speech by the Audubon Society's Frank Chapman to an audience of New Yorkers detailed his visit to a southern rookery in search of the highly prized plume birds. There he found a sportsmen's club whose warden had once been a plume hunter, but who now made more money as the game club's protector. The former "pot hunter" claimed he had taken as many as twenty-five hundred birds to market in one season. The white heron, once numerous in southern cypress swamps, the *New York Times* reported in 1907, was becoming an all-too-rare sight. The American Museum of Natural History acquired a few of the remaining, "practically exterminated" herons as specimens that year to be displayed for posterity.[24]

The heightened public focus on the plight of plume bird species highlighted the extent to which the conversion of birds into commodities had distanced consumers from the consequences of their fashion choices. The plume industry was internationally extensive, as events on Laysan Island demonstrated. Birds from a distant US territory in the Pacific, harvested by Japanese workers, sent to Paris millinery houses, and sold to New York City socialites were transformed by the modern market into abstract representations of wildness to consume. Although bird hat–wearing women valued the exotic, wild nature of the wings and feathers they donned, it was all too easy to ignore the elaborate economic processes that produced them and the resulting environmental consequences.[25] Pictures and descriptions like those from Laysan, however, made those connections explicit. Wearing plumes, argued Chapman to his urban audiences, did "infinitely more harm than hunters."[26] Wildfowl "slaughtered to extinction" in faraway places in America and abroad became a local crisis when their remains appeared in stores and on the streets.[27]

Other well-publicized extinctions or near extinctions happening in plain view underscored the effects of unrestricted consumptive wildfowl use.[28] The passenger pigeon was the most visible victim of consumerism's toll, and a variety of commentators drew the obvious parallels between the decimation of bison herds on the western plains and the destruction of the millions of passenger pigeons that had darkened the skies only a generation before. The birds' seemingly "strange and sudden disappearance" raised public curiosity and drove professional as well as amateur ornithologists into a vain search for survivors.[29] To explain the species' demise, Michigan businessman and sportsman William B. Mershon published an entire book on the passenger pigeon in 1907, including a compilation of the most recent pigeon sightings.[30] The frontispiece to Hornaday's *Our Vanishing Wild Life*, titled "The Folly of 1857 and the Lesson of 1912," paired an image of the last living passenger pigeon housed at the Cincinnati Zoo with a mid-nineteenth-century declaration from the Ohio legislature that "the passenger pigeon needs no protection." The bird's imminent extinction proved that no species could "withstand the systematic slaughter for commercial purposes."[31] Besides the celebrated case of wild pigeons, Hornaday provided a state-by-state catalog of recently exterminated bird and mammal species.[32] This piling up of modern examples of economic exploitation and repeated admonitions about the waste of the country's once-abundant natural resources made twentieth-century wildlife consumerism and conservation a matter for public debate.[33]

Market hunters responded to the widening public discussion about economic exploitation and extinctions with their own explanations for the unmistakable avian decline. All hunting had begun "to fall off very fast" in many areas after the turn of the century, and probable causes were not hard to find.[34] Some blamed the US Biological Survey's practice of poisoning pest species like crows and magpies for killing off game birds in the process.[35] Industrial pollution was a problem, too, since it destroyed bird habitats and disrupted migration and breeding habits.[36] Others claimed that the once-massive flocks of ducks and geese had not been killed off but had simply changed their flight patterns and gone elsewhere.[37] By far, though, commercial gunners most often cited agricultural development as the leading culprit of wildfowl destruction. Replacing prairie grasses with hayfields eliminated the prairie chicken nesting sites, they argued, while marshland drainage, field tiling, and levee creation dried up the landscapes that had previously drawn migratory waterfowl.[38] In this respect, market hunters advanced a nascent ecological perspective that was underdeveloped in the broader public conversation about wildlife conservation at the time.

Sport hunting ranked a close second to agricultural expansion in many market hunters' accountings of the damage caused to bird populations. As avian populations dwindled, urban sportsmen flocked to rural regions where birds could still be found, as Hoyt Elbert recalled of the large and well-financed Iowa hunting trips mounted in the late nineteenth century by prominent businessmen such as Cyrus McCormick and Jerome Case of threshing machine fame.[39] As wildfowl numbers declined further, large private shooting preserves, such as the one owned by Anheuser-Busch near St. Louis, monopolized the available birds, concentrating them in one spot to be shot by moneyed recreational consumers. Massive private preserves in Alabama and Louisiana, claimed market hunter William Brown, housed so many waterfowl that "you'd change your mind about their being gone."[40] Public condemnations of "gentlemen sportsmen" appeared regularly in California's Central Valley from the 1890s onward, with newspapers reporting on the latest "indignation meetings" organized by "city sports" who liked to rail against market hunters. Sportsmen's private shooting clubs, which they jealously guarded "against the encroachments of outsiders," were simply an "English system" that allowed "the hogging of duck-shooting lands."[41]

It was not just the sporting elite that posed a problem, though. More and more recreational shooters were taking to the field, and modern automatic weapons made "the average bozo" far more lethal than any market hunter of the past.[42] It was those sportsman-consumers who wanted improved, multiple-shot guns, argued Winchester Repeating Arms Company president T. S. Bennett, "not pot-hunters."[43] Automobiles only added to sportsmen's destructive capabilities. Ducks were "absolutely disappearing" along Iowa's Skunk River, noted the professional gunner Buck Lufkin, thanks to the newest transportation technology. With a car, hunters could easily invade locales that had formerly been difficult or even impossible to access.[44] According to professional wildfowlers, this growing cohort of sport hunting consumers, armed with ever-more-efficient machines, was a detrimental force far exceeding that of the market.

More than a few sportsmen were willing to concede this point. The market hunter was rapidly disappearing, *Forest and Stream* explained, and the "man who shoots purely for his own enjoyment" was now the greater threat to wildlife. The "new evil," agreed columnist Emerson Hough was "the over-hunting sportsman."[45] Gun clubs and sporting associations, however, were passive in the face of sport hunters' violations, and craven politicians were reluctant to do anything to anger this growing constituency.[46] Consequently, early-twentieth-century game codes retained high bag limits and long hunting seasons that allowed recreational hunters to "overshoot" wildfowl to an injurious extent.[47] Critics also maintained that "gunning from a car" was probably "destroying upland game birds faster than any other agency." Hunters could now sweep over far more territory than that "covered by the most energetic market shooter."[48] Iowa's shooters and gun club members were remarkable for their "lack of sportsmanship," wrote Leonidas Hubbard Jr. to *Outing* magazine in 1901. "In my humble opinion," he continued, "if Iowa were called in for judgment for her wretched sportsmanship, as two ancient cities of biblical fame were for their sins, the result would be a very small exodus of righteous ones, a shower of brimstone, and a lake as briny and bitter as the Dead Sea."[49] The acerbic Hough more simply proclaimed that "the American sportsman does not deserve any game" since he was so little interested in protecting it.[50]

Many observers blamed automobiles for wildfowl destruction in the early twentieth century. The new technology allowed recreational hunters to range over wider areas and to access wildfowl haunts more easily. Here a Utah hunter poses with his daughter beside a car and a variety of harvested waterfowl. Image from Compton's Studio Photograph Collection, courtesy of Special Collections & Archives, Merrill-Cazier Library, Utah State University, Logan, Utah

One could hope that moral suasion and appeals to sportsmanship would reform so-called game hogs and pot hunters, but many in sporting circles like Hubbard and Hough wished for more concrete action in the form of uniform wildfowl conservation laws. If the proposal sounded familiar, it was because many activist sportsmen had been advocating for alignment in state game laws for years.[51] Hallock's nineteenth-century "scheme for a system of game laws to cover the entire country" was endorsed by T. S. Palmer of the US Biological Survey, who characterized the former *Forest and Stream* editor's plan as "a step towards uniformity" in 1900.[52] The state ownership doctrine, however, gave rise to a hodgepodge of regulations that varied widely in subject, application, and complexity. State game laws often failed to distinguish between resident and migratory birds. Even for the same species, rules varied by state and sometimes by county. Legal protections for song and insectivorous birds were often missing from state codes concerned primarily with game birds, and sportsmen decried the people—usually youngsters or Italians, they claimed— who would "kill the tiniest songster whenever they get a chance."[53] Most important, all the game laws were subject to state-level politics and thus shifted with each new administration.[54]

Although the passage of the Lacey Act in 1900 established federal jurisdiction over the interstate game trade, it did not create legal uniformity. Activist sportsmen and conservationists generally agreed on an expanded federal role in providing migratory bird protections. This was, in their view, the best way to achieve statutory alignment and to combat shortsighted consumerist interests.[55] They also agreed that wildfowl regulations needed to extend beyond game and plume birds to include all avian species that had been commercially exploited or otherwise harmed by modernization.[56] Songbirds, for instance, were still being hunted, marketed, and consumed in some states well into the twentieth century. In Texas, just one dealer supplied over 120,000 robins to hotels in 1902.[57] As late as 1913, robins were killed for sport in Louisiana, Tennessee, Mississippi, Maryland, North Carolina, Virginia, and Florida, where "robin-potpie-loving inhabitants" abounded.[58]

To broaden public support for federal wildfowl legislation, conservationists began a campaign to advertise the tangible ecological and economic benefits that avian species provided and the costs to agriculture and human health when wild birds were recklessly destroyed. The vast majority of bird species, explained biologist Joseph Grinnell in 1914, "occupy an important position in maintaining the balance of nature" and served agricultural economic interests by acting "to check abnormal increase in plant-eating insects and excessive multiplication of weeds."[59] Nearly four hundred million dollars were "yearly sacrificed to insects" in the United States, according to 1901 estimates reported by the Massachusetts State Board of Agriculture, and insect-eating birds promised the best defense against such economic losses.[60] Predatory species such as hawks and owls were valuable as controls on rodent populations, while buzzards, vultures, and other scavenger types disposed of rotting carcasses, aiding public sanitation in the process. Swallows, martins, and other birds that fed on malaria-carrying mosquitoes also served the interests of public health.[61]

Activist sportsmen had little quarrel with conservation scientists about nongame bird protections crafted by biological experts. The AOU had blazed that path earlier, in the 1880s, when it put forward its "Model Law" for the protection of nongame birds. Although the AOU's recommendations for nongame birds were slow to be adopted, by 1903 twenty-nine states had revised their state wildlife regulations to conform with them.[62] Making rules for game bird preservation grounded in scientific principles, not sportsmen's predilections, however, was a much more contentious process. Wells W. Cooke's monumental bulletin, *Distribution and Migration of North American Ducks, Geese, and Swans*, published by the US Biological Survey in 1906, sought to set a scientific framework for "practical legislation in behalf of wild fowl." Toward that end, Cooke described breeding grounds, food habits, nesting seasons, and migration patterns for waterfowl species. "The problem of the legal protection of ducks, geese, and swans," he explained, "has two phases—protection during the breeding season and protection during migration and in winter." As activist sportsmen had long argued, Cooke advocated strict limitations on waterfowl marketing. But the ornithologist went further to insist that if there was any hope of preserving wildfowl for the future, consumptive recreational use had to be reined in as well.[63] While some sportsmen recognized the need for such restraint, many others were unwilling to believe that the costs of their consumption were in any way comparable to losses caused by market hunting.

Nationalizing Conservation

Calls for federal oversight from activists in both the sporting and scientific communities prompted congressional action on migratory birds in the first decades of the twentieth century. An initial effort introduced by Pennsylvania congressman George Shiras III in 1904 had not gone very far, but the simple premise underlying Shiras's bill, that the seasonal mobility of birds across state boundaries made state-level conservation legislation insufficient, would be repeated by proponents of federal regulation in future debates.[64] Both activist sportsmen and conservation scientists claimed credit for reviving the migratory bird issue in 1912 and spearheading the 1913 Weeks-McLean law. Key figures' differing recollections about the beginnings of the federal action reveal the emerging divisions between these two groups. According to *Forest and Stream* editor Grinnell, the campaign for a migratory bird law was led by the sportsmen's American Game Protective Association (AGPA) and its president, John Burnham, in 1912.[65] Their efforts to advance the cause during the House Agricultural Committee hearings, though, received little encouragement from Hornaday, who later complained that the "occasion was from beginning to end wholly a sportsman's affair, with 'migratory game protection' the shibboleth." Burnham reportedly refused to discuss songbird protections, and T. Gilbert Pearson, as founder of the National Association of Audubon Societies, was allowed only a brief statement on the subject.[66] Grinnell's recollections of the hearings diverged from Hornaday's. Representatives from both sportsmen's and bird protective associations attended the hearings, reported Grinnell, and no opposition accompanied the discussion.[67] After a lukewarm response from Congress, biologist Pearson took the lead in re-envisioning Shiras's bill more broadly to include nongame migratory species, thus widening its political appeal. Hornaday, however, claimed credit for promoting the idea at the National Conservation Congress in Indianapolis.[68]

When the redesigned migratory bird bill, now sponsored by Massachusetts representative John W. Weeks of Massachusetts and Senator George P. McLean of Connecticut, came up for debate in the House, it included protections for migratory songbirds and insectivorous species, as well as game birds. Opposition was limited but impassioned. Representative Franklin Wheeler Mondell of Wyoming took to the floor to argue against the "revolutionary" change conservationists proposed. Congress intended to eradicate "the people's sovereignty and control over a very important matter" and "was proposing an army of Federal officials, armed with power to make arrests anywhere . . . for trifling offenses." A tad hyperbolically, he declared that "no legislation so profoundly subversive of the fundamental principles of our Government ha[d] been suggested in the history of Congress." Mondell was convinced his constituents would never accede to the national government's authority to decide if, when, and how wildfowl could be killed.[69]

More measured discussion of bird conservation, however, dominated the proceedings. Some congressmen raised the specter of extinction, presenting the passenger pigeon as evidence. Others pointed to the discrepancies in state-level game legislation, the agricultural impact of avian destruction, and the consequences of unregulated commercial exploitation to argue for the bill's passage while drawing on the expert scientific opinions of conservationists such as Forbush and Hornaday to support their arguments.[70] As to the charge of federal overreach, Indiana's Representative Ralph W. Moss expressed frustration. "I am tired of hearing the Federal Constitution pleaded as a bar to rational and progressive legislation," he complained. All the states had legal protections for birds on the books, and the federal law was simply an attempt to put the national government "in active cooperation" with state efforts.[71] In the end, the "little wavelet of opposition" in the House failed to stop the Weeks-McLean measure, and it was signed into law (in part through Taft's inattention to the details in the agricultural appropriations bill to which it was appended) in March 1913.[72]

The new federal legislation, which put migratory birds "in the custody of the United States" and authorized the Department of Agriculture to develop wildfowl regulations, represented a dramatic shift in legal principles.[73] Less than twenty years earlier, *Geer v. Connecticut* had seemed to have settled the wildlife ownership question, establishing the individual states' control over wild animals. Weeks-McLean, however, overrode that precedent—but only for birds. The rationale for the change was mainly biological. As the debates over the national migratory bird law showed, politicians had begun to grasp the ecological realities that made state ownership problematic. Avian life habits, temperature zones, and migration patterns provided an ostensibly objective scientific backing for restrictions.[74] Activist sportsmen found much to like in the law, and *Forest and Stream* proclaimed it "the longest step in the direction of wild life protection yet taken in this country."[75]

Yet the Weeks-McLean Act was remarkably silent on the one issue that had most animated sportsmen. The new law did not abolish market hunting or sale of migratory wildfowl. Instead, the act was written in prohibitive wording; that is, all taking, possessing, and disposing of protected species were permitted unless specifically proscribed. The legislation did not address commercial sale, means of taking, or bag and possession limits. Rules issued following the law's passage defined migratory birds, eliminated night hunting, established a permanently closed season on insectivorous species, banned the hunting of certain other migratory birds for differing periods of time, and created two geographically defined zones for regulatory purposes: a northern zone where waterfowl bred and a southern zone where most birds wintered. Hunting seasons were set for each zone, and spring shooting was forbidden. A cadre of seven federal district inspectors and 172 state game wardens, working under the auspices of the US Biological Survey, would take charge of the law's enforcement. State laws prevailed in all other matters.[76]

The new, shorter federal open season for some migratory bird species indirectly affected the game market's operations, but they were not eliminated entirely.[77] State laws in most places outlawed commercial sale of all or certain kinds of game, but some significant exceptions persisted. Marketing of selected species (most commonly waterfowl) during the open season was still legal in some of the major game market states under certain conditions.[78] Illinois allowed the sale of out-of-state game after the federal migratory bird law was enacted and continued to do so until state law ended the practice in 1915. California's referendum on the Flint-Carey law reversed the state's no-sale provisions on ducks in 1914, *after* the passage of Weeks-McLean. Estimates from California's Fish and Game Commission showed that between 1911 and 1916 duck sales continued, though the numbers sold steadily declined over time. During the 1913–14 season—the first period in which the new federal regulations were in force—160,000 ducks made it into California's game markets. About the same number were sold the following season.[79] Market hunters operating in Louisiana killed tens of thousands of ducks in the years after Weeks-McLean, with some 80,000 killed on one shooting preserve during the 1916 season alone.[80] The fact that the "Sale" section in the Department of Agriculture's 1914 digest of game laws remained largely unchanged from earlier editions demonstrated how little the federal law affected the market in migratory birds.[81]

Sport hunting consumers, though, felt the effects of the new national regulations, and they rankled not a few sportsmen. The elimination of spring shooting—the law's key provision—was the main sticking point. Before the passage of Weeks-McLean, there had been little consistency in state laws on the matter of springtime gunning, and recreational hunters had long debated the detrimental effects of the practice.[82] Ornithologists argued that birds' nesting, food, and migration habits necessitated legal protection during spring breeding and migrating seasons.[83] Some sporting writers and game law officials subscribed to the same view.[84] It was the height of foolishness, insisted one game warden, to allow birds to be killed "when full of eggs en route to the nesting grounds."[85] Other sportsmen, however, denied that spring shooting had any significant influence on game bird populations. As one *Forest and Stream* contributor argued, the "hundreds of keen shooters" who engaged in the practice were "as a rule far better acquainted with the habits of ducks than most ornithologists." The experienced sportsman knew that "no harm [was] done by shooting at that season." Besides, the average recreational hunter's only chance at waterfowl usually came in the wetter spring season, when birds were more easily found.[86]

When the Weeks-McLean Act abolished spring shooting, vocal opposition immediately arose from many recreational hunters, sportsmen's associations, sporting goods retailers, and even some in game law enforcement. Sporting circles found their ranks split into pro– and anti–federal law factions. The law also upended sportsmen's position as the authorities in game protection and replaced them with faraway federal functionaries who, it was argued, possessed little of the local wildfowl knowledge required for sensible regulations. Sport hunters, long accustomed to being on the right side of wildfowl protection, now found their practices targeted and their intentions questioned. Hornaday haughtily dismissed the rising criticisms from his one-time allies, warning conservationists not to "take your cure from the sportsmen, especially regarding enactment of long close seasons!"[87] Game law uniformity, once the sportsmen's remedy for the market gunner, recoiled on the amateur hunter.

The conflict was not just a war of words. Sportsmen's opposition to Weeks-McLean spilled out of the pages of outdoor magazines to be played out in the fields and fought over in federal courtrooms. Within months of the implementation of the first federal hunting seasons, the spring shooting ban was put on trial. When Henry Shauver was arrested for shooting coots out of season in Arkansas, volunteer federal game warden (and local duck club president) Joseph Acklen saw an opportunity for a test case and promised to pay for Shauver's defense if he would agree to plead not guilty to the charges. When the case came before Justice Jacob Trieber in the US District Court for the Eastern District of Arkansas in May 1914, the judge found no basis for the national government's claims to legal authority over wild birds. Although

Trieber voiced his sympathy for the cause of wildfowl conservation, in his view, *Geer v. Connecticut* had settled that question in favor of state control, and he declared the federal migratory bird law unconstitutional.[88] Some, including Pearson and the AGPA, saw conspiracy at work in Acklen's actions and Trieber's decision.[89] Nevertheless, the district court's ruling was a clear signal that the legislation was on shaky legal ground.

While the *Shauver* case was waiting on the US Supreme Court's docket, duck hunters along the Missouri-Kansas border were mustering their own opposition to the spring hunting ban. The Interstate Sportsmen's Protective Association, formed in February of 1914 to counter the migratory bird law's provisions, argued that the elimination of spring shooting was detrimental to their interests as consumers. For these sportsmen, the spring shooting season offered the best, or even only, opportunities for hunting on the duck club lands for which they paid dues. Primed for confrontations with the federal law and its enforcement agents, many of these sportsmen took the 1914 *Shauver* decision as an invitation to continue their springtime hunting.[90] Their actions brought them into conflict with the newly appointed district inspector and US game warden, Ray P. Holland.

Holland, who gained modest notoriety as a freelance outdoor writer, had lobbied for the passage of Weeks-McLean, throwing his support behind "the anti–spring shooting crowd."[91] In 1914, he was named a federal game warden for a district covering seven midwestern states. Holland traveled extensively throughout his assigned territory, looking for game law violators, but also engaging in an educational campaign to convince shooters and sporting retailers to obey the new law. The outcome of the *Shauver* case turned his task into an uphill battle.[92] He found "two distinct elements, one for and one against the law." A number of hunters willingly admitted to having violated the spring shooting ban earlier that year, assuming from newspaper reports that the federal law was invalid and would not be enforced. Law-abiding hunters complained that others were ignoring the prohibitions against spring shooting and wanted more rigorous enforcement if they were to go on obeying voluntarily. Missouri hunters were "rabid against the law," wrote Holland in the fall of 1914, "because the violators got off scot free last spring." In some towns, many of them populated by former market hunters, Holland found a "general disregard for the law." In others, there was qualified support. The game warden hoped that he would find more compliance the next spring and toward that end, Holland placed notices in local newspapers informing readers that the federal law remained in effect and would be upheld despite the unfavorable court ruling.[93]

At the same time, Holland looked for another test case to bring before what he believed was the more sympathetic US district court in Kansas City. The April 1914 arrest of Galena, Kansas, sportsman and banker George McCullagh for shooting ducks out of season seemed a likely prospect, but Kansas game warden Lewis Dyche was less enthusiastic about its chances of success. Dyche was hard at work in the Kansas legislature, attempting to bring the state's hunting laws into conformity with the new federal seasons, though he doubted he would find much success there either. Despite the discouragement, Holland still held out hope that the case of *US v. McCullagh*, scheduled to be heard by Judge Pollock of the US district court in Kansas early in 1915, would turn things around.[94] Holland "worked the papers" to build support for the federal migratory bird law, but he found many were working actively against it.[95] Early in 1915, Holland met up with the "Social Target Club" in Kansas City, nearly all of whose members belonged to the Interstate Sportsmen's Protective Association as well. The federal warden found the group "awfully hard to talk to" and only interested in arguing about the constitutionality of the law.[96] The "Kansas City crowd" quickly became the bane of Holland's existence. He found that shooters in other parts of his district toed the line on the federal law and were willing to let the Kansas City sportsmen "take the lead in all the violating."[97]

By February, Holland's hopes for a clear judicial confirmation of the migratory bird law were dimming. Sportsmen in the state were confident that Judge Pollock would also declare the Weeks-McLean law unconstitutional in federal court, and Holland complained that Pollock had already "exposed himself as against this law." The attorney in charge of the case told Holland he had done all he could do and that he was "'too busy' to give the matter any further attention." It appeared Holland would have to do his own lawyering, and he worked tirelessly on crafting constitutional arguments.[98] But a few days later, as expected, Pollock ruled the Weeks-McLean provisions unconstitutional. As in the *Shauver* case, the decision relied on the findings in *Geer v. Connecticut*. Pollock concluded that once the state allowed wildlife "to come under the authority of the . . . national Constitution," the state's control of game no longer existed, "and its trust title for the common good of all the people of the state [was] cut off and destroyed."[99] Federal protection for migratory wildfowl was not simply supplemental to state game laws; it instead represented a major re-scission of state authority.

"Hell broke loose over in the bottoms," Holland wrote, when Pollock's decision was made public on March 20, 1915. Holland reported that "wires and telephone messages" flew across the Kansas-Missouri region, informing sportsmen of the decision and "telling them to shoot." The news interrupted a period of relative calm that spring during which Holland had only a few instances of illegal shooting. Now he was scrambling to correct reports about the federal law's status and to police emboldened duck hunters. When Holland got a line on a clear violation of the spring shooting ban in Missouri, the Kansas City attorney he consulted advised him not to bring any charges. It would be "ruinous to arrest" the culprit, the lawyer insisted, since the "Interstate [Sportsmen's Protective Association] would get ahold of him" and "bring a habeas corpus proceeding and immediate trial." The result would likely be a Missouri case decided "like the one across the river."[100]

Through the rest of the spring, Holland continued to police shooters, though he confined his work primarily to Missouri, where he worked in concert with state game officials to enforce Missouri license laws and Sunday shooting prohibitions while "spreading the news that [federal] law will be enforced if Pollock is reversed." The negative outcome in the *McCullagh* case, he believed, caused sportsmen in the state to "run wild." One Kansas City–area hardware store had shipped some half million shells—"duck loads"—to consumers in northern Missouri that season. A great many hunters and farmers took Pollock's decision as the final word on the federal law, Holland reported, though he still found support among others who had seen the law's benefits. As he waited for definitive word on the fate of Weeks-McLean, Holland sent out bulletins, gave newspaper interviews, wrote letters, and generally warned hunters that he would enforce the law in the interim.[101]

Two other court decisions at the state level in 1915, one in Kansas and another in Maine, reiterated the findings in *Shauver* and *McCullagh*. Federal authority to regulate interstate commerce did not apply to birds, maintained the Kansas Supreme Court, since "the natural flight of wild fowl from one point to another does not constitute 'commerce' unless that word be expanded beyond any significance heretofore given it."[102] The pile of unfavorable rulings proved to supporters that there were insurmountable flaws in the Weeks-McLean law. They allowed the *Shauver* case to be pushed back on the US Supreme Court's docket while conservationists worked on new ways to give wildfowl federal protection. The legal gutting of Weeks-McLean was met with varied responses from sportsmen.[103] Critics complained about overlegislating at both the state and national levels under "the present system of laws." The result, wrote George Klein to *Forest and Stream,* was that "our game is protected off the face of the earth." Lawmakers and bureaucrats without any knowledge of wildfowl habits and local conditions made the rules. Thousands of dollars were spent to make laws and convict hunters, "but not one dollar [was] spent to produce birds" for recreational consumers.[104]

Sport hunting opponents of Weeks-McLean argued that the law's focus on uniform hunting seasons ignored the market's ongoing depredations while targeting recreational hunters.[105] There was "absolutely no provision against possession after the Federal closed season nor bag limit shipment or sale" restrictions during the open season, complained a *Forest and Stream* correspondent in 1915.[106] Inconsistencies in state wildfowl sale restrictions also continued in the aftermath of Weeks-McLean. Under state law, Arkansas market hunters, for example, could harvest ducks from the flooded Chickasawba District during the open season and send them to Chicago, while nonresident sport hunters were prohibited from shooting there at all.[107] Market hunters raised no protest about the federal and state wildfowl laws, asserted critics, since they benefited from them, while the Missouri and Arkansas sportsmen who protested "were called 'enemies of wild life.'" In response, Missouri's sportsmen-led Fish and Game League had "expediently ignored the Federal seasons" and drafted their own bill "with further drastic and rigid provisions against all commercialism."[108]

New, state-level restrictions on the wildfowl market found more success than did the federal protections. In 1915, Illinois finally outlawed commercial sale of all game, except rabbits. This was "the great conservation movement of the season," according to *Forest and Stream.* Missouri's no-sale laws had been handicapped by its neighbor's liberal market policies that "afforded a great game market for the swamp hunters" in Missouri and Arkansas. With three of the greatest game market cities—Chicago, New York, and St. Louis— closed to wildfowl sales, the game shipper was fast approaching extinction in many parts of the country.[109] Even in California, where the state no-sale law had gone down in defeat, new daily bag limits of twenty-five birds per day on waterfowl squeezed game dealers, commission agents, and retailers whose operations required keeping larger amounts of game on hand. It was "an outrage for businessmen engaged in the Poultry & Game business . . . to be [declared] criminals" when they were simply "providing the general public with that much game that they are not able to kill themselves," one San Francisco merchant protested to the state's Fish and Game Commission.[110] Restrictions on commercial game sales were so pervasive by 1915, wrote a contributor to *Forest and Stream*, that the market hunter could no longer be blamed for the continuing decline in avian populations. "Just try to sell birds in the Midwest," he suggested, rather than "wear out pencils preparing articles on pot hunting and market hunting."[111] His was a minority view, however. For many sportsmen, market hunting continued to be the primary concern—a concern maddeningly unaddressed by the Weeks-McLean Act.

Holland's field work reflected the commercial limitations of Weeks-McLean. Although a portion of Holland's efforts included policing the wildfowl market, he could do so only under the auspices of state statutes, with the cooperation of state game authorities, and within the state's law enforcement system. Deputized into the state's warden ranks, Holland could investigate and prosecute no-sale, nonexport, and possession violations on railroads and in cold storage plants and restaurants.[112] For all his investigative efforts, Missouri market shooters promised to "fix [Holland] up the first time" he came across the river from Kansas. Convictions under state no-sale, nonexport, and possession laws remained elusive, however. The sheriff of Reno County, Kansas, who made "a business of trapping and selling ducks," told Holland that the state's game warden had tried to stop him several times, but "that he had beat him every time in court." The state attorneys general advised Holland to "let the game business go unless local citizens would swear out warrants."[113] When a porter discovered a bunch of wild ducks stored by area sportsmen in the Lincoln Hotel freezer in Nebraska, Holland was reluctant to pursue the matter. "I don't see he has any case under the state law," he reported, "and without a possession clause in the Federal Law we have no case." Much of the illegal wildfowl hunting, Holland concluded, was most likely "the natives" who were "simply killing a few to eat and not shooting for the market."[114]

As Holland's field diaries from 1914 onward show, his enforcement efforts were focused primarily on recreational hunters, not professional shooters. Typical was the example of "some men of means" who came in their private car to shoot black ducks out of season. These were "game law violators of the worst kind," according to Holland, and he produced "two A#1 cases" against the hunters.[115] This reversal of roles for sportsmen—from legitimate hunters to suspected game law violators—riled many. Even some wildlife officials had little use for federal intervention. Across the river from St. Joseph, Missouri, Holland found the federal warden for the Elwood, Kansas, district to have "absolutely no knowledge of game or shooting regulations." He insisted that "the law was intended for the pot-hunter" who sold birds commercially and not his friends who hunted in the spring for "a mess of ducks to eat." Kansas state game warden Dyche complained about the federal law to Holland, as well, blaming the shortened seasons for the reduction in license fees his office received from recreational hunters.[116] The Nebraska state game warden was likewise "not entirely in sympathy with the Federal law," Holland reported.[117]

One of the most vocal Weeks-McLean critics was former market hunter and Iowa fish and game commissioner E. C. Hinshaw. Like many other ex–professional wildfowlers, Hinshaw attributed the precipitous avian decline in the early twentieth century to habitat destruction rather than overhunting or

"Constable Examining Licenses—Hunting, ca. 1910–1915." By 1910, state and federal game wardens like Ray Holland were spending much of their time in the field policing sport hunters rather than market hunters. Many sportsmen found new conservation measures to infringe on what they saw as their right to consume wildfowl. Image from Bain News Service Photograph Collection, courtesy of Library of Congress Prints and Photographs Division, Washington, DC

spring shooting. Wetland drainage and intensive farming had driven the birds away. The agricultural value of insectivorous birds and warnings about species extinctions—key rationales for the federal legislation—were equally specious to Hinshaw. Quail and prairie chickens, classed among songbirds in the new law, did not eat enough insects to warrant the protections given them, the game warden claimed. Passenger pigeons had not been exterminated by gunners, and the loss of the species was no great injury. Modern Iowans had no more use of them than they did for "big herds of buffalo." The facts on the ground convinced Hinshaw that there was little use, and much harm, in closing the spring hunting season.[118] His open disdain for the Weeks-McLean law provoked Hornaday in 1914 to call publicly for his removal.[119] In response, Hinshaw claimed that all he wanted was "a square deal" for Iowa's sportsmen, in the form of adjusted open seasons, and the removal of quail, crows, and prairie chickens from the insect-eating bird category. In the meantime, he had "no intentions of changing his plans to suit a few cranks in New York" responsible for "bird daft propaganda."[120]

The migratory bird law (MBL) had been in effect a few years, but Hinshaw had not changed his tune. On a visit to Iowa in 1917, Holland found Hinshaw still wound up over the issue. "I doubt if this man will ever be in sympathy with the regulations of MBL or the Biological Survey," Holland reported. Spring shooting for ducks was the only shooting to be had in Iowa, according to the game warden, and the Weeks-McLean law was "worse than a joke." He claimed that more ducks had been killed in the spring since the law's passage than ever before and that the only people supporting Weeks-McLean were "a bunch of sentimental cranks." Holland found Hinshaw's views shared by other Iowa fish and game officials. At the state fish hatchery in Spirit Lake, he talked to a "most uncivilized crowd" of men who "took especial pains to show their contempt for the Biological Survey and all connected with it."[121]

Faced with such widespread opposition, Holland found himself spending much of his time educating hunters about the law and convincing them to obey it. Requests poured in for information about the federal regulations, especially after the constitutional challenges to the Weeks-McLean law.[122] When the expansion of the original, two-zone federal hunting season system created five new zones in 1916, Holland once again canvassed his district to educate hunters and seek their support.[123] Many, he found, were misinformed or "put out at the new regulations," and Holland expected an increase in out-of-season shooting as a consequence.[124] In Kansas, sportsmen were perturbed that the new season closed at the end of December, while neighboring Oklahoma's open season lasted until February 1. Wichita's sportsmen would be particularly intransigent, Holland predicted, since they had been "kept stirred up by the *Wichita Eagle*." The "Kansas City crowd" was also up in arms against the federal law and was reportedly "holding indignation meetings." Riled by the federal enforcement, one man on several occasions threatened to kill Holland. While not taking the claim too seriously, Holland nevertheless reported it, writing: "Should I get in a shooting scrape with him and shoot him, it might be well to make a note of this threat."[125]

While Holland worked to combat criticisms of the law on the ground, other prominent conservationists did the same in the broader public discussion. Pearson, as the head of the Audubon Society, attempted to downplay the revolutionary potential of the legal shift in bird protection. The federal statute "did not in itself change any of the existing game laws," he wrote in 1917, but simply "gave authority to certain functionaries to make regulations as they deemed wise, necessary, and proper." The regulations created by the committee of governmental experts appointed to the task were "remarkable for their clearness, directness, and fairness." The changes had brought "critical rumblings" from many sportsmen, as to be expected; but these were only the complaints of men accustomed to unsustainable privileges, Pearson maintained.[126]

The opposition to the Weeks-McLean law signaled the uphill fight its supporters would face in the following years as organized sportsmen repeatedly challenged the law's constitutionality. This, they insisted, was an intrusion on the principles of state ownership and states' rights previously espoused by the Supreme Court. The unfavorable court decisions at both the state and federal levels convinced the conservation minded that the rationale for a federal wildfowl law required a better constitutional argument. Leaving the *Shauver* case delayed on the Supreme Court's calendar, lawmakers and wildlife protection advocates began to craft new legislation to answer the Weeks-McLean critics. In the process, they pulled political control of wildfowl still further from sport hunters' grasp.

Waging a Duck War

The key to protecting migratory birds in the United States was to be found in international cooperation, an idea that had been floating around Congress since the first constitutional questions were raised about the Weeks-McLean

Act. The 1916 Convention on the Protection of Migratory Birds between the United States, Canada, and Great Britain was the first step in relocating the source of the federal government's authority over migratory wildlife in its treaty-making powers, rather than in more-nebulous assertions of regulatory control over interstate commerce or the "general welfare." After ratification of the convention treaty, its provisions and ensuing regulations were put into effect by enabling acts passed by Canada in the 1917 Migratory Birds Convention Act (MBCA) and by the United States in the 1918 Migratory Bird Treaty Act (MBTA). The passage of the MBTA rendered the still-pending *Shauver* case moot as the 1916 treaty superseded the previous US migratory bird law. The parties involved in *Shauver* allowed it to die in a quiet death, unresolved.[127]

The new tactic did not quiet the opposition, and the resulting dispute deepened the split among sportsmen between pro—and anti–federal legislation factions. Not a few sport hunting consumers raised familiar complaints. In imposing uniform hunting seasons in the MBTA, functionaries from Washington took aim at spring shooting again, and recreational hunters protested the lack of "some concession in the spring for the poor men" whose only opportunities to hunt were during the wet season.[128] *Forest and Stream*, on the other hand, supported the migratory bird law, insisting that commercial dealers and hotels were angling "for return of the good old days" when game markets flourished.[129] National wildfowl laws would ultimately put a stop to all market hunting, the magazine claimed, as more states revised their game codes to align with federal regulations. Only "the selfish element" opposed the sensible limits the MBTA imposed.[130] On this point, sportsman-conservationists had backing from conservation scientists. The problem with state game laws, field biologists in California pointed out, was that they shifted almost constantly in the political winds, caught between the "efforts of modern conservationists" and the immediate desires "of the more selfish element among the hunters." The result was legislation that ignored birds' ecological needs to satisfy recreational consumers. "The difficulty had always been," wrote the authors of *The Game Birds of California* in 1918, "that the political boundaries made use of in connection with the game laws, and the natural boundaries" that divided the "natural 'lifezones' in which the different kinds of game exist" rarely coincided.[131]

Lawmakers, too, were divided over federal conservation measures and their effects on wildfowl consumers. MBTA defenders in Congress offered economic rationales for the law, insisting that the legislation was not a tool to limit sportsmen but instead a weapon to control agricultural pests.[132] The migratory bird law's protections for insectivorous species were necessary "to prevent the extermination" of birds "valuable for purposes of food and now fast disappearing."[133] Opponents of the act, mainly midwestern politicians, took up the cause of the "true sportsman" against "well-intentioned but misinformed 'conservationists'" and "biological 'experts.'" The previous federal law had done nothing to curtail the market, Kansas representative Dudley Doolittle complained. Weeks-McLean had created a situation in which it was "eminently proper as a conservation argument" to allow professional shooters to ship birds to market "by the sackful," while the sportsman who killed a single bird in January was held accountable to federal law. National regulations profited only "the men who want to commercialize in game."[134] Representative Dorsey Shackleford of Missouri agreed. The "movement for Federal conservation of migratory birds" was merely a scheme for financial gain devised by large, eastern arms and ammunition manufacturers, and conservation groups like Hornaday's Permanent Wild Life Protection Fund were just fronts for Winchester, Remington, and other gun dealers.[135]

Despite the opposition, the MBTA was approved on July 3, 1918, and its passage was hailed in the *New York Times* with the headline: "New Law Bars Pot Hunting." The legislation did more than that, however.[136] In contrast to the Weeks-McLean Act's simple wording, which merely established federal custody of wild birds and federal authority to set hunting seasons, the MBTA contained a wider range of prohibitions. It was also written so that migratory birds were fully protected unless their use was expressly permitted—in contrast with the Weeks-McLean language that allowed wildfowl harvesting unless directly prohibited. This more comprehensive migratory bird law and the first set of regulations promulgated under its authority covered some 537 different species of migratory birds. Another 220 species, including quail, grouse, pelicans, flamingos, sparrows, and blackbirds, among others, were excluded from the law's provisions and thereby remained under state control. The new rules established closed seasons for treaty birds; prohibited all killing of insectivorous species; set restrictions on taking, possessing, and transporting birds; and, most significant, abolished the commercial sale of migratory wildfowl.[137]

The MBTA went beyond the Weeks-McLean's statement of proprietorship to begin with the extended stipulation that it was "unlawful at any time, by any means or in any manner to pursue, hunt, take, barter, offer to purchase, purchase, deliver for shipment, ship, export, import, cause to be shipped, exported, or imported, deliver for transportation, transport or cause to be transported, carry or cause to be carried, or receive for shipment, transportation, carriage, or export, any migratory bird, any part, nest, or egg of any such bird, or any product, whether or not manufactured."[138] Other sections of the law established varying bag and possession limits for different groups of birds. Though large by today's standards, the new limits made the game business unsustainable. The MBTA's foregrounding of commercial sale remedied one of sportsmen's key criticisms of the previous migratory bird law. The market in birds had been the largest drain on wildfowl supply, wrote the chief US game warden, George A. Lawyer. So long as markets remained anywhere, it was "profitable for the market hunter to continue his destructive vocation." In no uncertain language, the MBTA made the market in wild birds, and the market hunter, always illegal everywhere. With the market gone, "the incentive to slaughter [birds] in such large numbers" would vanish, Lawyer confidently asserted.[139]

A vocal segment of sport hunters, though, contended that the federal law was "the most detrimental to wild ducks in every imaginable manner."[140] These hunters' experiences told them something quite different from what the experts were claiming. While conservationists touted marvelous increases in wildfowl populations due to the new restrictions on recreational consumption, many hunters in the field saw the opposite. Conservationists' "ballyhoo" about spring shooting was nonsense, according to critics. The seasons should have been reversed, they argued, to allow hunting in the spring when there were ample water and food for ducks. Fall shooting only hurried the birds on to oil-polluted lakes farther south.[141] Other weaknesses of the MBTA included high bag limits and inconsistent enforcement.[142] Conservationists' attempts to use federal interventions to reduce wildfowl consumption smacked of the same futility as national prohibition.[143] Control and enforcement of the game laws would be better "turned over to the local sportsmen, the club men," some recreational hunters insisted. But, instead, sport hunters had been sidelined, unfairly placed "in the bootlegger and outlaw class" by outside authorities.[144]

It was true that enforcement of both federal migratory bird laws proved challenging for the Bureau of Biological Survey. In 1919, the appropriation for migratory bird work amounted to only fifty thousand dollars, and the bureau was able to employ only fifteen wardens for the entire country. State game wardens cooperated in policing hunters, but they mainly worked as informants, alerting federal wardens to possible violations.[145] Years into the MBTA's tenure, the lack of federal enforcers remained a difficulty. "The number of federal wardens in this country is ridiculously small," wrote Idaho's chief fish and game warden, "and it is an utter impossibility for the United States government to enforce the present federal laws."[146] By 1922 the number of US game wardens had risen to just twenty-eight.[147]

Working in one of the main hotbeds of sportsman dissent, US game warden Holland was well aware of the problems accompanying the implementation of the new federal law. The same organized forces that had derailed the Weeks-McLean law were no more pleased with its replacement. While Holland was gratified to find gunners voluntarily complying with some regulations, such as sunset laws, he knew that opposition to federal regulations remained and that there were other sportsmen willing to test it and him.[148] Months after the MBTA's passage, Holland discovered illegal shooting on the Kansas-Missouri border. "At least four boats have been shooting right along," he reported late in 1918: "the official police boat, a boat belonging to a city detective, Ed Costello's outfit, and [another hunter] from K.C." Holland was uncertain about bringing a case against the poachers, however. He was "confident the U. S. Attorney in Kansas would try to win" the case, but he was "afraid of the judge" in that district. "In Missouri, the court may be O.K.," Holland noted, but he was not sure that the US attorney would take on the sportsmen.[149]

Holland's fears were well founded. Early in 1919, Missouri's lawmakers directly challenged the federal legislation with a resolution reasserting the state's claim on all wildlife. The national government's intrusions into game regulation resulted in citizens being "grossly harassed, annoyed, and illegally deprived of their inherent rights to freely hunt game." The legislature thus instructed the state's attorney general "to enjoin the federal game inspectors" from interfering with residents' exercise of their state-granted hunting privileges.[150] The resolution, which was sponsored by the Interstate Sportsman's Protective Association, according to the *St. Louis Post-Dispatch*, was merely a suggested a course of action—one that the state's attorney general, Frank McAllister, was more than eager to pursue.[151]

This was just another salvo in Missouri's "Duck War on the United States."[152] A few days later, Holland arrested sport hunters George Samples and V. L. De Lapp for spring shooting violations. Defended by lawyers from the Interstate Sportsmen's Protective Association, the case seemed likely to become a constitutional test for the MBTA. But Attorney General McAllister was not willing to wait on a judicial decision, and he boasted that he would shoot ducks in the spring, the "unconstitutional law" and Holland be damned. McAllister was "continually in my hair," according to Holland.[153] An illegal duck hunt on Stultz Lake in western Missouri early in March 1919 brought confrontation between the two men. Warned that the game warden had been tipped off and was headed his way, McAllister reportedly replied: "Let him come. We know what we are doing."[154]

Arriving at the scene, Holland found the attorney general out waterfowl hunting with a well-connected group of sportsmen: the president and manager of the Kansas City Life Insurance Company, the mayor of Paris, Missouri, and a merchant from Macon.[155] The party had in their possession "41 pintails, 23 mallards, three green wing teals, and one blue goose," all of which Holland promptly seized.[156] In a retaliatory move, the sheriff of Clinton, Missouri, arrested Holland and his deputy for violating the state's laws against transporting game without a license as they returned with their confiscated waterfowl.[157] Although the state eventually dropped the illegal-transportation charges, McAllister quickly upped the ante on Holland. He filed for a restraining order against the federal warden to prevent him from enforcing the terms of the MBTA where they conflicted with the state's game laws.[158] McAllister's arrest was seen by some as purposeful—he had "offered himself up as a sacrifice" to provide a Supreme Court test case—but Holland and his allies denied that interpretation of events. The attorney general "was caught just like any other game law violator." The earlier De Lapp–Samples case was the obvious vehicle for a constitutional challenge, and McAllister had no reason to provide another.[159]

The legal wrangling in Missouri was clearly headed toward a constitutional showdown, but in the meantime, another test of the migratory bird law was making its way to federal court in the Eastern District of Arkansas. *United States v. Thompson* was set to go to the same judge who had ruled on the Weeks-McLean Act in *Shauver*: Justice Jacob Trieber. The defense relied heavily on the judge's previous ruling, arguing that the same conclusion must hold. But this time, Trieber took the plaintiff's side. The national treaty power was superior to any internal state authority, he concluded in June 1919.[160] This was "a hard blow" to "the enemies of the law" who had "pinned their faith on Judge Trieber." The AGPA celebrated the ruling, declaring it "second in importance only to a decision from the US Supreme Court."[161]

The *Thompson* decision presaged a favorable outcome for Holland and the MBTA in ever-contentious Missouri. The trial of the De Lapp–Samples case was scheduled in federal court early in July, and Holland, who had left his federal game warden position to take on the editorship of the *Bulletin of the*

American Game Protective Association in New York, was nevertheless keeping up on its progress through the judicial system. US district attorney Francis M. Wilson provided the new federal game warden, Jonathan Q. Holmes, a copy of the brief McAllister had filed to prove the law's unconstitutionality. Wilson "didn't think much of it, but McAllister was very proud of it." The judge seemed poised to support the federal law, Holmes wrote Holland, but was delaying his decision because he "hate[d] to break the news to Kansas City."[162] A month later, Judge Arba S. Van Valkenberg ruled the MBTA constitutional and refused McAllister's request to prevent US agents from enforcing federal wildfowl regulations.[163]

The Missouri fight won over some, including the District Attorney Wilson, to the conservationists' cause. "The old boy seems to be in dead earnest about bird conservation," one of Holland's informants told him.[164] For his part, Wilson was very pleased with his own performance in the Missouri case which he deemed "a fight from the heart." It was "quite an uphill fight," he wrote Holland, and "the other side felt sure they would win" on both the MBTA violation and the injunction against the federal game wardens. Yet the prosecutor's written and verbal arguments were "simply unanswerable," according to Judge Van Valkenberg.[165] Conservationists saw the case as significant because it came "from the very heart of the enemy's country"—a place where "the ever-troublesome Interstate Sportsmen's Association" held sway—and because it "convinced a confessedly antagonistic Federal Judge of the validity of the law." The result, wrote a correspondent to the *Auk*, was that "another long war [was] over and won."[166]

McAllister, though, was undeterred by the unfavorable state ruling. He immediately announced his intention to appeal the decision to the US Supreme Court, and the case of *Missouri v. Holland* was anxiously awaited by sportsmen and conservationists alike as the final word on federal wildfowl jurisdiction.[167] In July 1920, the constitutionality of the Migratory Bird Treaty of 1916 and the Migratory Bird Treaty Act of 1918 was upheld in a majority opinion written by Justice Oliver Wendell Holmes that settled the question of the government's legal authority over birds:

> Wild birds are not in the possession of anyone; and possession is the beginning of ownership. The whole foundation of the state's rights is the presence within their jurisdiction of birds that yesterday had not arrived, tomorrow may be in another state and in a week, a thousand miles away. . . . We see nothing in the

Constitution that compels the Government to sit by while a food supply is cut off and the protectors of our forests and crops are destroyed. It is not sufficient to rely upon the states. The reliance is in vain.[168]

Although the constitutional issue for Holmes centered on the nation's treaty power, the court's decision directly acknowledged the intersections of biology, economy, and law at stake in the case. Wild birds' fugitive natures, their value as economic and ecological resources, and the legalities of possession, ownership, and jurisdiction combined to produce both the controversy and its result.[169] Conservationists and allied sportsmen hailed the outcome. One widely circulated newspaper article trumpeted, "U.S. Supreme Court Saves the Birds."[170] An Iowa newspaper headline put it more bluntly: "Bird Hunters Must Obey the Law."[171] The conclusion of *Missouri v. Holland* understandably had a personal dimension for Holland. He received scores of copies of the court's decision to keep as mementos of his "only opportunity . . . to go down in history." Another prized keepsake from the affair was a framed photo of Missouri attorney general McAllister, which Holland hung on the wall of his office.[172]

The constitutional dispute over federal authority at the heart of *Missouri v. Holland* ultimately overshadowed the historical significance of the Migratory Bird Treaty Act: the legal end to the market in birds. It is easy to see why. The most visible controversies and most vocal opposition to the federal law came not from market hunters, game dealers, or shippers but from organized, elite, and politically powerful sportsmen. Preexisting and widespread state-level prohibitions on wildlife marketing made questions about the legitimacy of commercial sale seem almost quaint by twentieth-century standards. A uniform federal anti-marketing policy—and the elevation of sport hunting as the only legitimate form of wildfowl killing—looked inevitable from this perspective. Yet the consumer-capitalist market in wild avian commodities proved crucial to the revolutionary legal shift in the Migratory Bird Treaty Act. The modern wildfowl market first produced the conception of birds as interstate goods. The market was responsible for bringing far-flung avian resources in the form of meat, feathers, and eggs into a centralized public space. The market also placed an exchange value on the bodies of dead birds—as food, as feathers, and as collectibles—a value that was countered by sportsmen and conservationists arguing for different economic estimations of wild avian worth in the forms of profits derived from recreational consumers and increased agricultural production. The MBTA

Headline and illustrations for a widely reprinted story published in 1920 after the US Supreme Court issued its decision in *Missouri v. Holland*, featuring federal legislators Elihu Root and George Shiras III surrounded by ducks, songbirds, and plume birds. Image from the *Idaho Springs (CO) Siftings-News*, 2 July 1920. Courtesy of Chronicling America, Library of Congress, Washington, DC

was revolutionary in its absolute abolition of private ownership and commercial exchange for an entire category of environmental resources in a nation otherwise consumed by capitalism.

More important, biology and economy combined in the modern wildfowl market in ways that fundamentally reshaped American cultural attitudes toward wildlife. The federal migratory bird law marked a crucial shift in the language of wildlife conservation that originated in the short-lived modern market in birds. Turn-of-the-century game dealers not only sold food or other utilitarian commodities but also repackaged wildness for a new class of urban consumers. That wildness became the key issue in the market hunting debate, as activist sportsmen advanced a different and morally superior set of cultural values to be found in their recreational pursuit of wild birds. The sportsman's identity depended on these experiences with living wildness, and sport hunters' advocacy for game protection was motivated by the concern that unchecked commercial wildfowl exploitation would ultimately destroy the wildness experiences they treasured.

These kinds of abstract and esoteric estimations of wildfowl worth, most widely popularized by recreational hunters in the late nineteenth century, found new expression by the early twentieth century, as conservationists broadened their appeals to nonhunting audiences and elevated the cultural value of living wild birds to American consumers. The lines that had been first drawn in this debate between commercial and recreational wildfowl killing were replaced with new divisions separating those who saw greater value in living birds than in dead ones. Sportsmen, who had once taken the lead in wildfowl protection, found their ranks divided between those who supported conservationists' strict limits on recreational consumptive use and others who rejected them for their lack of "duck-sense."[173] In sum, the controversies provoked by the MBTA and played out in *Missouri v. Holland* reflected the transition from wildlife consumption to a modern and multifaceted wildlife consumerism in America.

There were some who realized a significant change had happened. "The greatest step forward in protection of waterfowl was when Federal law came in and stopped the sale of ducks," wrote J. S. Hunter of the California State Fish and Game Commission in the 1920s.[174] State laws alone failed to stop waterfowl commercialization, and Hunter estimated, somewhat generously, that between a half million and a million ducks were sold yearly before the game market was outlawed. Game dealers had used every influence to allow them "to have as many ducks as they could get hold of in their possession."[175] Even though a remnant of commercial wildfowling continued illegally in the 1920s after the MBTA, Hunter maintained that in California, it amounted to less than "1% of the sale in olden days."[176] *Forest and Stream* likewise noted the salutary environmental effects of uniform anti-market policies. In the past, wrote Archibald Rutledge, men "made it their business to kill ducks for the market" along the Santee River in South Carolina. But in 1920, legislation and "the passing of most of the lands on the delta into the hands of a good sportsman's club" had stopped this kind of exploitation.[177] Another contributor credited the combination of federal and state laws for "ending the days of the market hunters." If not for the timely federal intervention, more wild birds would have followed the passenger pigeon's route to extinction.[178]

Conservationists and their activist sportsmen allies pointed to the recent history of wildfowl commercialization and profits made by some individuals from wildlife held in the public's trust to explain why national legislation was necessary. In the wake of the first federal migratory bird law in 1915, *Forest and Stream* offered a warning to readers complacent about the market's effects on birds. The warning came in the form of excerpts taken from Merritt's 1904 book. These were the "Confessions of a Market Hunter" that revealed "the horror of butchery" inherent in commercial gunning. Two pages of quotations from *The Shadow of a Gun* followed, detailing the hundreds of birds Merritt shot at a time and the tens of thousands Merritt shipped in a season. If younger sport hunters chafed at the stringent new laws, like those prohibiting spring shooting, they needed only to read Merritt's recollections or peruse the back issues of *Forest and Stream* to understand why wildlife conservation was necessary. The nonsale of game was the magazine's signal achievement, one that "put Merritt and his class out of business." His "confessions" proved that market hunters had "stolen outright" the people's property in wildfowl. The game dealer's greed resulted in "a crime equally great" perpetrated on "the present generation," whose legacy in wild birds and wildness had been irreparably diminished.[179]

☙

By the time *Forest and Stream* passed its judgment on Merritt and *The Shadow of a Gun*, the old game dealer had been dead for nearly a decade. Three years before he died, the *Kewanee Star Courier* offered a much more favorable account of Merritt's book. It was "full of local reference" and a "complete history of the game business" that would likely interest the present generation. Included, the newspaper noted, was poetry "written on a high plane of thought and voice" as well as "an account of the game litigation in which Merritt was involved" and his characterization of "the motives which . . . inspired the charges brought." The book sold for $1.60 to $1.75 per copy.[180]

About a month after the publication of *The Shadow of a Gun*, Merritt's wife of more than forty years, Sarintha, passed away. Sarintha's niece Jennie Hainds came afterwards to work for Merritt as his housekeeper. He had mostly given up the game business, citing the lack of birds and the pestering presence of game wardens, but retained much of his substantial business holdings, including the Merritt Block and the hotel. Years before the market was officially outlawed, Merritt had predicted its demise, declaring that "the hunter for profit has had his day." In the end, the old market hunter's regrets were not so much for the birds killed as for a way of life lost to the present generation of young men, eager to distinguish themselves "in the field with dog and gun."[181]

Merritt's retirement was hardly a quiet one, though. A younger generation closer to home brought him new legal troubles. His sons, Clarence and Herbert, filed suit against him late in 1906, claiming that he had hidden from them an inheritance bequeathed by their grandfather Nathan Merritt years earlier. They painted an unflattering portrait of their father as a selfish man who allowed his sons "a slight education" and no funds "for personal needs or pleasures." Told they would have to work for their money, the sons were forced to support their father's business: Clarence traveled to buy game birds while "Herbie," who had "always been physically weak," worked in the cold storage plant when he was able. Rather than turn over the inheritance to his sons, Merritt was alleged to have used the money on Jennie's divorce and her three children. According to Clarence and Herbie, their father now spent "most of his time love making, writing poetry, and conspiring with Mrs. Hainds [about] how to cut off the rights of the complainants."[182]

Shortly after the news of the lawsuit went public in 1906, Merritt married the much younger Hainds in a small ceremony in Chicago. A year later, Merritt died suddenly of apoplexy at the age of seventy-six. He was the "best known man of the city," the *Kewanee Weekly Star* proclaimed, with real estate holdings in Kewanee that included the Merritt Block, the Cliff House, and a three-story building formerly known as Merritt's Flats. All told, Merritt's worth was valued at some $300,000, much of it derived from the avian capital that *Forest and Stream* would later claim had been stolen from posterity. His substantial estate was left entirely to his new wife, who was listed, along with her children, as Merritt's survivors in the newspaper obituary. Clarence and Herbie were not.[183]

The Culture of Conservation

I have killed a good many ducks in my life. I wish that I could bring
some of them back to life again.

HENRY KLEINMAN, FORMER MARKET HUNTER

Twenty years after Henry Clay Merritt's passing, Kewanee's newspaper interviewed his son about the town in its early days. At sixty-eight, Herbie Merritt
was a local landmark, and his reminiscences carried the weight of lifelong experience in the region. His recollections painted a picture of a landscape
brimming with avian wildness. He remembered a time when "prairie chickens were everywhere" and flocks of "thousands of pigeons might be found in
cultivated fields." The nearby Annawan swamp drew a "surfeit of ducks" in
those years, and "the waters of the Mississippi were congregated with them."
The flights of birds seemed endless. "As one flock was destroyed, new ones
came." Not surprisingly, many of Herbie's wildfowl memories involved his
father, the "famous hunter of his day." Of particular note to the article's author
were the details of Merritt's game-shipping business and his properties, including his Spoon River icehouse and the Merritt Block in Kewanee. The poultry and game storehouse, from which "choice game was shipped to the east,"
still stood "near the old Merritt residence."[1]

Herbie's feelings about his father appeared to have softened in the years
since Henry's death. The article revealed none of the conflicts that had roiled
the Merritt family in its patriarch's final years: the game dealer's court battles,

his sons' lawsuit for their inheritance money, Merritt's remarriage to his wife's niece, and his exclusion of Herbie and Clarence from his will. The only hint of family discontent came from Herbie's critical observation about his father's penchant for poetry. An inscription on one of Merritt's buildings—a quotation, taken from a line of Lord Byron's satiric poem *Don Juan*—"was a waste of time and money," according to the son.[2] Otherwise, it seemed that Herbie was willing to set aside his personal grievances, at least publicly, to memorialize a way of life that had disappeared.

Nostalgia kept Merritt's reputation alive in Kewanee. He was "an outstanding figure" of the late nineteenth century, wrote E. A. Lincoln in 1934, "one of the greatest shippers of wild game in the middle west and probably in the United States." His "famous cold storage plant" held "many thousands of wild fowl" that were sent to "the lucrative markets of New York, Boston, Philadelphia, and other large eastern cities" to be purchased by "dealers and consumers at fancy prices."[3] More than just a businessman in these retellings, Merritt was, according to local newspaper stories in the 1930s and 1940s, a "lover of nature and of hunting and the world lying out-of-doors."[4] His memoir revealed him to be "a genuine wild-life philosopher," a true "lover of the wild things of the woods and streams" whose nature writing was "reminiscent of Henry D. Thoreau."[5]

Market hunters of the previous century such as Merritt were eyewitnesses to an avian environment the likes of which would never be seen again. Their testimonies were invaluable as records of America's wild heritage, and many sport hunters and conservationists were eager to document the details of the natural world that commercial hunters had inhabited and exploited. One of those was Des Moines realtor and banker Frederick O. Thompson, who recaptured market hunting's past in a series of interviews with a number of longtime professional and amateur wildfowlers (including several of his hunting buddies) in the early 1930s. Thompson's collection of memoirs chronicled the turn-of-the-century world of market hunters, "duck special" trains, and game law evasions.[6] Two of Thompson's informants, Fred Gilbert and Richard Harker, appeared in midwestern newspapers in the 1920s and 1930s as current target-shooting champions with market hunting pasts. The *Des Moines Register* acknowledged that Gilbert, known locally as the "Wizard of Spirit Lake," had started out as a commercial gunner, though that was "a perfectly legitimate vocation in those days."[7] Minnesota's Harker was likewise identified as "a market hunter in the old days," but he was better known in the 1930s as "one of the greatest wing shots in the middle west," according to the

Minneapolis Star. Outdoor columnist Jimmy Robinson explained that "thirty or 40 years ago, market hunting was not considered anything terrible" and the market hunter "was recognized as a hero in his section of the country."[8] The reminiscences of Louisiana ex–market hunter Captain Theodore Johnson constituted a significant "document in American wildlife history," noted the US Fish and Wildlife Service in a 1941 press release, important enough to warrant the interior secretary's, as well as the public's, attention.[9]

Though culpable for the destruction of wildfowl in years past, many of these old-timers had been transformed into "our most rabid game conservationists."[10] Since former market hunters could attest to the efficacy of state and federal wildlife policies, their conservationist conversions lent an air of legitimacy to such measures and support for new ones. Louisiana's Captain Johnson provided just this kind of ammunition to wildlife authorities. It was "a darn good thing the Federal Government started regulatin' the bag limits and seasons," he reportedly said to US Fish and Wildlife representatives, "or else there wouldn't be many ducks and geese left to look at today."[11] These kinds of twentieth-century transformations from market hunter to game warden, bird sanctuary creator, or some other "disciple of conservation" were clear evidence that a sea change in environmental consciousness had occurred.[12]

To be sure, some things remained the same. Despite the legal abolition of commercial wildfowling and a federal migratory bird law, avian populations continued to shrink in the first half of the twentieth century. Widespread drought conditions and land reclamation practices in the 1930s contributed to the decline as migratory waterfowl numbers in North America plummeted from 150 million in 1900 to less than 30 million by 1934.[13] In Iowa, conservationist Aldo Leopold described the once-waterfowl-rich landscape as a "veritable cemetery of dead lakes and ducking grounds" in the 1930s.[14] The situation grew so dire that some proposed a one-year moratorium on waterfowl hunting in 1935 to offset the effects of "legal hunting, commercialization, drought, drainage, and other causes" on "the plight of ducks."[15]

Adding to the problems were unreformed market hunters who continued to ply their time-honored trade surreptitiously in many areas of the country.[16] Known as "duckleggers" in the 1920s and 1930s, illicit hunters like California's "Dillinger of Duck Hunting," Howard "Bluejay" Blewett, sold their game on the sly to locals, stores, and "swanky restaurants" and faced incarceration when they were caught.[17] Major wildfowl "bootlegging" activities persisted well into the postwar era and were answered by some equally major game law enforcement operations. A two-year probe into a loosely organized market hunting network in 1956 along the upper Texas gulf coast, led by undercover US Fish and Wildlife Service agent Tony Stefano, resulted in fifty-three indictments for violations of the migratory bird law. Posing as a wholesale jeweler and syndicate game buyer near Houston, Stefano purchased 3,000 ducks and geese destined for sale at local restaurants and nightclubs. More than fifty agents from eight states took part in the raids and mass arrests, which brought in rice farmers, oil field workers, ranch hands, restaurant owners, and even a town constable to face federal charges.[18] Stefano's 1958 operation against market hunters based in East Peoria, Illinois, resulted in arrests in Michigan, Wisconsin, and Illinois and broke up a "ducklegger ring" that was handling about 250,000 ducks annually.[19] Observers declared that "no penalty [was] heavy enough" for these "commercial killers."[20] Roundly criticized as "low-lifes," twentieth-century market hunters could no more escape the public blame for wildlife destruction than could their nineteenth-century predecessors.[21]

Sportsmen, too, received their share of the blame as the battle between "real conservationists" and "shotgun conservationists" raged in the first half of the twentieth century. Conflicts over public shooting grounds in federal refuges and bag limit reductions fractured and realigned wildlife interest groups in unexpected ways in the 1920s and 1930s.[22] Conservation purist William Hornaday campaigned against game refuges, while the National Association of Audubon Societies opposed reductions in federal bag limits on migratory birds.[23] The US Biological Survey stood behind efforts to establish public shooting grounds, whereas the Emergency Conservation Committee opposed them.[24] The editor of *Outdoor Recreation* threw stones at the "double-crosser" and "four-flushing" Will Dilg, editor of the Izaak Walton League's publication, *Outdoor America*, for his "betrayal of the Shooting Grounds Bill."[25] Manly Harris of California's Wild Life Development League challenged federal authority to lower duck bag limits, while Emerson Hough, once a champion of the sport hunter–conservationist alliance, called for a halt to all sport hunting, claiming that, "of the alleged true sportsmen of this country, not 10 percent have practiced the creed which hypocritically they profess."[26] Sport-hunting commercialism had replaced game market commercialism, critics argued, with similar destructive consequences for wildfowl.

National Audubon Society president T. Gilbert Pearson tried to steer a middle course between the moral purity of the radical conservationists and the proprietary attitudes of activist sportsmen. Fearing "the extremists on both sides," he sought to rebuild the conservation coalition of earlier years. Enlisting the more "level-headed crowd"—which included Ray Holland, now editor of *Field and Stream*—Pearson promoted "a new doctrine of game administration," hoping to "stem the battle between bird-lovers and sportsmen."[27] But his brand of "conservative conservation" (labeled "compromised conservation" by detractors) ran headlong into rival versions of wildlife use and value.[28] At the heart of the conflict over these differing visions of proper human-wildlife relations was, in Hornaday's words, "the scourge of guns and commercialism."[29] Sportsmen and game conservation organizations had entered into unholy alliances with gunmakers, ammunition suppliers, guide services, and other recreational hunting businesses.[30] Conservationists like Hornaday and Hough grew disillusioned with what seemed a fatal disconnect between sportsmen's high-minded rhetoric and their profligate actions. Preaching self-restraint, hunters all too often "practice[d] self-indulgence."[31]

Advocating alternative forms of nonconsumptive engagement with living wildlife, anti-hunting conservationists touted a whole host of intangible advantages modern Americans derived from wildfowl. The mere "contemplation of birds as possessors of pleasing form and plume, cheerful manner, and attractive song" brought "refreshment to the mind" and made them a "community asset."[32] Birds, especially, offered great aesthetic value to wilderness-starved urbanites, who could encounter avian wildness in their backyards, in public parks, or along waterfronts.[33] Wildfowl also contributed "inherent value" to children's education, supplying lessons on ethics, compassion, and nature appreciation.[34] According to educators, there was "no subject better adapted to develop the moral or humane features of a child's character than the subject of birds."[35] Young and old could find pleasure equally from hearing "the early song of the robin on the lawn, or cooling, delicious notes of the canyon wren in a wild ravine of a mountain desert." Both experiences taught that, "despite human cares," there could be happiness and "relief in the return to Nature."[36]

Faced with competing estimations of wildlife values and growing anti-hunting sentiments, organized sportsmen began to craft a defense of hunting in twentieth-century terms. Their reformed image of sport hunting stressed the more esoteric values of recreational wildlife consumerism. Companionship, well-being, and increased efficiency were important by-products of hunters' encounters with wildness. The "healthful recreation of hunting furnishes relief from nervous strain" for adults, asserted Manly Harris, president of the United Duck Hunters of California in 1927. A bit of waterfowl gunning could also "divert attention of youth away from less desirable pursuits" (including premarital sex)—an outcome that clearly "offset the costs of killed ducks." Without "open sport" to serve "as an outlet for their young desire for expression," boys would be lost in a "wave of youthful delinquency." Indeed, as Harris concluded, if only "Loeb and Leopold had a little duck shooting they would never have conceived 'the perfect crime.'"[37]

Sportsmen's expansive vision of "constructive conservation," however, went beyond abstract values to characterize recreational hunting "as a commercial asset as well."[38] Hunting clubs, sportsmen argued, served an important ecologic purpose by maintaining, with private funds, wildfowl habitats threatened by agricultural and industrial development.[39] The sportsman paid for the birds he killed in license fees, club dues, guns, and ammunition. Unlike the market hunter of old, the sport hunting consumer valued wild creatures for the wildness experiences they provided, and the money he spent funded important conservation work. Anti-hunting forces saw such arguments as merely defenses of "killing privileges" for a profit, hardly different from the market hunter's objective.[40] Moral purists insisted that wildlife conservation would always fail "when one commercialized motive comes into its thought."[41]

Despite the in-fighting among competing wildlife protectors, federal conservation measures expanded in the 1930s. Chief among the new provisions were those aimed at increasing avian populations through habitat conservation, especially along the major and minor continental flyways that migratory birds traversed.[42] The Migratory Bird Conservation Act of 1929 established refuges for North American birds, and the Migratory Bird Stamp Act of 1934 used sport hunter fees for avian conservation and wetland preservation.[43] The Pittman-Robertson Act of 1937 funneled monies raised from excise taxes on arms and ammunition into habitat restoration and scientific wildlife management.[44] Although waterfowl reached historically low population levels in the mid-1930s, these new measures gave conservationists cause for hope. A significant shift in wildlife protection policies was occurring, one that Aldo Leopold described as a turn away from the "negative and prohibitory" enactments that had characterized the movement in its early phases toward a "positive and affirmatory ideology" focused on environmental health.[45] For avian activists, the wildfowl refuge victories of the 1920s and 1930s proved that "at last the God of Nature and the wild places and wild things WON."[46]

These pockets of avian refuge, however, were only the barest reminders of the natural landscapes that had once nurtured the market hunter and were now a distant memory. The evidence of loss could be found in the original survey plats "of some dried-up township," Leopold wrote, where surveyors had drawn sloughs and marshes in "quaint water color arabesques of blue and brown." Vast grasslands once alive with wildfowl now spread "lifeless from horizon to horizon, in final surrender to the legions of corn."[47] Or one could remember New York's "virgin timber . . . interlaced with wild grape vines" that fed innumerable grouse in days gone by.[48] The almost-incomprehensible numbers of birds those environments supported and the immense quantities harvested for the market represented remarkable natural wealth and, at the same time, deplorable national waste.

The market hunter of old occupied a different cultural landscape, as well. To explain the market hunter's late-nineteenth-century existence, modern writers pointed to wildfowl use for frontier subsistence, the lack of game laws, and the nearly universal belief in wildlife inexhaustibility in arguing that utilitarian values had ruled the day. Market hunting was "the most feasible means of harvesting Nature's resources" in early America, and even the most visionary could not have foreseen the need for conservation, especially for birds.[49] The public's awakening to the tragedy would come slowly. Flickers of concern at the end of the nineteenth century grew into flames with the passing of Martha, the last surviving passenger pigeon, in a Cincinnati zoo in 1914. "The period of wanton slaughter began to taper off" as a new era of enlightened management imposed strict terms for wildfowl use and, in the process, slated the market hunter for extinction. "The wasteful years had come to an end."[50]

That was the story, at least. The twentieth-century triumph of wildlife conservation was an unambiguous epic of market hunting villains, sportsman-conservationist heroes, and wildfowl victims. This narrative forms the historical foundations of what has come to be called the North American Model of Wildlife Conservation. A distinctively American construct according to its proponents, the North American Model (NAM) comprises seven core tenets:

1. Wildlife resources are a public trust.
2. Markets for game are eliminated.
3. Allocation of wildlife is effected by law.
4. Wildlife can be killed only for a legitimate purpose.
5. Wildlife is considered an international resource.

6. Science is the proper tool for discharging wildlife policy.
7. Democracy in hunting is standard.

Though not formally articulated until 1995, these principles nevertheless drove modern wildlife conservation policymaking, according to some researchers, ultimately producing "the greatest environmental success story of the twentieth century," and perhaps even "one of the greatest achievements of North American culture."[51] Wildlife experts employ the model both descriptively and prescriptively, offering it as a schematic for conservation's historical and political evolution, as well as a road map for future directions in wildlife management. The past successes of sportsmen-led wildlife conservation campaigns that brought populations of migratory birds and other terrestrial creatures back from the brink, NAM advocates suggest, offer lessons in how incentivizing sustainable, nonmarket wildlife consumerism can work.[52]

Other researchers, however, have in recent years challenged the North American Model's dominance in wildlife management circles. Some take issue with the prohibitions against privatization and commercialization of wildlife—a position that reflects the widely held notion that "conservation and business are natural enemies."[53] This anti-capitalist ideology, they argue, is a necessarily limiting vision for wildlife policy and does not account for conservation successes in parts of the world where private ownership and wildlife markets are the norm.[54] The NAM also fails to fully acknowledge the ways commercialism is implicated in its own paradigm. Sport hunting, especially, is market driven. Gun and ammunition companies, recreational outfitters, and other businesses are economically vested in the maintenance of certain wildlife populations.[55] Indeed, a major argument for sport hunting's legal preservation has been its ability "to pay its own way" not only through license fees devoted to conservation projects but also in increased commercial activity.[56]

For many critics, the model's key weakness is its selective privileging of sport hunters in both current wildlife policies and in wildlife conservation history. NAM principles serve the interests of sport hunters, inspiring wildlife management strategies to increase game populations often at the expense of nongame species. Democratic access to hunting opportunities means little to nonhunters—the majority of Americans—whose participation in conservation policy decisions is consequently minimized by this emphasis. The public trust doctrine on which the model rests does little to address the legalities of private landownership and the concerns of landowners whose relationships with wildlife are different from those of recreational hunters. Limiting financial and ethical investment in wildlife conservation to a dwindling cohort of sport hunters, NAM detractors insist, is hardly a prescription for either environmental justice or sustained wildlife conservation.[57]

This controversy has split wildlife conservation science into opposing disciplinary factions, spawning the separate field of conservation biology to counter the NAM's game management emphasis. Ironically, the fundamental sticking point is not in the present but in the past. The model's current authority relies on a particular progressive historical narrative, one that pits rapacious market hunters against enlightened sportsmen in a transformative battle that ultimately ended an age of free-taking profligacy.[58] Skeptics dispute this celebratory, sport hunter-centric historical explanation of the conservation movement, contending that it attributes conservation successes almost exclusively to elite, white males; overlooks the contributions of nonhunters to significant environmental advances; ignores preexisting indigenous patterns of wildlife use and management; and underplays conservation failures.[59] In short, the model hangs on a dangerously oversimplified history.

For a historian, saying the past is complicated is more a dodge than a revelation. But in the case of wildlife conservation, the past may really be more complex than even the critics have allowed. NAM detractors and proponents alike have tended to interpret market hunting one-dimensionally, as simply an early foil for wildlife activists. Consequently, they have yet to fully come to terms with the ways consumer capitalism shaped and continues to shape the culture of conservation. If, as some have provocatively claimed, an "inadequate history" is to blame for the shortcomings of current wildlife conservation policies, then an "adequate history" that addresses those limitations must consider the ways the modern market in birds produced the consumer-capitalist language of wildlife value we use today.[60]

This book is a crucial chapter in what I hope is a more than adequate wildlife history. That history reveals that the market hunter stood not at the end of an era, but at the beginning of one. He was part of a larger economic, legal, and cultural transformation set in motion by the market in birds, a transformation that redefined the value of North American wildlife for a twentieth-century consumer culture. Sport hunters, the subjects of so much heated scholarly and popular debate, were not the vanguard of a new effort to reconcile nature with modern society. The market was the key mechanism of change. It was here that wildlife's cultural worth was debated and decided in terms the market set. The turn-of-the-century market in birds ushered in the transition from wildlife consumption to wildlife consumerism in North America.

The modern wildfowl market was revolutionary because it commodified animal wildness. The game dealer's stall, the specimen catalog, and the millinery shop sold not just meat, eggs, and feathers but a host of esoteric qualities that distinguished wild creatures from tame. What else could explain consumers' continued cravings for wildfowl, asked the New York Emergency Conservation Committee in 1929, when it was "not only the most expensive meat to obtain" but also often "the poorest?"[61] The answer could be found in the creatures themselves. Birds remained recognizable as wildlife in the marketplace, and consumers desired them because they were authentically wild. The commercial production of wildness, like wilderness, promised consumers some relief from the innervating effects of industrial capitalism and the curse of modern artificiality.[62] But unlike wilderness, which was ultimately tied to a geographic space separate from consumers' urbanized environments, the wildness that was built into avian natures could be packaged, transported to cities, and sold. This was the pivotal innovation of the modern market in birds, and it set the direction for what was to follow.

Commodified avian wildness was at once both biological and economic. Diversity and mobility, biological traits of living birds, were crucial to wildfowl marketability. The late-nineteenth-century market exploited these characteristics, using the variety of avian species and their seasonal migrations to shape consumers' desires. As dead commodities, birds remained surprisingly mobile. Modern technologies and commercial networks designed to move ever-greater numbers of wildfowl from the countryside to the cities broke down biophysical barriers and expanded birds' ranges into places from which they were normally scarce or absent. The market removed birds from their ecological contexts, transported them across space, and (quite literally) froze them in time. This spatial and temporal separation between production and consumption, between living-bird environments and dead-bird markets, is the quintessence of capitalist commodification.

Yet avian commodities did not entirely conceal their environmental origins or the natural relationships that produced them. They were not subject to the same sort of human forgetting that masked the socio-natural relations contained in other commodities.[63] The wildfowl market sold wild celery–fed Maryland canvasbacks, the plumage of forest-dwelling Carolina parakeets, or the collectible eggs of the Florida red-shouldered hawk. The market in birds linked consumers to distant environments and gave them a vested interest in avian species both near and far. Through their purchases, buyers could reconnect in some small way to natural patterns and rhythms, assuaging their estrangement from nature in the process.

The conservation movement drew on these market-mediated engagements with wildness to generate public support. The market transformed wildfowl into a national resource—the process that Merritt once described as "the benefit of one State" becoming "the happy heritage of all"—and nationalized conservation as a consequence.[64] Thus, Royal Phelps of the NYPGA could make midwestern prairie chickens an East Coast issue when they were served in a big-city restaurant, and Frank Chapman of the Audubon Society could expose the snowy egret's plight by counting hats in a New York City streetcar rather than counting birds in a Louisiana swamp. Little wonder that the foundational US Supreme Court decision in the public trust doctrine for wildlife, *Geer v. Connecticut*, and the first federal wildlife conservation laws concerned marketed, migratory birds.[65] In the context of the market, migration was not just a biological fact. It was also a political act.

For Merritt, that "happy heritage of all" was a product of the wildfowl market. Turning creatures into commodities was a democratizing act, giving the public access to American avian wildness in the marketplace. Sport hunting consumers had to counter this capitalist argument head on to craft a democratic definition of recreational hunting for a modern and increasingly urban consumer culture in which hunting was no longer a necessity. Assigning cultural value to the experience of living wildness, sport hunters drew a bright line between their killing and the market's, elevating their wildlife consumption above other commercialized forms. They asserted that those experiences were available to all even as they advocated for restrictive game laws and privatized their shooting preserves. The sportsman's insistence on the superiority of democratic hunting over the democracy of the market enshrined in the North American Model of Conservation, was a direct outcome of the turn-of-the-century debate about the market in birds.[66] For a while, sportsmen's analyses of wildfowl worth dominated the popular discourse about conservation and guided wildlife law making. The anti-commercial conservation vision sportsmen advanced, however, left them in a conflicted position with modern American consumer culture. Decrying wildlife commercialism as morally, culturally, and ecologically bankrupt, sport hunters failed to recognize the extent to which their experiences with wildness were commodified ones.

Modern wildness was, in many ways, a market invention. The wildlife market fundamentally redefined wildness not as an environmental condition but as an intrinsic attribute of individual animal identities and an object of consumer desire. Birds were particularly compelling exemplars of wildness based on the freedom of the Hunted. Highly mobile, migratory avian species perfectly embodied the free-roaming natures of truly wild creatures. Avian entrepreneurs capitalized on that wild identity, pricing the wild duck differently than the domestic fowl or even the game farm–raised bird. In the modern market, wild birds' symbolic worth to the consumer surpassed their utilitarian value.

The market's construction of individuated wildness, and its explicit monetary valuation of it, shaped the contours of the debate over wildfowl commercialization and conservation at the twentieth century's turn. The key participants in that debate—the Market Hunter, the Game Dealer, the Sportsman, and the Conservationist—came to define their respective identities by their personal ethical relationships to avian wildness. Each claimed special moral authority to speak for birds, and each claimed to know the true value of wildness. Commodification of wildness threatened the social identities and cultural authority of sport hunters and conservationists. Conservation of wildness delegitimized and destroyed the identities of game dealers and market hunters. All prized wildness, but in their own ways, and they sought to convince the public to do the same. In the end, the contest between wildlife commercialization and wildlife conservation turned not on capitalist logic or scientific principles, but on the value of wildness to consumers in a modern world.[67]

Wildness is a slippery concept. Like its wilderness sibling, wildness is a cultural invention bound up in a material nature.[68] We have trouble defining it, but we know it when we see it. "Our ability to perceive quality in nature," wrote Aldo Leopold in his portrait of a crane marsh, was capable of growth "to values yet uncaptured by language." The sandhill cranes' qualities belonged to that higher realm, "beyond the reach of words." In the crane, Leopold saw "wildness incarnate."[69] But in the act of assigning human value to this nebulous quality of nonhuman wildness, we catch ourselves in a bind. The higher the value we accord it, the greater the bind. The dilemma was most obvious in the wildfowl market, where it was no paradox, only a simple case of supply and demand. As consumers' desires for avian wildness grew, there was less of it to be found. In the conservation movement, where preservation of wildness was the goal, the contradiction was far less clear. Leopold saw it, though, in the marsh and in the wildness of the cranes. "All conservation of wildness is self-defeating," he wrote, "for to cherish we must see and fondle, and when enough have seen and fondled, there is no wilderness left to cherish."[70] The culture of conservation was itself a culture of consumption.

❧

Herbert Merritt died in 1946 at the age of eighty-eight, the "last of a well-known local family."[71] With his death passed away Kewanee's last living connection to the nineteenth-century world of the market hunter. The visible remnants of Henry Clay Merritt's business in birds had slowly faded. Merritt himself had torn down his icehouse in 1901, using the bricks to build a new block on Chestnut Street.[72] The Merritt home followed in 1937, leveled to make way for an oil station.[73] Merritt's ice pond was sold, closed, and filled in with rubbish. His hotel and old storehouse with its curious inscription were razed for a parking lot in the 1950s, after decades of litigation posthumously resolved the final dispute between Merritt's sons and their stepmother, Jennie.[74]

So too had disappeared the environment that, Merritt once claimed, "held more game than any land of its size in the world."[75] The saturated landscape of Henry County that had drawn countless waterfowl in the nineteenth century was wrung dry in the twentieth with the aid of modern engineering and significant capital investment. The Green River, "a stream formerly sluggish from its tortuous windings," was straightened and drains were laid. The "rescued soil" from the river now grows corn and hogs instead of cattails and teal.[76] People can still hunt birds along the edge of the old Great Willow Swamp north of Kewanee, "a sportsman's paradise" that used to contain "one of the most concentrated and varied wildlife populations in the central part of North America," but now they hunt planted pheasants by permit in a state-managed program.[77]

Some traces of the past remain. A 230-acre patch of the old Annawan wetlands endures as Mineral Marsh Nature Preserve. The Kewanee Historical Society occupies the Butterwick Building, where Merritt used to store game birds in the basement, and a local restaurant serves a custom-brewed "Prairie Chicken Ale" in a nod to the old Kewanee Brewery's flagship beverage of more than a century ago (advertised as "the beer that makes the smile that won't come off").[78] Still, the most detailed record of those bird-filled times rests on a shelf in the Kewanee Public Library. Merritt's loosely organized recollections paint a picture of an unfamiliar world, one that bears little resemblance to our

Waterfowl viewing area on a dammed lake in Johnson-Sauk Trail State Recreation Area, north of Kewanee, Illinois. The recreation area occupies a fragment of the formerly vast wetlands hunted by Henry Clay Merritt in the nineteenth century. Once drawing incredible numbers and varieties of wildfowl, it now attracts recreational wildlife consumers: hunters and birdwatchers. Author photo

own, either environmentally, economically, legally, or culturally. It seems a forgotten time, this "day of the professional hunter and plentiful game," and Merritt, a forgotten representative of that time.[79]

Appearances, however, can be misleading. Though not exactly a household name, Henry Clay Merritt has hardly been consigned to the dustbin of history. Like clockwork, once every decade or so, *The Shadow of a Gun* is rediscovered by local history buffs, nostalgic sport hunters, and historians. Others of his ilk have found their places in historical memory, too. John Curl, E. M. Brubaker, and Dale Hamm are among those one-time professional wildfowlers whose stories make up what could be called "the last of the market hunters" literature. With his image reshaped in the retrospectives that began to appear from the 1920s onward, the character of the market hunter has been reclaimed by sportsmen as part of their genealogy. Viewed with a mixture of admiration, fascination, and disgust, both legal and "outlaw" market hunters serve as a reminder of a time when the value of wildlife was measured on a narrow utilitarian scale, before Americans learned to properly value "wildness incarnate."[80]

Although there is no equivalent body of literature featuring "the last of the game dealers" or the "outlaw cold storage man"—perhaps for obvious reasons—the late-nineteenth-century wildfowl market has had its chroniclers too. But the histories of the market hunter and the market have existed largely in parallel. The former focuses on colorful individuals engaged in an extinct way of life; the latter is an institutional analysis of political economy.[81] To those two narrative threads, some environmental historians have added a strand of cultural analysis tying the market consumer to changing constructions of nature.[82] Each line of inquiry is valuable. A more integrated history of the market in birds, however, unravels the connections between economy, ecology, law, and culture that these separate investigative paths leave ambiguous. It reveals, too, the interplay of avian biology, individual personalities, organized group actions, and broader social forces—subjects more often treated in isolation—that nationalized the public's interest in birds in the twentieth century.

These are not merely academic considerations. Market hunting's history is obviously deeply implicated in current conservation paradigms, and it explains why impassioned controversies over different versions of wildlife use continue without clear resolutions. The killings of Zimbabwe's "Cecil the Lion" in 2015 and of "Pedals," New Jersey's upright-walking bear, in 2016 generated heated public outrage but failed to produce serious public attention to the realities of human-wildlife conflicts and conservation funding reliant on sport hunting revenues.[83] Today, a shrinking cohort of hunters struggles to make the economic case for recreational, consumptive wildlife use to a much larger nonhunting population that imagines a more affective connection to wild animals they rarely encounter.[84] Wise-use supporters, who argue that productive relations with nature are superior to those based in consumption, clash with elite environmentalists claiming to represent the public's interests in wildlife.[85] Current efforts to privatize wildlife by introducing game ranching, commercial harvesting, and the sale of hunting privileges in North America produce swift backlash from conservationists convinced that markets in dead wildlife would destroy the existing system of wildlife conservation.[86] And, on the one hundredth anniversary of the Migratory Bird Treaty Act, the Trump administration made good on its effort to "clip the wings" of the groundbreaking federal wildlife legislation by removing penalties for companies that incidentally kill or injure migratory birds. Opposition forces have called for immediate congressional action in response, insisting that "America is being robbed of its birds and wildlife" to benefit corporations.[87]

It all seems clear-cut, this struggle between economics and environmental ethics. That is why the market hunter's story is so compelling. The market in birds neatly illustrates the morality tale that is all too often construed as the history of wildlife preservation. In that narrative, market hunters and game dealers like Merritt were callous exploiters of nature without the higher sensibilities of wildlife protectors who appreciated the ineffable qualities of avian wildness. It was Merritt's loss. "History could have read: H. Clay Merritt, hunter, naturalist, conservationist," wrote an Illinois columnist in 1964. "Only a point of view made the difference."[88]

Merritt's story and the history of the market in birds can tell us so much more, though. To imply that the Kewanee market hunter and game dealer had no appreciation for wild birds beyond profits would be a mistake. His depiction of a jacksnipe's spring migration northward as "in the nature of a romance," for example, hardly seems a coldhearted financial view:

> For many months he has fed at the Gulf, he may have traversed the Isthmus. He hears the beating of the Pacific; he covers the footsteps of Balboa. Behind him the Summer, before him the screech of the wild fowl opening up the frozen fastness of the North. Enwrapped in the clouds of the night he plunges into the airy abyss. Headland and Mainland vanish before him. Turbulent rivers and mountain ranges disappear as if by magic. Rising cities throw out a line of light. To-day he feeds in the cotton fields, to-night he will be beyond the black belt of the bondman. The constellations with their jeweled fingers open up the gateway of the Mississippi, he enters the vortex of commerce, he follows the coastline.[89]

We want to imagine wildlife like Merritt's jacksnipe: roaming free in nature, separate from the urban-industrial world below. We believe we can save wildness by holding it apart from the economy with the right ethics, the right science, and the right laws.[90] Merritt's point of view, though, offers a glimpse of the fundamental problem at the heart of human-wildlife coexistence in modern America. The bird in Merritt's romance escaped "the vortex of commerce." The birds in Merritt's freezer did not. Before we can think about how to prevent wildness from disappearing under the weight of the market, we need to escape the vortex of commerce ourselves.

Acknowledgments

This book would not have been possible without the assistance of a great many wildlife researchers, historians, librarians, archivists, and others who provided valuable insights, located hard-to-find sources, corrected the inevitable errors, and read all or parts of the manuscript in its various forms. In addition to the numerous individuals and institutions listed in the foreword to this book, I would like to include the librarians and archivists at the California State Archives, Sacramento; California State Library, Sacramento; San Joaquin County Historical Society and Museum, Lodi, California; New York State Archives and Library, Albany; Olin Library, Special Collections and Archives, Wesleyan University, Middletown, Connecticut; and the State Archives of Iowa, Des Moines.

Assistance with illustrations came from the staffs at the Chesapeake Bay Maritime Museum, the California Academy of Sciences Library, the Minnesota Historical Society, the Oshkosh (Wisconsin) Public Museum, the Renton (Washington) Historical Museum, the Merrill-Cazier Library at Utah State University, the US Fish and Wildlife Service Conservation Library, the Kewanee Historical Society, the Chicago History Museum, the Special Collections Department at the University of Maryland Baltimore County, Newspapers.com, and the *Baltimore Sun*.

My graduate research assistants at Northern Illinois University, Chester "David" Carlson and Amanda Allen, provided valuable aid in the early stages of the project. David also supplied sorely needed moments of parenthetical hilarity along the way.

Clait E. Braun, Richard McCabe, and the anonymous reviewer for Johns Hopkins University Press deserve thanks for reading, commenting, and generally improving the manuscript. Jim Schmidt, my first reviewer and partner in all things, knows how much his help means to me. Acquisitions editor Laura Davulis, assistant editor Esther Rodriguez, managing editor Juliana

McCarthy, and copyeditor Steven B. Baker have my eternal gratitude for bringing this long-awaited project to its conclusion.

Finally, the support of Merilyn Reeves, Julie Burke, and the rest of the Reeves family has been unwavering. I thank them for introducing me to my fellow Illinoisan Henry Clay Merritt, and for the chance to tell these stories.

Notes

Abbreviations

CT *Chicago Tribune*
EAP Edward Alders Papers, San Joaquin County Historical Society, Lodi, California
FS *Forest and Stream*
FTP Frederick Thompson Papers on Hunting in Iowa and Illinois, 1930–1937. Special Collections, State Historical Society of Iowa, Des Moines.
JSHP Natural Resources, Fish and Game Division, Game Management, J. S. Hunter Papers, California State Archives, Sacramento
NYAPG New York Association for the Protection of Game Papers, New York State Library, Albany
NYT *New York Times*
RPHP Ray P. Holland Papers on Enforcement of the Migratory Bird Treaty, Olin Library, Wesleyan University, Middletown, Connecticut

Prologue

1. Henry Clay Merritt, *The Shadow of a Gun* (Chicago: F. T. Peterson Company, 1904), 11.

2. Henry L. Kiner, *History of Henry County, Illinois* (Chicago: Pioneer Publishing Company, 1910), 1:62.

3. "Breeding Parks and Game Preserves," *Brooklyn Daily Eagle*, 1 June 1871, 2.

4. "Game Dealers Checked," *San Francisco Daily Call*, 16 January 1896, 10.

5. "Breeding Parks and Game Preserves," *Brooklyn Daily Eagle*, 1 June 1871, 2.

6. James L. Huffman, "Speaking of Inconvenient Truths: A History of the Public Trust Doctrine," *Duke Environmental Policy & Law Forum* 18, no. 1 (Fall 2007):100.

7. See James B. Trefethen, *Crusade for Wildlife: Highlights in Conservation Progress* (Harrisburg, PA: Stackpole Books, 1961); Frank Graham Jr., *Man's Dominion: The Story of Conservation in America* (New York: M. Evans, 1971); John F. Reiger, *American Sportsmen and the Origins of Conservation* (New York: Winchester Press, 1975); Thomas R. Dunlap, "Sport Hunting and Conservation, 1880–1920," *Environmental History* 12, no. 1 (Spring 1988): 51–60; and Thomas R. Dunlap, *Saving America's Wildlife: Ecology and the American Mind, 1850–1990* (Princeton, NJ: Princeton University Press, 1988), 5–46. Market hunting is highlighted in James A. Tober, *Who Owns the Wildlife? The Political Economy of Conservation in Nineteenth Century America* (Westport, CT: Greenwood Press, 1981) and Robin W. Doughty, *Feather Fashions and Bird Preservation: A Study in Nature Protection* (Berkeley: University of California Press, 1975). Legal histories addressing market hunting include Michael J. Bean, *The Evolution of National Wildlife Law* (Washington, DC: Council on Environmental Quality, 1977) and Thomas A. Lund, *American Wildlife Law* (Berkeley: University of California Press, 1980).

8. See Karl Jacoby, *Crimes against Nature: Squatters, Poachers, Thieves, and the Hidden History of Conservation* (Berkeley: University of California Press, 2001); Dorceta E. Taylor, *The Rise of the American Conservation Movement: Power, Privilege, and Environmental Protection* (Durham, NC: Duke University Press, 2016); Louis S. Warren, *The Hunter's Game: Poachers and Conservationists in Twentieth-Century America* (New Haven, CT: Yale University Press, 1997); Steven Hahn, "Hunting, Fishing, and Foraging: Common Rights and Class Relations in the Postbellum South," *Radical History Review* 26 (1982): 37–64.

9. See Michael Blumm and Lucas Ritchie, "The Pioneer Spirit and the Public Trust: The American Rule of Capture and State Ownership of Wildlife," *Environmental Law* 50 (2005): 845–51; Ann-Marie Szymanski, "Wildlife Protection and Centralized Governance in the Progressive Era" in *State Building from the Margins: Between Reconstruction and the New Deal*, ed. Carol Nackenoff and Julie Novkov (Philadelphia: University of Pennsylvania Press, 2014), 141; Theodore W. Cart, "The Lacey Act: America's First Nationwide Wildlife Statute," *Forest History* 17, no. 3 (October 1973): 4–13; Kurk Dorsey, *The Dawn of Conservation Diplomacy: US-Canadian Wildlife Protection Treaties in the Progressive Era* (Seattle: University of Washington Press, 2009), esp. 165–237.

10. On early American principles of unrestricted "free taking," see Cecilia Campbell-Mohn, Barry Breen, and J. William Futrell, *Environmental Law: From Resources to Recovery* (St. Paul, MN: West Group, 1993), 203; Eric Pearson, *Environmental and Natural Resources Law*, 4th ed. (New York: Matthew Bender, LexisNexis, 2012), 280; and Thomas R. Dunlap, *Saving America's Wildlife: Ecology and the American Mind, 1870–1990* (Princeton, NJ: Princeton University Press, 1991), 5–7. On the inevitability of market hunting's decline, see David Kimball and Jim Kimball, *The Market Hunter* (Minneapolis: Dillon Press, 1969), iii.

11. See Philip Shabecoff, *A Fierce Green Fire: The American Environmental Movement* (Washington, DC: Island Press, 2003), 160–62; Roderick Nash, *The Rights of Nature: A History of Environmental Ethics* (Madison: University of Wisconsin Press, 1989), 33–54; Taylor, *Rise of the American Conservation Movement*, 24–49; Miles A. Powell, *Vanishing America: Species Extinction, Racial Peril, and the Origins of Conservation* (Cambridge, MA: Harvard University Press, 2016), 64–66; Warren, *Hunter's Game*, 55–59.

12. Kimball and Kimball, *Market Hunter*, iii.

13. Regional studies include R. K. Sawyer, *Texas Market Hunting: Stories of Waterfowl, Game Laws, and Outlaws* (College Station: Texas A&M University Press, 2013); Terry Grosz, *Slaughter in the Sacramento Valley: Poaching and Commercial-Market Hunting—Stories and Conversations* (Denver: Johnson Books, 2008); Roy E. Walsh, *Gunning in the Chesapeake: Duck and Goose Shooting on the Eastern Shore* (Centreville, MD: Tidewater, 1960); and C. John Sullivan, *Waterfowling on the Chesapeake, 1819–1936* (Baltimore: Johns Hopkins University Press, 2003). Popular works include, Kimball and Kimball, *Market Hunter*; Dale Hamm with David Bakke, *The Last of the Market Hunters* (Carbondale: Southern Illinois University Press, 1996); Harry M. Walsh, *The Outlaw Gunner* (Atglen, PA: Schiffer, 2008); and John N. Davis, *The Life and Times of Fred Kimble* (Fayetteville, NC: Old Mountain Press, 2005).

14. See, for example, Jon T. Coleman, *Vicious: Wolves and Men in America* (New Haven, CT: Yale University Press, 2004); Andrew C. Isenberg, *The Destruction of the Bison: An Environmental History* (New York: Cambridge University Press, 2000); Dan Flores, *American Serengeti: The Last Big Animals of the Great Plains* (Lawrence: University Press of Kansas, 2016); and Tara Kathleen Kelly, *The Hunter Elite: Manly Sport, Hunting Narratives, and American Conservation, 1880–1925* (Lawrence: University Press of Kansas, 2018).

15. Mark V. Barrow Jr., *A Passion for Birds: American Ornithology after Audubon* (Princeton, NJ: Princeton University Press, 2000); Carolyn Merchant, *Spare the Birds! George Bird Grinnell and the First Audubon Society* (New Haven, CT: Yale University Press, 2016); Thomas R.

Dunlap, *In the Field, among the Feathered: A History of Birders and Their Guides* (New York: Oxford University Press, 2011); Michael Edmonds, *Taking Flight: A History of Birds and People in the Heart of America* (Madison: Wisconsin Historical Society Press, 2018); Robert M. Wilson, *Seeking Refuge: Birds and the Landscapes of the Pacific Flyway* (Seattle: University of Washington Press, 2010); Joel Greenberg, *A Feathered River across the Sky: The Passenger Pigeon's Flight to Extinction* (New York: Bloomsbury, 2014).

16. See, for example, Clayton Koppes, "Efficiency, Equity, and Esthetics: Shifting Themes in American Conservation," in *The Ends of the Earth: Perspectives on Modern Environmental History*, ed. Donald Worster (New York: Cambridge University Press, 1989), 230–51; Stephen R. Kellert, "American Attitudes toward and Knowledge of Animals: An Update," *International Journal for the Study of Animal Problems* 1 (1980): 87–119; and Lisa Mighetto, *Wild Animals and Environmental Ethics* (Tucson: University of Arizona Press, 1991). On utilitarian versus aesthetic values in conservation, see Stephen R. Fox, *The Conservation Movement: John Muir and His Legacy* (New York: Little, Brown, 1981); Michael P. Cohen, *The Pathless Way: John Muir and American Wilderness* (Madison: University of Wisconsin Press, 1984); and Roderick Nash, "John Muir, William Kent, and the Conservation Schism," *Pacific Historical Review* 36, no. 4 (November 1967): 423–33.

17. Studies of cultural valuations of wildlife make up their own interdisciplinary subfield. See Darla S. Deruiter, "A Qualitative Approach to Measuring Determinants of Wildlife Value Orientations," *Human Dimensions of Wildlife* 7, no. 4 (2002): 251–71; Michael B. Usher, "Wildlife Conservation Evaluation: Attributes, Criteria, and Values," in *Wildlife Conservation Evaluation*, ed. Michael B. Usher (Dordrecht, Netherlands: Springer, 1986), 1–44; Ken G. Purdy and Daniel J. Decker, "Applying Wildlife Values Information in Management: The Wildlife Values and Attitudes Scale," *Wilderness Society Bulletin* 17, no. 4 (Winter 1989): 494–500; and M. J. Manfredo, L. Sullivan, A. W. Don Carlos, A. M. Dietsch, T. L. Teel, A. D. Bright, and J. Bruskotter, *America's Wildlife Values: The Social Context of Wildlife Management in the U.S.*, national report of the research project "America's Wildlife Values," Department of Human Dimensions of Natural Resources, Colorado State University, Fort Collins, 2018.

18. John Smith, *A Description of New England, or Observations and Discoveries in the North of America in the Year of Our Lord 1614* (1616; facsimile reprint, Boston: William Veazie, 1865), 57–58.

19. Colonial laws controlling commercial exploitation of wildlife were limited. Authorities, for example, tried to restrict southeastern deerskin hunters' actions because they threatened farmers' possessions and destroyed a useful species. Andrea L. Smalley, *Wild by Nature: North American Animals Confront Colonization* (Baltimore: Johns Hopkins University Press, 2017), 163–70.

20. "Game Protection in Congress," *FS*, 5 February 1898, 107.

21. On changing spatial dimensions of migratory bird protection, see Robert M. Wilson, "Directing the Flow: Migratory Waterfowl, Scale, and Mobility in Western North America," *Environmental History* 7, no. 2 (April 2002): 247–66.

22. For catalogs of the agricultural, health, and social benefits of wildfowl, see, for example, Massachusetts State Board of Agriculture, *Forty-Eighth Annual Report of the Secretary of the Massachusetts State Board of Agriculture* (Boston: Wright & Potter Printing Co., 1901), xv; C. E. Pleas, "The Turkey Buzzard," *Collector's Monthly*, February 1891, 14; "A Mosquito Brief," *Scientific American*, 1 September 1906, 154; Joseph Grinnell, "Bird Life as a Community Asset," *California Fish and Game* 1, no. 1 (October 1914): 20.

23. Grinnell, "Bird Life as a Community Asset," 20.

24. Texas Agricultural Experiment Station, *Bulletin: 1583–1623* (College Station: Texas Agricultural and Mechanical College System, 1888), 39.

25. Colorado, State Game and Fish Commission, *Biennial Report of the Fish Commissioner of the State of Colorado, 1870–80* (Denver: Tribune Publishing Company, 1881), 31.

26. Ernest Schaeffle, "Some Notes on the Non-sale of Game," *California Fish and Game* 1, no. 1 (October 1914): 3.

27. "Why Some Men Are Hunters," *FS*, January 1922, 17.

28. Merritt, *Shadow of a Gun*, 234–35, 240, 242.

29. Merritt, *Shadow of a Gun*, 234–35.

30. T. Gilbert Pearson to Ray Holland, 8 January 1926, Migratory Bird Conservation, 1913–1937, Box 16, RPHP.

31. See James Morton Turner, "From Woodcraft to 'Leave No Trace': Wilderness, Consumerism, and Environmentalism in Twentieth-Century America," *Environmental History* 7, no. 3 (June 2002), 463; Matthew W. Klingle, "Spaces of Consumption in Environmental History," *History and Theory* 42, no. 4 (December 2003): 93.

32. Merritt, *Shadow of a Gun*, 226.

33. For contemporary debates, see Terry L. Anderson and Peter J. Hill, eds., *Wildlife in the Marketplace* (Lanham, MD: Rowman & Littlefield, 1995); Patrick F. Noonan and Michael D. Zagata, "Wildlife in the Market Place: Using the Profit Motive to Maintain Wildlife Habitat," *Wildlife Society Bulletin* 10, no. 1 (Spring 1982): 46–49.

CHAPTER ONE: The Hunter

1. Merritt, *Shadow of a Gun*, 13.

2. William J. Blake, *A History of Putnam County, N.Y.* (New York: Baker & Scribner, 1849), 268–269.

3. Merritt, *Shadow of a Gun*, 13, 15–16, 19.

4. Merritt, *Shadow of a Gun*, 14, 20, 25–26, 30–31, 35–36.

5. Merritt, *Shadow of a Gun*, 36–37.

6. "H. Clay Merritt Suddenly Dies," *Kewanee Weekly Star Courier*, 4 December 1907, 1; Merritt, *Shadow of a Gun*, 14, 20, 25–26, 30–31, 35–37, 40.

7. Merritt, *Shadow of a Gun*, 40–41.

8. Smalley, *Wild by Nature*, 11–32; Tober, *Who Owns the Wildlife?* 99.

9. The *Oxford English Dictionary* identifies James G. Cooper's "Zoological Report," based on his explorations of Washington Territory, as the first published US example of "market hunter." See J. G. Cooper and George Suckley, *The Natural History of Washington Territory and Oregon* (New York: Balliere Brothers, 1859), 225.

10. See, for example, 1880 US Census, Middletown, Jefferson Co., Kentucky, Enumeration District 090, Benjamin W. Winn, Roll 421, 76A; 1880 US Census, Lincoln, Placer Co., California, Enumeration District 071, Old Man Mike, Roll 70, 298C; 1900 US Census, Suisun, Solano Co., California, Enumeration District 0147, Abraham Hitchcock, Roll 1240113, 1, all available on Ancestry.com.

11. Robert S. Miller, interview, 22 February 1930, hunting compilation typescript, Folder BR T36, 188, FTP; George L. Hopper, "Ducking on the Susquehanna Flats, Past and Present," in *Tales of Duck and Goose Shooting*, ed. William Hazelton (Chicago: Eastman Brothers, 1916), 33.

12. Smalley, *Wild by Nature*, 203–6.

13. David W. Cartwright and Mary F. Bailey, *Natural History of Western Wild Animals and Guide for Hunters, Trappers, and Sportsmen* (Toledo, OH: Blade Printing and Paper Co., 1875), 3, 5.

14. "Arkansas—Improving Conditions of the State," *Sacramento Daily Union*, 25 January 1866, 2.

15. H. B. Roney, "Efforts to Check the Slaughter," *American Field*, 11 January 1879, reprinted in *The Passenger Pigeon*, ed. William B. Mershon (New York: Outing Publishing Company, 1907), 84.

16. "Hunting in the Provinces: Toils, Trials, Profits, and Pleasures of the Professional Hunter," *New York Times*, 31 December 1865, 2. On sportsmen's market hunting, see also William B. Mershon, *Recollections of My Fifty Years of Hunting and Fishing* (Boston: Stratford Company, 1923), 138; A. C. Connor, interview, 22 September 1930, 556, FTP.

17. W. W. Sawyer, "Market Hunters of the Eighties," n.d., 16; E. C. Hinshaw, interview, 23 February 1930, 171; Judd Brownlee, interview, 25 April 1930, 290–92, all in FTP.

18. Glenda Riley, *The Life and Legacy of Annie Oakley* (Norman: University of Oklahoma Press, 1994), 11.

19. Lon Main, interview, 22 February 1930, 260–61, FTP.

20. "Game along the Sacramento," *(San Francisco) Daily Alta*, 6 December 1850, 2; Herbert Gardner, *Come Duck Shooting with Me* (New York: Knickerbocker Press, 1917), 3–5; John N. Davis, *The Life and Times of Fred Kimble* (Fayetteville, NC: Old Mountain Press, 2005), 33.

21. Eugene V. Connett, ed., *Wildfowling in the Mississippi Flyway* (New York: D. Van Nostrand, 1949), 64, 66.

22. James J. Dinsmore, *A Country So Full of Game: The Story of Wildlife in Iowa* (Iowa City: University of Iowa Press, 1994), 155; "Personal," *Spirit Lake (IA) Beacon*, 23 August 1889, 5; Richard Harker, interview, 22 February 1930, 214–16, FTP.

23. David Kimball and Jim Kimball, *The Market Hunter* (Minneapolis: Dillon Press, 1969), 51.

24. Murray Crowder, "The Golden Years," *Outdoors in Illinois*, 1955, xxiv.

25. For example, Harker, interview, 219, FTP.

26. Forrest Crissey, "The Great Shotgun Syndicate: The Modern Nimrod Who Hunts by Wire and Slays by Companies," *Saturday Evening Post*, 10 October 1903, 8.

27. W. G. [pseud.], "Marsh Scenes in Louisiana: A Hunter's Funeral," *Spirit of the Times* 26, no. 21 (1872): 322.

28. Roney, "Efforts to Check the Slaughter," 89; T. Gilbert Pearson, *Adventures in Bird Protection* (New York: D. Appleton-Century Company, 1937), 117.

29. Hinshaw, interview, 171, FTP.

30. John Mortimer Murphy, *American Game Bird Shooting* (New York: Orange Judd Company, 1882), 282.

31. The 1870 census lists 719 white "hunters," but this number includes men listed as "cattle hunters." In the 1900 census, 7 men are identified as either a "market hunter" or "hunter for the market." The 1910 census lists 20 "market hunters," all residing in the same St. Charles, Louisiana, ward, and 1 "hunter for the market" in California. The missing 1890 census leaves a particularly large gap in these records, falling centrally in the market hunting era. See 1870 US Census, 1900 US Census, and 1910 US Census, Ancestry.com.

32. US Census Office, *Report on the Statistics of Agriculture in the United States at the Eleventh Census, 1890* (Washington, D.C.: Government Printing Office, 1894), 37.

33. 1900 US Census, Speedwell, St. Clair Co., Missouri, Enumeration District 0152, Raleigh W. Zener, Roll 1240886, 9, Ancestry.com.

34. Ralph E. Eshelman and Patricia A. Russell (Eshelman & Associates), "Historic Context Study of Waterfowl Hunting Camps and Related Properties within Assateague Island National Seashore, Maryland and Virginia," 21 July 2004, for Assateague Island National Seashore, National Park Service, US Department of Interior; 1900 US Census, Chincoteague Island, Accomack Co., Virginia, Enumeration District 0014, Daniel T. Whealton, Roll 1241697, 6, Ancestry.com; 1910 US Census, Chincoteague Island, Accomack Co., Virginia, Enumeration District

0004, Daniel T. Whealton, Roll 1375632, 1A, Ancestry.com; 1920 US Census, Chincoteague Island, Accomack Co., Virginia, Enumeration District 4, Daniel T. Whealton, Roll T625_1877, 1A, Ancestry.com.

35. "About the State," *Ashland (WI) Weekly News*, 2 October 1889, 6; "One Day's Hunting Bagged 160 Ducks," *(Madison) Wisconsin State Journal*, 4 November 1962, 41; 1880 US Census, Marquette, Green Lake Co., Wisconsin, Enumeration District 070, John E. Ades, Roll 1241791, 357C, Ancestry.com; 1900 US Census, Marquette, Green Lake Co., Wisconsin, Enumeration District 0057, John E. Ades, Roll 1241791, 1, Ancestry.com.

36. Joel Barber, *Wild Fowl Decoys* (1934; repr., Lyon, MS: Derrydale Press, 1989), 18; Hopper, "Ducking on the Susquehanna Flats," 32.

37. Gardner, *Come Duck Shooting*, 3–4.

38. Gary Richter, *Heron Lake: 100 Years of Good Living; for 100 Years of Progress in Heron Lake, Jackson County, Minnesota, and Including the Surrounding Townships* (Marceline, MO: Walsworth, 1983), 15.

39. Alexander Hunter, "Sport: Past, Present, and Future, Part II," *Outing: An Illustrated Monthly Magazine of Recreation*, January 1889, 326.

40. C. B. W., "San Joaquin Market Hunters," *FS*, 12 May 1892, 445.

41. Witmer Stone, *Bird Studies at Old Cape May: An Ornithology of Coastal New Jersey*, vol. 1 (1937; repr., New York: Dover, 1965), 407–8.

42. Hoodoo (pseud.), "A Word about Market Hunters," *FS*, 8 August 1889, 46.

43. Tober, *Who Owns the Wildlife?* 54.

44. George H. Tinkham, *History of San Joaquin County: With Biographical Sketches of the Leading Men and Women Who Have Been Identified with Its Growth and Development from the Early Days to the Present* (Los Angeles: Historic Record Company, 1923), 1183; Edward Alders, Record of Birds Killed and Shipped, Folders 76-589-1 and 76-589-4, EAP.

45. Merritt, *Shadow of a Gun*, 47.

46. Murphy, *American Game Bird Shooting*, 242.

47. Franklin Satterthwaite, "Snipe Shooting on the American Prairies," *Outing*, March 1887, 552.

48. Roney, "Efforts to Check the Slaughter," 81.

49. A. W. Schorger, *The Passenger Pigeon: Its Natural History and Extinction* (Madison: University of Wisconsin Press, 1955), 147.

50. George Bird Grinnell, "American Wildfowl and How to Take Them," *FS*, 8 September 1901, 186.

51. Edward H. Forbush, *A History of Game Birds, Wild-Fowl, and Shore Birds of Massachusetts and Adjacent States* (Boston: Mass. Board of Agriculture, 1912), 513.

52. Tinkham, *History of San Joaquin County*, 1183.

53. Kimball and Kimball, *Market Hunter*, 51.

54. Harker, interview, 218, FTP.

55. Merritt, *Shadow of a Gun*, 212.

56. Tom Ditto, interview, 23 August 1930, 482, FTP.

57. E. T. Martin, "Among the Pigeons—A Reply to Prof. Roney's Account of the Michigan Nestings of 1878," *American Field*, 25 January 1879, reprinted in *Passenger Pigeon*, 97.

58. Quoted in Harry M. Walsh, *The Outlaw Gunner* (Atglen, PA: Schiffer, 2008), 134.

59. Eugene V. Connett, ed., *Duck Shooting along the Atlantic Tidewater* (1947; repr., New York: William Morrow, 1956), 127.

60. William Brown, interview, 19 August 1930, 514, FTP.

61. 1880 US Census, Irondale, Cook Co., Illinois, Enumeration District 194, Henry Kleinman, Abraham Kleinman, George Kleinman, Roll 199, 148D, 149A, Ancestry.com.

62. 1880 US Census, St. Charles, Louisiana, Enumeration District 150, Anthony Tragle, Roll 468, 358C, Ancestry.com.

63. Merritt, *Shadow of a Gun*, 20, 63–65.

64. William N. Smith, *Marsh Tales: Market Hunting, Duck Trapping, and Gunning* (Centreville, MD: Tidewater, 1985), 134.

65. Percy Cushing, "A Market Gunner's Diary," *Outing*, March 1913, 689. Other examples abound. In 1865, "George Randall, a market hunter living near San Pablo, accidentally shot himself dead with his own shot gun yesterday." Untitled item, *(Olympia) Washington Standard*, 1 April 1865, 3.

66. Harold Umber and Craig Bihrle, "The Early Days," *North Dakota Outdoors*, June 1989, 92.

67. Spencer Fullerton Baird, Thomas Mayo Brewer, and Robert Ridgway, *A History of North American Birds: Land Birds* (Boston: Little, Brown and Company, 1874), 1:121.

68. "An Odd City of Birds: A New and Perilous Industry on the Farallon Islands," *(Rochester, NY) Democrat and Chronicle*, 8 September 1895, 10; "Aid Carried to Marooned Egg Hunters by the *Call*'s Stanch Tug *Reliance*," *San Francisco Call*, 12 July 1899, 1; "Marooners' Simple Story," *San Francisco Examiner*, 13 July 1899, 14.

69. Chas. Warren Stoddard, "With the Egg-Pickers on the Farollones," *Victorian Review*, 1 March 1882, 552–54.

70. "Rumored Accident at the Farallones," *(San Francisco) Daily Alta California*, 25 May 1858, 2.

71. "Odd City of Birds," 10.

72. "Telegraphic," *Sacramento Bee*, 5 June 1863, 3; "Out in the Fog: A Boating Adventure on the Pacific," *San Francisco Examiner*, 25 September 1869, 1.

73. Stoddard, "With the Egg-Pickers."

74. Gardner, *Come Duck Shooting*, 9.

75. "Serious Accident," *New York Tribune*, 11 August 1857, 6.

76. Merritt, *Shadow of a Gun*, 45–47.

77. Merritt, *Shadow of a Gun*, 46, 48–51, 56.

78. Forbush, *History of Game Birds*, 512.

79. Smith, *Marsh Tales*, 34.

80. William L. Finley, "The Grebes of Southern Oregon," *Condor* 9 (July–August 1907): 97–98.

81. Roney, "Efforts to Check the Slaughter," 89–90.

82. H[enry] T. Phillips to W. B. Mershon, 2 November 1904, reprinted in *Passenger Pigeon*, 117–18.

83. Gardner, *Come Duck Shooting*, 3–4.

84. See, for example, Wilmer M. Ely, *The Boy Chums in the Forest or Hunting Plume Birds in the Florida Everglades 1910* (New York: A. L. Burt Company, 1910), 57.

85. "Hunting in the Provinces: The Toils, Trials, Profits, and Pleasures of the Professional Hunter," *New York Times*, 31 December 1865, 2.

86. Jack Miner, *Jack Miner and the Birds: Some Things I Know about Nature* (Toronto: Ryerson Press, 1923), 6–7.

87. Merritt, *Shadow of a Gun*, 10–11.

88. Connett, *Wildfowling in the Mississippi Flyway*, 206–7.

89. Buck Lufkin, interview, 1 March 1930, 141, FTP; "Gunning from a Car," *FS*, March 1920, 120.

90. Jennie Holliman, *American Sports, 1785–1835* (Durham, NC: Seeman Press, 1931), 52–53.

91. "Pete and His Cannon," *Sacramento Daily Union*, 18 January 1891, 5.

92. Merritt, *Shadow of a Gun*, 95.

93. Harker, interview, 230, FTP.

94. Daybreak (pseud.), "South-Western Ohio," *FS*, 17 January 1878, 454.

95. Sawyer, "Market Hunters," 10.

96. Merritt, *Shadow of a Gun*, 282; Henry T. Finck, "Los Angeles," *Nation*, 27 October 1887, 330.

97. "Game in the Market of San Francisco, *(San Francisco) Daily Alta California*, 29 November 1857, 1.

98. Harold C. Bryant, "A Survey of the Breeding Grounds of Ducks in California in 1914," *Condor* 16, no. 5 (September–October 1914): 233; Scott Herring, *Opening Days: Classic Waterfowling in California* (Davis, CA: Greenwing Books, 2006), 49–50.

99. Holliman, *American Sports*, 27.

100. "Hunting in the Provinces."

101. Merritt, *Shadow of a Gun*, 50, 229–30.

102. T. Gilbert Pearson, "Canada Goose," *Bird-Lore*, 1 October 1921, 271; "Gunmen Four Participate in a Remarkable Goose Shoot," *San Francisco Call*, 20 March 1902, 8.

103. Louis Bagger, "Reminiscences of Havre de Grace," *FS*, 5 May 1887, 321.

104. Van Campen Heilner, *A Book on Duck Shooting* (1939; repr., New York: Alfred A. Knopf, 1945), 65–66; Connett, *Duck Shooting*, 172, 186–87.

105. J. B. White, "Canvasback Ducks," *FS*, 14 November 1889, 324; Sullivan, *Waterfowling on the Chesapeake*, 466; "Duck Shooting," *Baltimore Sun*, 3 November 1891, 6.

106. "The Ducking Season: Many Devices to Lure and Kill the Wild Fowl of the Chesapeake," *Baltimore Sun*, 8 October 1886, 6. See also, "A Ducker's Danger," *Baltimore Sun*, 19 November 1903, 9.

107. Alexander Wilson, *American Ornithology, or The Natural History of the Birds of the United States*, 9 vols. (Philadelphia: Bradford & Inskeep, 1808–14), 7:85, 113; Witmer Stone, "The Birds of New Jersey: Their Nests and Eggs," in *Annual Report of the New Jersey State Museum*, ed. Silas R. Morse (Trenton, NJ: John L. Murphy Publishing Company, 1908), 111.

108. John James Audubon, *Birds of America* (1840–44; repr., New York: Dover, 1966), 6:357.

109. "Eggs and Their Foes," *Saturday Review*, 21 April 1894, 412.

110. George Sennett, "Eggers of the Texas Coast," *Audubon Magazine*, March 1887, 34–35.

111. "Mud Hen Eggs for Chicago: Industry But Little Known a Money Producer for Hunters of Bureau County," *Garden City (KS) Telegram*, 30 July 1900, 3.

112. "The Farallon Rookeries—The Breeding Place of Thousands of Sea Birds off San Francisco," *San Francisco Chronicle*, reprinted in *Ornithologist and Oologist*, August 1891, 127.

113. Samuel C. Clarke, "Western Field Sports in Early Days," in *Shooting on Upland, Marsh, and Stream: A Series of Articles Written by Prominent Sportsmen*, ed. Bruce C. Leffingwell (Chicago: Rand McNally & Co., 1890), 264.

114. Stone, "Birds of New Jersey," 302; "Eagle Isle," *Minneapolis Star Tribune*, 24 December 1900, 2.

115. Barbara Mearns and Richard Mearns, *The Bird Collectors* (San Diego: Academic Press, 1998), 202.

116. "Eggs and Their Foes."

117. Mark V. Barrow, "The Specimen Dealer: Entrepreneurial Natural History in America's Gilded Age," *Journal of the History of Biology* 33, no. 3 (August 2000): 497; Spencer F. Baird, "Report of Assistant Secretary," in *Annual Report of the Board of Regents of the Smithsonian Institution, 1858* (Washington, DC: Government Printing Office, 1859), 57–62.

118. B. B. Haines, "Long-Billed Marsh Wren," *Ornithologist and Oologist*, 7 January 1883, 8.

119. Elliott Coues, *Field Ornithology: Comprising a Manual of Instruction for Procuring, Preparing, and Preserving Birds, and a Check List of North American Birds* (Salem, MA: Naturalists' Agency, 1874), 27.

120. Roney, "Efforts to Check the Slaughter," 80–81; Peter Steinhart, "A Common Possession," in *The National Audubon Society: Speaking for Nature; A Century of Conservation*, ed. Les Line (Southport, CT: Hugh Lauter Levin Associates, 1999), 47.

121. Jacob P. Giraud Jr., *Birds of Long Island* (New York: Wiley & Putnam, 1844), 109, 118, 127, 132, 138.

122. See W. W. Greener, *The Gun and Its Development*, 9th ed. (1881; repr., New York: Cassell & Company, 1910); Sullivan, *Waterfowling on the Chesapeake*, 61–70.

123. Gardner, *Come Duck Shooting*, 4.

124. Robin W. Doughty, *Wildlife and Man in Texas: Environmental Change and Conservation* (College Station: Texas A&M University Press, 1983), 98.

125. Merritt, *Shadow of a Gun*, 19, 45, 76, 173, 428; John H. Walsh, *Encyclopedia of Rural Sports* (Philadelphia: Porter & Coates, 1867), 25–27.

126. Gardner, *Come Duck Shooting*, 5.

127. Charles Askins, *Modern Shotguns and Loads: Together with a Treatise on the Art of Wing-Shooting* (Marshallton, DE: Small-Arms Technical Publishing Company, 1929), 268–69.

128. Walsh, *Outlaw Gunner*, 36; Sawyer, *Texas Market Hunting*, 51–52.

129. Walsh, *Outlaw Gunner*, 91.

130. W. H. "Chip" Gross, "Throwback Thursday: History of Punt Guns," *NRA Family*, 16 May 2019, https://www.nrafamily.org/articles/2019/5/16/throwback-thursday-history-of-punt-guns/.

131. Smith, *Marsh Tales*, 76.

132. Walsh, *Outlaw Gunner*, 105.

133. Many sporting authors used the term "battery" synonymously for the sinkbox.

134. Eric J. Dolin and Bob Dumaine, *The Duck Stamp Story: Art, Conservation, History* (Iola, WI: Krause, 2000), 14–15, 39.

135. Merritt, *Shadow of a Gun*, 284.

136. Jane E. Townsend, *Gunners' Paradise: Wildfowling and Decoys on Long Island* (Stony Brook, NY: Museums at Stony Brook, 1979), 13.

137. Smith, *Marsh Tales*, 190.

138. William B. Cronin, *The Disappearing Islands of the Chesapeake* (Baltimore: Johns Hopkins University Press, 2005), 12–13; "Ducking on the Flats," *Baltimore Sun*, 3 November 1890, 6.

139. Henry Kleinman, "Hints and Points on Ducks," *FS*, 18 June 1890, 431–32.

140. Caroline C. Fisher, Allison W. Elterich, Bernard L. Herman, and Rebecca J. Siders, *Marshland Resources in the Delaware Estuary, 1830–1950: An Historic Context* (Newark: University of Delaware Center for Historic Architecture and Engineering, 1993), 79.

141. "Duck Shooting," *Baltimore Sun*, 3 November 1891, 6; White, "Canvasback Ducks," 324; "Hartford County—Complete," *Baltimore Sun*, 3 October 1850, 2; Tober, *Who Owns the Wildlife?* 78; Walsh, *Outlaw Gunner*, 113, 130–32; Sullivan, *Waterfowling on the Chesapeake*, 466; Elisha J. Lewis, *The American Sportsman, Containing Hints to Sportsmen, Notes on Shooting, and the Habits of the Game Birds and Wild Fowl of America*, 3rd ed. (Philadelphia: J. B. Lippincott & Co., 1879; 1st ed., 1856), 284–85; Townsend, *Gunners Paradise*, 13.

142. Tober, *Who Owns the Wildlife?* 78.

143. Arrow (pseud.), letter to the editor, *FS*, 6 May 1879, 315. Knapp's gun was considered the largest of its kind in Illinois at the time.

144. Walsh, *Outlaw Gunner*, 35–36. See also William T. Hornaday, *Wild Life Conservation in Theory and Practice* (New Haven, CT: Yale University Press, 1914), 32.

145. See, Connett, *Wildfowling in the Mississippi Flyway*, 205; R. I. Brasher, "Sniping on the South Side of Long Island," *Outing* 17, no. 2 (November 1890): 83; Edwyn Sandys and T. S. Van Dyke, *Upland Game Birds* (New York: MacMillan Company, 1904), 321; William J. Mackey Jr., *American Bird Decoys: The Authoritative Story of Our Classic Bird Decoys and Their Makers* (New York: E. P. Dutton, 1965), 174.

146. Walsh, *Outlaw Gunner*, 132; "Greatest Duck Shooters," *Baltimore Sun*, 19 January 1904, 9.

147. See, for example, "Three Generations of Duck Shooters," *Indianapolis Journal*, 24 January 1904, 7; and "Maryland's Crack Duck Shot," *New York Sun*, 7 February 1904, 24.

148. Davis, *Fred Kimble*, 45–46; Kenneth H. Smith, "Illinois," in *Wildfowling in the Mississippi Flyway*, 267–; Joseph W. Long, *American Wild-Fowl Shooting* (New York: J. B. Ford and Company, 1874), 180.

149. "A Family of Good Shots," *Sydney (Australia) Morning Herald*, 18 February 1881; untitled item, *(Decatur, IL) Daily Review*, 34 February 1883, 4.

150. Adam H. Bogardus, *Field, Cover, and Trap Shooting*, 3rd ed. (New York: Forest & Stream Publishing Co., 1891), 409.

151. "Personal," *Spirit Lake (IA) Beacon*, 23 August 1889, 5; "A Short Sketch of the Wizard of Spirit Lake," *Spirit Lake Beacon*, 23 March 1916, 2; untitled item, *Spirit Lake Beacon*, 4 September 1896, 5; untitled item, *Milford (IA) Mail*, 11 November 1909, 5. In the 1910 census, Gilbert's occupation was listed as "wing shot." Under the "Industry" column, "champion" was noted. 1910 US Census, Center Grove, Dickinson Co., Iowa, Enumeration District 0039, Fred Gilbert, Roll 1374409, 8A, Ancestry.com.

152. See, for example, "Capt. Bogardus' Patent Glass Ball Trap and Rough Balls" (advertisement), *FS*, 16 August 1877, 40.

153. Bogardus, *Trap Shooting*; Marksman (pseud.), *The Dead Shot, or Sportsman's Complete Guide: Being a Treatise on the Use of a Gun*, 5th ed. (London: Longman, Green, and Company, 1882), 381.

154. Riley, *Annie Oakley*, 27–30.

155. Greener, *Gun and Its Development*, 256, 258; Davis, *Fred Kimble*, 106, 118–19, 168; Rollin B. Organ, "An Anecdote of Fred Kimble," in *Ducking Days: Narratives of Duck Hunting, Studies of Wildfowl Life, and Reminiscences of Famous Marksmen on the Marshes and at the Traps*, ed. William Chester Hazelton (Chicago: W. C. Hazelton, 1918), 109.

156. See, for example, "The New Dittmar Powder: The Champion Powder of the World" (advertisement), *FS*, 23 August 1877, 58; "Trade Topics: Eastern Handicap Winnings" (advertisement), *Western Field*, September 1907, 155.

157. Maurice Thompson, "Marvin and His Boy Hunters," *St. Nicholas: An Illustrated Magazine for Young Folks*, June 1884, 646–47.

158. Harry Castlemon [Charles Austin Fosdick], *The Young Wild-Fowlers* (Philadelphia: Porter & Coates, 1885), 5, 7–8, 11, 274–75, 305–7.

159. Thompson, "Marvin and His Boy Hunters," May 1884, 564.

160. Maurice Thompson, "Marvin and His Boy Hunters," in *The Boys' Book of Sports and Outdoor Life*, ed. Maurice Thompson (New York: Century Company, 1886), 42.

161. Sawyer, "Market Hunters," 16.

162. Harker, interview, 229, FTP.

163. "Gunmen Four Participate in a Remarkable Goose Shoot."

164. Hoyt Elbert, interview, 19 January 1930, 70, 74, FTP.

165. Miller, interview, 185, 187–89, FTP. See also William F. Gram, interview, 28 April 1930, 506, FTP.

166. Elbert, Interview, 63–64, 68, FTP.

167. Robert B. Roosevelt, *The Game-Birds of the Coasts and Lakes of the Northern States of America* (New York: Carleton, 1866), 271.

168. "Pot-Hunting," *Rutland (VT) Herald*, 30 November 1875, 4.

169. Merritt's occupation was listed as "hunter" in both the 1863 draft registration and the US census of 1870. See Consolidated Enrollment Lists, 1863–1865 (Civil War Union Draft Records), NAI 4213514, Vol. 2, 474; 1870 US Census, Kewanee, Henry Co., Illinois, H. C. Merritt, Roll M593_230, 47B, Ancestry.com.

170. Merritt, *Shadow of a Gun*, 78–79, 97, 191.

171. Merritt, *Shadow of a Gun*, 122–23.

172. Merritt, *Shadow of a Gun*, 128.

173. Merritt, *Shadow of a Gun*, 125–52.

174. Merritt, *Shadow of a Gun*, 161–64.

175. Merritt, *Shadow of a Gun*, 160.

CHAPTER TWO: The Dealer

1. Merritt, *Shadow of a Gun*, 47, 50, 54, 80, 109–10.

2. Merritt, *Shadow of a Gun*, 113.

3. Merritt, *Shadow of a Gun*, 109–11.

4. Merritt, *Shadow of a Gun*, 10, 45, 109–11.

5. Merritt, *Shadow of a Gun*, 52–54, 73, 109–12.

6. Merritt, *Shadow of a Gun*, 53–54, 76, 167–70, 196.

7. Merritt, *Shadow of a Gun*, 77.

8. Merritt, *Shadow of a Gun*, 109.

9. Dave Hall and Brian Cheramie, "Louisiana Champagne," *Ducks Unlimited* (July/August 1985), quoted in Stephen P. Havera, *Waterfowl of Illinois: Status and Management* (Urbana: Illinois Natural History Survey, 1999), 25; Gay M. Gomez, *A Wetland Biography: Seasons on Louisiana's Chenier Plain* (Austin: University of Texas Press, 1998), 115.

10. Crissey, "Great Shotgun Syndicate." *Saturday Evening Post*, 10 October 1903, 8.

11. Tober, *Who Owns the Wildlife?* 79.

12. Merritt, *Shadow of a Gun*, 10.

13. Kimball and Kimball, *Market Hunter*, 108.

14. C. Anne Wilson, *Waste Not, Want Not: Food Preservation from Early Times to the Present Day* (Edinburgh: Edinburgh University Press, 1991), 87; Sue Shephard, *Pickled, Potted, and Canned: How the Art and Science of Food Preserving Changed the World* (New York: Simon & Schuster, 2006), 190–91.

15. Audubon, *Birds of America*, 6:47.

16. John Latham, *A General Synopsis of Birds*, vol. 3, bk. 2 (London: Leigh & Sotheby, 1785), 445.

17. A. D. Livingston and Helen Livingston, *Edible Plants and Animals: Unusual Foods from Aardvark to Zamia* (New York: Facts on File, 1993), 143.

18. William and Robert Chambers, *Chambers's Encyclopedia: A Dictionary of Useful Knowledge*, new ed., vol. 4 (Philadelphia: J. B. Lippincott Company, 1889), 718.

19. Henry William Herbert, *Frank Forester's Field Sports of the United States and British Provinces*, 2 vols. (New York: Stringer & Townsend, 1849), 1:167.

20. Lewis, *American Sportsman*, 440.

21. Harold F. Duebbert, ed., *Wildfowling in Dakota, 1873–1903: Old-Time Duck and Goose Shooting on the Dakota Prairies* (Bismarck, ND: Windfeather Press, 2003), 299–300.

22. Michael R. Miller and Frederick W. Hanson, *Waterfowl Decoys of the Pacific Coast* (Davis, CA: MBF, 1989), 229.

23. Merritt, *Shadow of a Gun*, 45–46.

24. Most early-nineteenth-century hunters and consumers believed that wild game should be chilled rather than frozen. See, for example, Murphy, *American Game Bird Shooting*, 347; James Troubridge Critchell and Joseph Raymond, *A History of the Frozen Meat Trade: An Account of the Development and Present Day Methods of Preparation, Transportation, and Marketing of Frozen and Chilled Meats* (London: Constable, 1912), 276–77.

25. Thomas Cuthbert, Plaintiff's Points on Demurs, Royal Phelps v. J. H. Racey, District Court in the City of New York for the Sixth Judicial District, 12 March 1874, Box 1, Folder: Cases: Phelps v. Joseph H. Racey, NYAPG.

26. Jonathan Rees, *Refrigeration Nation: A History of Ice, Appliances, and Enterprise in America* (Baltimore: Johns Hopkins Press, 2013), 3–4; Lydia Bjornlund, *How the Refrigerator Changed History* (Minneapolis: Abdo, 2015), 51. A similar severing of domestic livestock production from natural barriers occurred in this period. See William Cronon, *Nature's Metropolis: Chicago and the Great West* (New York: W. W. Norton, 1991), 231–48.

27. Harker, interview, 214–15, FTP; Jack Musgrove, "Market Hunting in Northern Iowa," *Annals of Iowa* 26 (1945): 174.

28. Merritt, *Shadow of a Gun*, 204.

29. Dean Karau, "Merritt's Pond: The Rediscovery of Weathersfield's First Swimming and Skating Pond," *Dusty Roads: Viewing Kewanee's Past from the Present*, 1 March 2019, 1–2.

30. Gris [pseud.], "Nebraska Prairie Chickens," *FS*, 17 July 1890, 512.

31. Merritt, *Shadow of a Gun*, 204.

32. Merritt, *Shadow of a Gun*, 287.

33. G. T. Ferris, "How a Great City Is Fed," supplement, *Harper's Weekly*, 22 March 1890, 232.

34. "Christmas in the Markets," *Harper's Weekly*, 5 January 1878, 14.

35. Richard O. Cummings, *The American Ice Harvests: A Historical Study in Technology, 1800–1918* (Berkeley: University of California Press, 1949), 171.

36. Merritt, *Shadow of a Gun*, 206.

37. Merritt, *Shadow of a Gun*, 206.

38. Merritt, *Shadow of a Gun*, 50.

39. Heilner, *Book on Duck Shooting*, 65–66; Connett, *Duck Shooting*, 172, 186–87.

40. Connett, *Duck Shooting*, 150.

41. David C. Holly, *Tidewater by Steamboat: A Saga of the Chesapeake* (Baltimore: Johns Hopkins University Press, 1991), 65.

42. Miller and Hanson, *Waterfowl Decoys*, 273.

43. Albro Martin, *Railroads Triumphant: The Growth, Rejection, and Rebirth of a Vital American Force* (New York: Oxford University Press, 1992), 176, 277; Cronon, *Nature's Metropolis*, 70, 92–93.

44. Merritt, *Shadow of a Gun*, 44–45; "Railroad Construction in the Near Future," *Bradstreet's Weekly*, 11 February 1888, 91–92; Thomas Curtis Clarke, *The American Railway: Its Construction, Development, Management, and Trains* (New York: Skyhorse, 2012), 430–32.

45. A. Hogeland, comp., *Centennial Report of the Mineral and Agricultural Resources of the State of Kentucky* (Louisville: Courier-Journal, Job Rooms, 1877), 19.

46. "Freaks of Railroad Transportation," *American Railroad Journal and Mechanics' Magazine*, 15 June 1842, 364–65.

47. "New York and Erie Railroad," *American Railroad Journal and Mechanics' Magazine*, May 1844, 156.

48. Sherman Strong Hayden, *The International Protection of Wild Life* (New York: Columbia University Press, 1942), 80.

49. Herbert, *Frank Forester's Field Sports*, 1:36.

50. See, for example, Charles Hallock, *The Sportsman's Gazetteer and General Guide*, 5th ed. (New York: Forest & Stream Publishing Co., 1880).

51. William C. Harris, *The Sportsman's Guide to the Hunting and Shooting Grounds of the United States and Canada* (New York: The Anglers' Publishing Company, 1888), 6, 130. Thirty-nine out a total forty-eight advertisers in Harris' guide were railroad companies.

52. John C. Phillips, *American Game Mammals and Birds: A Catalogue of Books, Sport, Natural History, and Conservation, 1582–1925* (Boston: Houghton Mifflin, 1930), 630–35.

53. Elbert, interview, 67–68, FTP; Billy Finnicum, interview, 23 February 1930, 3, FTP; "The Old Shooting Territory: D. M. Division Conductor 'Bill' Finnicum Has Interesting Experience," Folder: Miscellaneous Hunting Articles, FTP; James J. Dinsmore, *A Country So Full of Game: The Story of Wildlife in Iowa* (Iowa City: Iowa University Press, 1994), 162–63.

54. Anthony Arnold, *Suisun Marsh History: Hunting and Saving a Wetland* (Marina, CA: Monterey Pacific, 1996), 101–2.

55. "Pigeons," *(Kilbourn, WI) Mirror*, 6 May 1871, 4.

56. Miller and Hanson, *Waterfowl Decoys*, 319.

57. Daniel Giraud Elliot, "The 'Game Birds' of the United States," in US Department of Agriculture, *Report of the Commissioner of Agriculture for 1864* (Washington, DC), 383.

58. "The Chicago Game Trade," *Memphis Daily Appeal*, 12 December 1868, 1.

59. "Great Central Route. 'Blue Line'" (advertisement), *Railroad Gazette*, 4 February 1871, 450.

60. Merritt, *Shadow of a Gun*, 122.

61. George H. Mackay, "Observations on the Knot (*Tringa Canutus*)," *Auk*, January 1893, 29.

62. H[enry] T. Phillips to W. B. Mershon, 2 November 1904, in Mershon, *Passenger Pigeon*, 109.

63. M. R. B., "Game Protection" and "Treatment of Stool Pigeons," *American Field* 21 (1884): 227, 395–96, quoted in Schorger, *Passenger Pigeon*, 154.

64. Merritt, *Shadow of a Gun*, 120.

65. Merritt, *Shadow of a Gun*, 60, 192.

66. Mershon, *Passenger Pigeon*, 110.

67. T. M. Metcalf, Statistics of Minnesota for 1877 (St. Paul, 1878), 226 in Evadene Burris Swanson, *The Use and Conservation of Minnesota Wildlife, 1850–1900* (MA thesis, University of Minnesota, June 1940) (St. Paul: Minnesota Department of Natural Resources, 2007), 21.

68. Iowa Board of Railroad Commissioners, *Thirteenth Annual Report of the Board of Railroad Commissioners for the Year Ending June 30, 1890* (Des Moines, IA: G. H. Ragsdale, State Printer, 1891), 14.

69. Henry Oldys, "The Game Market of To-day," in *USDA Yearbook of Agriculture, 1910* (Washington, DC: Government Printing Office, 1911), 249.

70. Merritt, *Shadow of a Gun*, 78.

71. David E. Blockstein and Harrison B. Tordoff, "Gone Forever: A Contemporary Look at the Extinction of the Passenger Pigeon," in *American Birds* 39, no. 5 (Winter 1985), 845–51.

72. Crissey, "The Great Shotgun Syndicate," 8.

73. Merritt, *Shadow of a Gun*, 264.

74. Ferris, "How a Great City Is Fed," 232.

75. George Laycock, *The Hunters and the Hunted* (New York: Outdoor Life, 1990), 26.

76. "Prairie Chickens," *Buffalo (NY) Weekly Express*, 14 August 1860, 3.

77. "Chicago Daily Market," *Chicago Tribune*, 19 November 1864, 2.

78. "Centre Market," *(Baltimore) Daily Exchange*, 28 February 1859, 1.

79. A. L. Heerman, "Report upon Birds Collected in the Survey," in *Reports of Explorations and Surveys, to Ascertain the Most Practicable and Economical route for a Railroad from the Mississippi River to the Pacific Ocean*, 12 vols. (Washington, DC: Government Printing Office, 1855–61), 10:62; "San Francisco Grain and Produce Market," *Red Bluff Independent*, 31 October 1862, 2; "San Francisco Grain and Produce Market," *Santa Cruz Weekly Sentinel*, 27 October 1866, 2.

80. Merritt, *Shadow of a Gun*, 204.

81. "Chicago as a Game Market," *American Poultry Journal*, November 1889, 337.

82. See, for example, "Market Report," *(Milwaukee) Daily Commercial Letter Price Current*, 22 October 1879, 1; "Country Produce," *St. Louis Daily Market Reporter Merchants' Exchange Price Current*, 20 January 1886, 44, "Game in Market," *FS*, 3 September 1874, 58.

83. Miller and Hanson, *Waterfowl Decoys*, 16.

84. Alders, Record of Birds Killed and Shipped, EAP, 76-589-4.

85. Merritt, *Shadow of a Gun*, 114–15.

86. "Wildfowl Shooting in Maryland—The Great Guns at Night," *Wilkes' Spirit of the Times*, 12 March 1864, 28; Sawyer, *Texas Market Hunting*, 53–55.

87. Merritt, *Shadow of a Gun*, 126, 159–60.

88. Roney, "Efforts to Check the Slaughter," 109.

89. Murphy, *American Game Bird Shooting*, 347

90. Audubon, *Birds of America*, 5:342.

91. Lewis, *American Sportsman*, 446.

92. Harker, interview, 216, FTP.

93. Herring, *Opening Days*, 51.

94. "Kept in Cold Storage," 3.

95. Walsh, *Outlaw Gunner*, 105, 140.

96. Jack Dudley, *Carteret Waterfowl Heritage* (Morehead City, NC: Coastal Heritage Series, 1993), 40.

97. Cushing, "A Market Gunner's Diary," 686–88.

98. Tom Horton, *An Island Out of Time: A Memoir of Smith Island in the Chesapeake* (New York: W. W. Norton, 1996), 132.

99. Smith, *Marsh Tales*, 93, 185. See also advertisements in *Furniture Trade Review and Interior Decorator* 15 (1894–95): 7, 60, 444, 465, 616, 1004, 1080.

100. Wilson, *American Ornithology*, 8:55, 57.

101. George Mackay, *Shooting Journal of George Henry Mackay*, ed. Henry M. Reeves (Whitefish, MT: Kessinger, 2008), 284; New York Department of Labor, *Sixth Annual Report of the Bureau of Labor Statistics . . . for the Year 1888*, vol. 6, pt. 3 (Albany, NY: Troy Press Company, 1889), 131–32.

102. "Only a Feather," *Millinery Trade Review* (NY) 1, no. 11 (November 1876): 131.

103. "Game and Fish Notes," *New York Sun*, 8 January 1893, 4.

104. "A New Cold Storage Warehouse," *Ice and Refrigeration*, June 1893, 475

105. Merritt, *Shadow of a Gun*, 78.

106. Ferris, "How a Great City Is Fed," 232; "Cold Storage," *Ice and Refrigeration Illustrated*, June 1890, 472.

107. "Special Mention," *(Wellington, KS) Sumner County Press*, 1 January 1880, 3.

108. Untitled item, *(Independence) South Kansas Tribune*, 6 November 1878, 3; "At It Again" (advertisement), *Independence Kansan*, 22 November 1877, 3.

109. Untitled item, *Gibson City (IL) Courier*, 11 January 1884, 5.

110. "Melvin," *Gibson City Courier*, 21 December 1883, 4.

111. Mackay, *Shooting Journal*, 174.

112. Pearson, *Adventures in Bird Protection*, 118.

113. Murphy, *American Game Bird Shooting*, 344.

114. US Department of Agriculture, *Report of the Commissioner of Agriculture for the Year 1870* (Washington, DC, 1871), 246.

115. Ferris, "How a Great City Is Fed," 232.

116. US Department of Agriculture, *Report of the Commissioner of Agriculture for the Year 1870*, 246.

117. George O. Hill and US Department of Agriculture, *Farmers' Bulletin No. 62: Marketing Farm Produce* (Washington, DC: Government Printing Office, 1903), 14.

118. Miller and Hanson, *Waterfowl Decoys*, 320.

119. Advertisements, *Daily (Des Moines) Iowa State Register*, 12 December 1874, 2.

120. US Department of Agriculture, *Report of the Commissioner of Agriculture for the Year 1870*, 246.

121. "Form Express Company," *Sacramento Union*, 23 November 1909, 4; "Wild Birds of Southern Oregon," *Sacramento Union*, 4 August 1907, 4; Joseph Grinnell, Harold Child Bryant, and Tracy Irwin Storer, *Game Birds of California* (Berkeley: University of California Press, 1918), x.

122. See, for example, Henry G. Langley, comp., *The San Francisco Directory for the Year Commencing December 1869* (San Francisco: Henry G. Langley, 1869), 53, 59, 75, 446; *A. N. Marquis & Co.'s Handy Business Directory of Chicago, 1886–1887* (Chicago: A. N. Marquis & Company, 1886), 603.

123. See, for example, Isaac Costa, comp., *Gopsill's Philadelphia Business Directory for 1869* (Philadelphia: James Gopsill, 1869), 19; "Edward Frank: Dealer in Fish, Turtle and Game" (advertisement), *(New Orleans) Times-Picayune*, 17 February 1885, 9.

124. "The New York Game Laws," *FS*, 1 January 1896, 476; *Illustrated New York: The Metropolis of To-Day* (New York: International Publishing Co., 1888), 144.

125. "Christmas in the Markets," 14.

126. *Origin, Growth, and Usefulness of the Chicago Board of Trade* (New York: Historical Publishing Co., 1885–86), 265.

127. "End of the Game Season," *(Chicago) Daily Inter Ocean*, 1 February 1899, 10.

128. John S. Hittell, *The Commerce and Industries of the Pacific Coast of North America* (San Francisco: A. L. Bancroft & Co., 1882), 341.

129. Mershon, *Passenger Pigeon*, 146; Merritt, *Shadow of a Gun*, 183; Otto Widmann, "A Preliminary Catalog of the Birds of Missouri," *Transactions of the Academy of Science of St. Louis* 17 (1907), 85.

130. Greenberg, *Feathered River across the Sky*, 82.

131. Forbush, *History of Game Birds*, 453; Schorger, *Passenger Pigeon*, 145.

132. Greenberg, *Feathered River across the Sky*, 90; Henry K. Coale, "Notes on *Ectopistes migratorius*," *Auk* 39 (1922): 255.

133. Mershon, *Passenger Pigeon*, 223–24.

134. Tober, *Who Owns the Wildlife?* 81, 108.

135. "Seafowl Eggs: A Curious Way of Obtaining a Livelihood," *(Winston-Salem, NC) People's Press*, 16 June 1887, 4.

136. T. S. Palmer, "A Review of Economic Ornithology in the United States," in *Yearbook of the United States Department of Agriculture, 1899* (Washington, DC: Government Printing Office, 1900), 271–72.

137. "Court Proceedings," *(San Francisco) Daily Alta California*, 10 November 1863, 1; "Conditions of Wounded Italians," *Daily Alta California*, 6 June 1863, 1; "The Farallones War— Arrests for Murder," *Daily Alta California*, 6 June 1863, 1.

138. "The Albatross May Disappear," *CT*, 24 June 1900, 42.

139. Joseph Rosenthal advertisement, *Millinery Trade Review*, December 1899, 32.

140. Louis Stern advertisement, 7 September 1899, reprinted in T. S. Palmer, "Protest against the Collection of Plume Birds through Postmasters," *Bird Lore* 2 (April 1900): 66.

141. William Dutcher, "The Destruction of Bird-Life in the Vicinity of New York," *Science* 7, no. 160 (26 February 1886): 198.

142. George Bird Grinnell, "The Destruction of Small Birds," *FS* 23 (7 August 1884): 24.

143. Clarence M. Weed and Ned Dearborn, *Birds in Their Relations to Man: A Manual of Economic Ornithology for the United States and Canada* (Philadelphia: J. B. Lippincott Company, 1903), 261.

144. Mark V. Barrow, "The Specimen Dealer: Entrepreneurial Natural History in America's Gilded Age," *Journal of the History of Biology* 33, no. 3 (August 2000): 493–98, 502–8.

145. F. T. Corless, advertisement, *Natural Science News* (Albion, NY), 4 May 1895, n.p.

146. Ernest H. Short, advertisement, *Natural Science News*, 4 May 1895, 100. See also William Gallaher, advertisement, *The Oologist*, October 1904, 162; and T. Adolph LeGasse, advertisement, in Fred H. Clifford, *In the Maine Woods* (Bangor, ME: Bangor and Aroostook Railroad, 1905), 163.

147. Merritt, *Shadow of a Gun*, 177–78.

148. Kiner, *History of Henry County*, 1:593.

149. Karau, "Merritt's Pond," n.p.; James W. Beasley, comp., *Beasley's Kewanee Directory for 1876-7* (Princeton, IL: Republican Steam Book and Job Printing House, 1876), 39, 42–43, 47, 52, 59, 63, 68 72, 87, 90, 93; Merritt, *Shadow of a Gun*, 190; "Death of Henry C. Parker," *Kansas City (MO) Journal*, 2 December 1898, 7; *Portrait and Biographical Album of Henry County, Illinois* (Chicago: Biographical Pub. Co., 1885), 490, 611; Kiner, *History of Henry County*, 1:593.

150. Merritt, *Shadow of a Gun*, 223–25.

151. Merritt, *Shadow of a Gun*, 31, 189, 222–23.

152. Merritt, *Shadow of a Gun*, 223–25.

153. Merritt, *Shadow of a Gun*, 120–23.

154. Merritt, *Shadow of a Gun*, 177–78.

155. Herbert, *Frank Forester's Field Sports*, 1:36.

156. Ferris, "How a Great City Is Fed," 232.

157. "Our Sportsmen," *CT*, 23 November 1872, 3.

158. Thomas F. De Voe, *The Market Assistant, Containing a Brief Description of Every Article of Human Food Sold in the Public Markets of the Cities of New York, Boston, Philadelphia, and Brooklyn* (New York: Hurd & Houghton, 1867), 145.

159. John Strong Newberry, "Report on the Birds of California and Oregon" (1859), *Pacific Railroad Reports*, vol. 6, pt. 6, quoted in Elliott Coues, *Birds of the Northwest: A Hand-Book of Ornithology of the Region Drained by the Missouri River and Its Tributaries* (Washington, DC: Government Printing Office, 1874), 533; Grinnell, Bryant, and Storer, *Game Birds of California*, 587.

160. Quoted in Walsh, *Outlaw Gunner*, 59.

161. "In the Markets," *New York Evening Post*, 23 October 1897, 23.

162. W. Mackay Laffan, "Canvas-back and Terrapin," *Scribner's Monthly*, November 1877, 2.

163. "Where Some Game Goes To," *FS*, 17 March 1881, 123.

164. Dutcher, "Destruction of Bird-Life," 197.

165. "Where Some Game Goes To," 123.

166. "Washington Market: Preparing for the Holidays," *Frank Leslie's Illustrated Newspaper*, 27 December 1879, 291.

167. Daniel Giraud Elliot, *The Wildfowl of the United States and British Possessions, or The Swan, Geese, Ducks, and Mergansers of North America* (New York: Francis P. Harper, 1898), 152.

168. Patricia Fleming and Thomas Carpenter, eds., *Traditions in Wood: A History of Wild-fowl Decoys in Canada* (Ontario: Camden House, 1987), 135.

169. W. Ross King, *The Sportsman and Naturalist in Canada, or Notes on the Natural History of the Game, Game Birds, and Fish of That Country* (London: Hurst & Blackett, 1866), 152.

170. Palmer, "Review of Economic Ornithology," 268.

171. Oldys, "Game Market of To-day," 253–54.

172. C. B. R. Kennerly, "Report on the Birds Collected on the Route" (1859), *Pacific Railroad Reports*, vol. 10, pt. 6, no. 3, p. 34, quoted in Grinnell, Bryant, and Storer, *Game Birds of California*, 288.

173. "Game in the Market of San Francisco," *(San Francisco) Daily Alta California*, 29 November 1857, 1.

174. "City Intelligence," *Daily Alta California*, 8 February 1851, 2.

175. *Langley's San Francisco Directory for the Year Commencing May 1890* (San Francisco: Geo. B. Wilbur, 1890), 1559; *The San Francisco Directory . . . 1869* (San Francisco: Henry G. Langley, 1869), 734.

176. Missouri Fish Commissioners, *Third Biennial Report of the Fish Commission of the State of Missouri* (Jefferson City, MO: Tribune Printing Company, 1885), 12.

177. *Gould's St. Louis Directory for 1880* (St. Louis: David B. Gould, 1880), 1218.

178. "Game Galore," *St. Louis Post-Dispatch*, 28 December 1881, 5.

179. "In the Game Market: St. Louis the Largest Market in the United States," *St. Louis-Post Dispatch*, 15 September 1880, 7.

180. "Game Galore," 5.

181. H. H. Kohlstaat, "The Greatest Game Market in the World," *Saturday Evening Post*, 26 April 1924, 72.

182. *Edwards' Fourteenth Annual Directory of the City of Chicago for 1871* (Chicago: Richard Edwards, Publisher, 1871), 1068; *A. N. Marquis & Co.'s Handy Business Directory of Chicago, 1886–7* (Chicago: A. N. Marquis & Company, 1886), 76, 202, 255, 574–75, 578, 603, 631.

183. "Chicago's Game Market: It Is the Greatest One in the Civilized World," *CT*, 22 September 1889, 3.

184. "Oysters, Fish, and Fowl: The Season for These Never Ends in Chicago," *CT*, 8 June 1890, 34; "Chicago as a Game Market," *American Poultry Journal*, October 1889, 337.

185. James Fenimore Cooper, *Notions of the Americans as Picked Up by a Travelling Bachelor*, 2 vols. (London: Henry Colburn, 1828), 1:182, 188.

186. Ferris, "How a Great City Is Fed," 232.

187. De Voe, *Market Assistant*, 111.

188. Junius Henri Browne, *The Great Metropolis: A Mirror of New York* (Hartford, CT: American Publishing Company, 1869), 408.

189. "Washington Market, New York City," *American Agriculturalist*, December 1860, 364; "The New York Game Market," *FS*, 3 January 1884, 454.

190. Ferris, "How a Great City Is Fed," 232.

191. Ferris, "How a Great City Is Fed," 232.

192. "Washington Market: Preparing for the Holidays," 291.

193. Merritt, *Shadow of a Gun*, 221.

194. Ferris, "How a Great City Is Fed," 232.

195. See, for example, Louis Stern, advertisement, 7 September 1899, reprinted in Palmer, "Protest against the Collection of Plume Birds," 66.

196. "San Francisco Market Review," *(Rocklin, CA) Placer Herald*, 17 June 1876, 7.

197. "Canvas-Back Duck Trust," *NYT*, 5 February 1888, 10.

198. "The New York Law," *FS*, 24 February 1887, 81.

199. "Game Protection: The Coming Convention of Sportsmen and Game-Dealers," *CT*, 2 January 1882, 2; "Game Protection: The Chicago Sportsmen and Game Dealers Effect an Organization," *CT*, 6 January 1882, 12; "A Western Game Association," *NYT*, 6 January 1882, 1.

200. "Sportsmen and Game Dealers: Their Association Disbanded," *(Chicago) Inter Ocean*, 17 October 1883, 3.

201. "Millinery Trade at War," *NYT*, 15 November 1903, 24.

202. Emerson Hough, "Chicago and the West," *FS*, 2 April 1898, 266; "Game Dealers Elect Officers," *St. Louis Post-Dispatch*, 5 May 1903, 16; "Organizing and New Game Association," *New York Produce Review and American Creamery*, 18 September 1907, 881; "Incorporated," *San Francisco Examiner*, 26 March 1877, 3.

203. "A Call to Ohio Sportsmen," *Recreation*, February 1904, 130; "Organizing a New Game Association," *New York Produce Review and America Creamery*, 18 September 1907, 881.

204. Herring, *Opening Days*, 50.

205. Merritt, *Shadow of a Gun*, 70.

206. Merritt, *Shadow of a Gun*, 119, 125.

207. Roney, "Efforts to Check the Slaughter," 81.

208. Kohlstaat, "Greatest Game Market," 72.

209. US Department of Agriculture, *Report of the Commissioner of Agriculture for the Year 1870*, 246.

210. Merritt, *Shadow of a Gun*, 195.

CHAPTER THREE: The Hunted

1. Merritt, *Shadow of a Gun*, 248–53.

2. Merritt, *Shadow of a Gun*, 301.

3. Merritt, *Shadow of a Gun*, 204.

4. Merritt, *Shadow of a Gun*, 204, 293, 301.

5. Merritt, *Shadow of a Gun*, 204, 293. See also US Department of the Interior, "Old-Time Market Hunter Reminiscences: Says Protection Is Bringing Game Back," Fish and Wildlife Service press release, 21 January 1941, 4.

6. Merritt, *Shadow of a Gun*, 121.

7. Elsa G. Allen, *The History of American Ornithology before Audubon*, Transactions of the American Philosophical Society 3 (Philadelphia: American Philosophical Society, 1951), 401–14.

8. Christopher Columbus, *Personal Narrative of the First Voyage of Columbus to America*, trans. Samuel Kettell (Boston: Thomas B. Wait & Son, 1827), 55–56; Arthur Barlowe, "The first voyage made to the Coasts of America . . . ," in *The Principall Navigations, Voyages, Traffiques & Discoveries of the English Nation*, comp. Richard Hakluyt, ed. Edmund Goldsmid (Edinburgh: E. & G. Goldsmid, 1884), 13:284.

9. Samuel Clarke, comp., *A True and Faithful Account of the Four Chiefest Plantations of the English in America* (London: Robert Clavel, Thomas Passenger, William Cadman, William Whitwood, Thomas Sawbridge, and William Birch, 1670), 12, 35–36, available at Early English Books Online Text Creation Partnership, accessed June 13, 2021, name.umdl.umich.edu/A33345.0001.001.

10. John Lawson, *A New Voyage to Carolina* (London: n.p., 1709), 135–37, available at Documenting the American South, accessed 13 June 2021, https://www.docsouth.unc.edu/nc/lawson/menu.html.

11. Allen, *American Ornithology before Audubon*, 465. On Catesby, see George Frederick Frick and Raymond Phineas Stearns, *Mark Catesby: The Colonial Audubon* (Urbana: University of Illinois Press, 1961).

12. Wilson, *American Ornithology*; Allen, *American Ornithology before Audubon*, 552–53; John James Audubon, *The Birds of America, from Original Drawings* (London: privately printed, 1827–38).

13. North American Bird Conservation Initiative, "The State of North America's Birds, 2016," Environment and Climate Change, Ottawa, Ontario, www.stateofthebirds.org; Charles Gald Sibley and Burt Leavelle Monroe, *Distribution and Taxonomy of Birds of the World* (New Haven, CT: Yale University Press, 1990).

14. W. L. McAtee, *Local Bird Refuges*, US Department of Agriculture Farmer's Bulletin 1644 (Washington, DC: Government Printing Office, 1931), 2; J. W. Aldrich, R. C. Banks, T. J. Cade, W. A. Calder, F. G. Cooch, S. T. Emlen, G. A. Greenwell, T. R. Howell, J. P. Hubbard, D. W. Johnston, R. F. Johnston, and L. R. Mewaldt, "Report of the American Ornithologists' Union Ad Hoc Committee on Scientific and Educational Use of Wild Birds," supplement, *Auk* 92, no. 3 (1975): 1A–27A; John L. Trap, "How Many Birds Are There?" *Birds Etcetera*, 23 July 2002, http://birdstuff.blogspot.com/2002/07/how-many-birds-are-there.html. Worldwide estimates of bird population run from 200 to 400 billion individuals. See Kevin J. Gaston and Tim M. Blackburn, "How Many Birds Are There?" *Biodiversity and Conservation* 6 (April 1997): 615–25.

15. Julian P. Hume, *Extinct Birds*, 2nd ed. (London: Bloomsbury, 2017), 60–61, 153–55, 166–68, 215–16.

16. Elliott Coues and D. Webster Prentiss, *Avifauna Columbiana, Being a List of Birds Ascertained to Inhabit the District of Columbia, with the Times of Arrival and Departure of Such as are Non-residents, and Brief Notices of Habits, Etc.* (Washington, DC: Government Printing Office, 1883), 18, 53, 91.

17. Thomas Froncek, *An Illustrated History of the City of Washington* (New York: Wings Books, 1992), 168.

18. Clark, "Western Field Sports," 264.

19. Kohlstaat, "The Greatest Game Market in the World," 72.

20. Greenberg, *Feathered River across the Sky*, 53.

21. North American Bird Conservation Initiative, US Committee, *The State of the Birds, United States of America, 2009* (Washington, DC: US Department of Interior, 2009), 3; George G. Williams, *Geographical Factors in the Distribution of American Birds: Evolutionary Aspects of Migration* (Berkeley: University of California Press, 1958), 4–5.

22. John H. Rappole, *The Avian Migrant: The Biology of Bird Migration* (New York: Columbia University Press, 2013), 5–7.

23. A little over 18.5 percent of all avian species are long-distance migrants. See Jonathan Rolland, Frédéric Jiguet, Knud Andreas Jønsson, Fabien L. Condamine, and Hélène Morlon, "Settling Down of Seasonal Migrants Promotes Bird Diversification," *Proceedings of the Royal Society B: Biological Sciences*, 7 June 2014, https://doi.org/10.1098/rspb.2014.0473.

24. Stephen N. G. Howell and Sophie Webb, *A Guide to the Birds of Mexico and Northern Central America* (Oxford: Oxford University Press, 1995), 38; Rappole, *Avian Migrant*, 180–81.

25. Frederick C. Lincoln, "Bird Banding," in *Fifty Years' Progress in American Ornithology, 1883–1933*, ed. Frank Chapman and T. S. Palmer (Lancaster, PA: American Ornithologists Union, 1933), 65–87; Paul Johnsgard, *Wings over the Great Plains: Bird Migrations in the Central Flyway* (Lincoln, NE: Zea Books, 2012), 15; "The U.S. Response," in *Flyways: Pioneering Waterfowl Management in North America*, ed. A. S. Hawkins, R. C. Hanson, H. K. Nelson, and H. M. Reeves (Washington, DC: Government Printing Office, 1984), 2; Robert M. Wilson, *Seeking Refuge: Birds and Landscapes of the Pacific Flyway* (Seattle: University of Washington Press, 2011), 73–74, 97.

26. Johnsgard, *Wings*, 15–16; Frederick C. Lincoln, *The Migration of North American Birds*, United States Department of Agriculture Circular No. 363 (October 1935), 40.

27. Lincoln, *Migration of North American Birds*, 4–5.

28. Rappole, *Avian Migrant*, 102.

29. Lincoln, *Migration of North American Birds*, 55; Rappole, *Avian Migrant*, 6.

30. Hugh Dingle, *Migration: The Biology of Life on the Move* (Oxford: Oxford University Press, 2014), 135–36.

31. M. C. Witmer, D. J. Mountjoy, and L. Elliot, "Cedar Waxing," *Birds of North America* 8, no. 309 (1997), available at Birds of the World, Cornell Lab of Ornithology, https://birdsoftheworld .org/bow/species/cedwax/cur/introduction; Rappole, *Avian Migrant*, 6; Dingle, *Migration*, 344, 358–51; Vicky J. Meretsky, Jonathan W. Atwell, and Jeffrey B. Hyman, "Migration and Conservation: Frameworks, Gaps, and Synergies in Science, Law, and Management," *Environmental Law* 41, no. 2 (Spring 2011): 454–56.

32. Coues and D. Webster Prentiss, *Avifauna Columbiana*, 18, 91; Coues and Prentiss quoted in Paul Russell Cutright and Michael J. Brodhead, *Elliott Coues: Naturalist and Frontier Historian* (Urbana: University of Illinois Press, 1981), 254.

33. Lewis, *American Sportsman*, 262.

34. Charles W. Townsend, cited in Arthur Cleveland Bent, *Life Histories of North American Shore Birds: Order Limicole, Part 1*, United States National Museum Bulletin 142 (Washington, DC: Government Printing Office, 1927), 330.

35. W. A. A., "Coweens on the Niagara," *FS*, 24 March 1887, 177.

36. "Grouse of the Prairies and Plains," *American Agriculturalist*, August 1889, 387; Witmer Stone, "Economic Ornithology—Birds in Relation to Man," *Turf, Field, and Farm*, 9 April 1897, 569.

37. Palmer, "Protest against the Collection of Plume Birds," 66.

38. Grinnell, Bryant, and Storer, *Game Birds of California*, 288.

39. Grinnell, Bryant, and Storer, *Game Birds of California*, 587; Walter E. Bryant, *Birds and Eggs from the Farallon Islands*, extract from the *Proceedings of the California Academy of Sciences*, 2nd ser., 1 (19 December 1887): 35.

40. On migration and the repeated concentrations of birds available to market hunters, see National Research Council, "Wildlife Values in a Changing World," in *Land Use and Wildlife Resources* (Washington, DC: National Academies Press, 1970), 38, https://doi.org/10.17226/9576.

41. Hugo Ziemann and Mrs. F. L. Gillette, *The White House Cook Book: A Comprehensive Cyclopedia of Information for the Home* (1887; repr., New York: Saalfield Publishing Co., 1915), 473–77.

42. Roney, "Efforts to Check the Slaughter," 83; Gideon Lincecum, "The Nesting of Wild Pigeons," *American Sportsman*, reprinted in *(Elkton, MD) Cecil Whig*, 4 July 1874, 1.

43. Greenberg, *Feathered River across the Sky*, 136–37; Errol Fuller, *The Passenger Pigeon* (Princeton, NJ: Princeton University Press, 2015), 62.

44. Mershon, *Passenger Pigeon*, 125.

45. J. Kemp Bartlett Jr., "Chesapeake Bay," in Connett, *Duck Shooting*, 127.

46. Pearson, *Adventures in Bird Protection*, 118–19. Recent research on avian movement ecology has found that ducks and other avian species which migrate in the autumn are influenced primarily by wind conditions. Temperature, precipitation, and cloud cover are also contributing factors. See Benjamin J. O'Neal, Joshua D. Stafford, Ronald P. Larkin, and Eric S. Michel, "The Effect of Weather on the Decision to Migrate from Stopover Sites by Autumn-Migrating Ducks," *Movement Ecology* 6, no. 23 (2018), https://doi.org/10.1186/s40462-018-0141-5; Anders P. Tøttrup, Kalle Rainio, Timothy Coppack, Esa Lehikoinen, Carsten Rahbek, and Kasper Thorup, "Local Temperature Fine-Tunes the Timing of Spring Migration in Birds," *Integrative and Comparative Biology* 50, no. 3 (September 2010): 293–304, https://doi.org/10.1093/icb/icq028.

47. Merritt, *Shadow of a Gun*, 295.

48. G. L. R., "Shooting Mallards in Ice Holes," *FS*, 20 February 1897, 146.

49. Ed. W. Sandys, "Rod and Gun," *Outing*, June 1899, 312.

50. "Delaware River Rail Tides," *FS*, 5 September 1896, 186; J. B. Shannon & Sons (advertisement), *Sporting Life*, 10 November 1906, 15.

51. Hallock, *Sportsman's Gazetteer and General Guide*, 181.

52. Lewis, *American Sportsman*, 322–23.

53. George Ross Starr Jr., *Decoys of the Atlantic Flyway* (New York: Winchester Press, 1974), 88; Robert T. Morris, "Again the Woodcock," *FS*, 19 March 1910, 457; Lewis, *American Sportsman*, 187–88; John J. Audubon, *Ornithological Biography, or, An Account of the Habits of the Birds of the United States of America*, 5 vols. (Edinburgh: Adam & Charles Black, 1831–39), 1:323.

54. Lewis, *American Sportsman*, 306.

55. "Shooting," *Outing*, June 1886, 356.

56. "Street Produce Markets," *Daily Commercial Bulletin*, 14 September 1885, 2.

57. H. Albert Hochbaum, *Travels and Traditions of Waterfowl* (Minneapolis: University of Minnesota Press, 1955), 7–11; John C. Phillips, *Shooting Stands of Eastern Massachusetts* (Cambridge, MA: Riverside Press, 1929); "Camping in the Woods," *Farm Implement News*, November 1886, 11; *History of Sandusky County, with Portraits and Biographies of Prominent Citizens and Pioneers* (Cleveland, OH: H. Z. Williams & Bro., 1882), 727.

58. George Bird Grinnell, *American Duck Shooting* (New York: Forest and Stream Publishing, 1901), 425.

59. Ralph Payne-Gallwey, *Wildfowl and Wildfowl Shooting with Shotgun and Punt Gun* (1896; repr., Redditch, UK: Read Books Limited, 2017), n.p.; Wilfred K. Merrill, *The Hunter's Bible: An Illustrated Book of Outdoor Advice for the Sportsman* (New York: Doubleday, 1986), 115.

60. Audubon, *Ornithological Biography*, 5:138; Sandys and Van Dyke, *Upland Game Birds*, 353; Lewis, *American Sportsman*, 201; Grenville Mellen, ed., *A Book of the United States Exhibiting Its Geography, Divisions, Constitution, and Government* (Hartford, CT: H. F. Sumner & Co., 1839), 220.

61. William Rohn, interview, 1930, 416–17, FTP; Richard Harker, interview, 22 February 1930, 224, 226–28, FTP.

62. Mershon, *Passenger Pigeon*, 80.

63. Andrews Wilkinson, "Christmas Week among the Lagoons of Lower Louisiana," *Outing*, December 1897, 216; Frank Dodd, interview, 16 December 1930, 535, FTP; R. K. Sawyer, *A Hundred Years of Texas Waterfowl Hunting: The Decoys, Guides, Clubs, and Places* (College Station: Texas A & M University Press, 2012), 37. Many market hunters developed and manufactured their own duck calls, including J. T. Beckhart of Buckspoint, Arkansas; Charles Ditto of Keithsburg, Illinois; and Fred Allen, of Monmouth, Illinois. Allen supposedly developed the first modern instrument in 1863, and his commercially produced, nickel-plated call was infamous for freezing onto hunters' lips. Matt Young, "Great Moments in Waterfowling," *Ducks Unlimited*, accessed 15 June 2021, https://www.ducks.org/hunting/duck-hunting-stories/great-moments-in-waterfowling; Gary Koehler, "Legends of the Call," *Ducks Unlimited*, accessed 15 June 2021, https://www.ducks.org/hunting/duck-calling/legends-of-the-call.

64. William C. Hazelton, *Duck Shooting and Hunting Sketches* (Chicago: n.p., 1916), 2.

65. "Redhead: Life History," *All About Birds*, Cornell Lab of Ornithology, accessed 15 June 2021, https://www.allaboutbirds.org/guide/Redhead/lifehistory#.

66. Kimball and Kimball, *Market Hunter*, 108.

67. Parker Gillmore, *Prairie and Forest: A Description of the Game of North America, with Personal Adventures in Their Pursuit* (New York: Harper & Brothers, 1874), 237.

68. Mershon, *Passenger Pigeon*, 20, 219.

69. Henry Jay Case, "The Making of a Refuge," *Outing*, October 1913, 16; O. P. Brownlow to Chief of Patrol, 16 February 1933, Box 3, Folder F3735:592, B, Misc. 1926–36, 48, JSHP.

70.   Gasper Howland, *The Complete Sportsman: A Manual of Scientific and Practical Knowledge Designed for the Instruction and Information of all Votaries of the Gun* (New York: Press of Forest and Stream Publishing Company, 1893), 76; Luc Bélanger and Jean Bédard, "Hunting and Waterfowl," in *Wildlife and Recreationists: Coexistence through Management and Research*, ed. Richard L. Knight and Kevin Gutzwiller (Washington, DC: Island Press, 1995), 243–56.

71.   Charles H. Ditto to F[red] O. Thompson, 26 August 1930, FTP, 497–98; Matthieu Guillemain, Jean-Maxime Bertout, Thomas Kjær Christensen, Veli-Matti Väänänen, Patrick Triplet, Vincent Schricke, and A. D. Fox, "How Many Juvenile Teal *Anas crecca* Reach the Wintering Grounds? Flyway-Scale Survival Rate Inferred from Wing Age-Ratios," *Journal of Ornithology* 151, no. 1 (January 2009): 51–60.

72.   Bertie J. Weddell, *Conserving Living Natural Resources in the Context of a Changing World* (New York: Cambridge University Press, 2002), 37; Robin W. Doughty, "San Francisco's Nineteenth-Century Egg Basket: The Farallons," *Geographical Review* 61, no. 4 (October 1971): 561; Forbush, *History of Game Birds*, 143, 363, 413.

73.   Maria R. Audubon, *Audubon and His Journals, With Zoological and Other Notes by Elliott Coues*, vol. 1 (New York: Charles Scribner's Sons, 1897), 374; Janet Foster, *Working for Wildlife: The Beginning of Preservation in Canada*, 2nd ed. (Toronto: University of Toronto Press, 1998), 184; Sveinn Are Hanssen and Kjell Einar Erikstad, "The Long-Term Consequences of Egg Predation," *Behavioral Ecology* 24, no. 2 (March–April 2013): 567, https://doi.org/10.1093/beheco/ars198.

74.   Katharine C. Parsons and Terry L. Master, "Snowy Egret (*Egretta thula*)," version 1.0, in *Birds of the World*, ed. A. F. Poole and F. B. Gill (Ithaca, NY: Cornell Lab of Ornithology, 2020), https://doi.org/10.2173/bow.snoegr.01.

75.   Grinnell, *American Duck Shooting*, 589–90.

76.   Merritt, *Shadow of a Gun*, 188.

77.   Merritt, *Shadow of a Gun*, 178.

78.   Neltje Blanchan, *Birds That Hunt or Are Hunted: Life Histories of One Hundred and Seventy Birds of Prey, Game Birds, and Water-Fowls* (New York: Doubleday & McClure, 1898), 106.

79.   Merritt, *Shadow of a Gun*, 301.

80.   "A Novel Egg Farm," *The Friend*, 9 July 1891, 382; "Arrived from the Farallones," *Sacramento Daily Union*, 17 June 1851, 2.

81.   "Seafowl Eggs: One Curious Way of Obtaining a Livelihood," *(Winston-Salem, NC) People's Press*, 16 January 1887, 4; C. H. Thompson, "Egg-Hunting on the South Farallon," *Frank Leslie's Popular Monthly*, November 1896, 594.

82.   Ernest, Peixotto, *Romantic California* (New York: Charles Scribner's Sons, 1911), 178.

83.   Glen Chilton, "Labrador Duck," in *The Birds of North America*, no. 307 (Philadelphia: Academy of Natural Sciences and American Ornithologists' Union, 1997), 7.

84.   "Hints about Geese and Fish," *NYT*, 10 February 1878, 9.

85.   De Voe, *Market Assistant*, 159.

86.   Merritt, *Shadow of a Gun*, 46.

87.   Tober, *Who Owns the Wildlife?*, 79.

88.   *New York Evening Post*, 26 April 1851, quoted in Schorger, *Passenger Pigeon*, 147.

89.   Schorger, *Passenger Pigeon*, 149.

90.   Roney, "Efforts to Check the Slaughter," 90–91.

91.   E. T. Martin, *American Field*, 25 January 1879, reprinted in Mershon, *Passenger Pigeon*, 103. On passenger pigeons and the principles of supply and demand, see Schorger, *Passenger Pigeon*, 155.

92.   "Cold Storage of Game: Medical and Gastronomical Objections to It," *New York Sun*, 22 January 1899, 3.

93. Daniel Giraud Elliot, *The Gallinaceous Game Birds of North America* (New York: Francis P. Harper, 1897), 114.

94. "Kept in Cold Storage: Poultry and Game Just as Good as Ever after Six Months," *New York Tribune*, 16 May 1897, 3.

95. P. C. Beaman, "Cold Storage and the Game Laws, *FS*, 12 March 1898, 207.

96. Kohlstaat, "Greatest Game Market in the World"; "Game in Market," *FS*, 8 October 1874, 139; Oldys, "Game Market of To-day," 245–46.

97. For example, "San Francisco Grain and Produce Market," *Santa Cruz Weekly Sentinel*, 27 October 1866, 2.

98. "Game in Market," *FS*, 1874–76 issues. Data on prices per pound calculated by Henry M. Reeves, "Avian Entrepreneurs: The Economic Exploitation of North American Migratory Birds" (unpublished manuscript, 1 February 2014), 263–69. Thomas De Voe's 1867 assessment of the woodcock's market value aligned with Hallock's from nearly a decade later. De Voe wrote that this was a "highly prized bird" that garnered "the highest price of any bird brought to market." See De Voe, *Market Assistant*, 164.

99. Tober, *Who Owns the Wildlife?* 79.

100. "City Intelligence," *(San Francisco) Daily Alta California*, 8 February 1851, 2.

101. Oldys, "Game Market of To-day," 253.

102. S. H. Greene, "Mallard Shooting at Shoalwater," *FS*, 21 May 1891, 348.

103. Robert Ridgeway, *The Ornithology of Illinois*, vol. 2, pt. 1 (Springfield, IL: H. W. Rokker, 1895), 160–61.

104. "The Texas Game Market: From the Galveston News," *Recreation*, July 1897, 47.

105. "Force-meat" was a ground mixture of meats with fats and flavorings. Maria Parloa, "The Household Market Basket: A Practical Family Provider," *Good Housekeeping*, September 1893, 101–102.

106. "The Chase in Louisiana: The Game Birds in New Orleans Markets," *New Orleans Republican*, 17 November 1872, 2.

107. "Georgia Gossip," *Atlanta Constitution*, 11 April 1883, 2.

108. Murphy, *American Game Bird Shooting*, 282.

109. "Miscellaneous City News: Game, Poultry, and Fruit," *NYT*, 5 August 1883, 5; "The Retail Markets," *NYT*, 29 April 1883, 13; "Features of the Market: Prices Asked for Provisions—The Supply of Fish," 31 July 1881, 9.

110. Sullivan, *Waterfowling on the Chesapeake*, 72.

111. Fred. Chapman Mathews, "A Plea for Our Game," *Lippincott's Monthly Magazine*, April 1897, 573.

112. Elisha J. Lewis, *Hints to Sportsmen: Containing Notes on Shooting; The Habits of the Game Birds and Wild Fowl of America; The Dog, the Gun, the Field, and the Kitchen* (Philadelphia: Lea & Blanchard, 1851), 237.

113. "A Game Dinner," *FS*, 6 December 1877, 352.

114. Garnett Laidlow Eskew, *Willard's of Washington: The Epic of a Capital Caravansary* (New York: Coward-McCann, 1954), 14, 64, 108, 129–30.

115. Connett, *Duck Shooting*, 127.

116. Calvin Dill Wilson, "The Land of the Epicure," *Cosmopolitan*, October 1895, 661, 666.

117. Michael Batterberry and Ariane Batterberry, *On the Town in New York: The Landmark History of Eating, Drinking, and Entertainments from the American Revolution to the Food Revolution* (New York: Routledge, 1998), 64.

118. Cleveland Amory, Helen Bullock, Lucius Beebe, and Helen McCully, *The American Heritage Cookbook and Illustrated History of American Eating and Drinking*, Part 1 (New York: American Heritage, 1964), 332.

119. "Tempting Hotel Menus: Fine Christmas Dinners at Many Hostelries," *NYT*, 26 December 1890, 8.

120. Charles Ranhofer, *The Epicurean: A Complete Treatise of Analytical and Practical Studies on the Culinary Art . . . Making a Franco-American Culinary Encyclopedia* (New York: Charles Ranhofer, 1894), 637–38, 643, 647, 654–55, 659, 671–73. Chef Alexander Filippini, another alumnus of Delmonico's and cookbook author, created numerous signature game bird entrees. See Alexander Filippinni, *The International Cook Book* (New York: Doubleday, Page & Co., 1911), 824, 981, 1006.

121. "Chitchat on New York and Philadelphia Fashions for November," *Godey's Lady's Book*, November 1865, 458.

122. T. S. Palmer, *Legislation for the Protection of Birds Other than Game Birds*, Bulletin No. 12, U.S. Department of Agriculture, Division of Biological Survey (Washington, DC: Government Printing Office, 1900), 25. Wild bird feathers were also used by anglers for tying flies. Grouse, ibis, widgeon, wild turkey, and blue jay were a few of the species employed. See S. K. Putnam, "Amateur Fly-Tying," *American Angler*, January 1892, 166–71.

123. Robert Welker, *Birds and Men* (Cambridge, MA: Harvard University Press, 1955), 196.

124. "Chitchat on Fashions for December," *Godey's Lady's Book*, December 1874, 574; "A Chat about Feathers," *Godey's Lady's Book*, February 1874, 191.

125. Helen Rowe, "Family Fashions and Fancies," *Good Housekeeping*, 1 September 1888, 204.

126. Charles Dixon, *Stray Feathers from Many Birds, Being Leaves from a Naturalist's Note-Book* (London: W. H. Allen & Co., 1890), 26.

127. Frank M. Chapman, "Birds and Bonnets," *FS*, 25 February 1886, 84; Tober, *Who Owns the Wildlife?* 80.

128. "Destruction of Birds for Millinery Purposes," *Science* 26 (February 1886): 197.

129. "Fashionable Women Wear Bird Hats," *El Paso Times*, 19 November 1901, 7.

130. Sue Ann Prince, "Introduction," in *Stuffing Birds, Pressing Stuffing Birds, Pressing Plants, Shaping Knowledge: Natural History in North America, 1730–1860*, ed. Sue Ann Prince (Philadelphia: American Philosophical Society, 2003), 1–3, 6–8.

131. Elliott Coues, *Key to North American Birds* (Boston: Estes & Lauriat, 1890), 10.

132. Natural history museums comprised a specialized segment of the collector economy, acquiring enormous quantities of bird, egg, and nest specimens from suppliers ranging from professional hunters to explorers, scientific collectors, other museums, and game markets. The Smithsonian Institution alone held more than eighty thousand specimens at the end of the 1880s. See Charles Coleman Sellers, *Mr. Peale's Museum: Charles Wilson Peale and the First Popular Museum of Art and Natural History* (New York: W. W. Norton, 1980), 206, 262–63; William B. Krohn and Marilyn R. Massaro, "Fur into Feathers: Manly Hardy and His Collection of North American Birds," in *New England Collectors and Collections: Dublin Seminar for New England Folklife, Annual Proceedings 2004*, ed. Peter Benes (Boston: Boston University, 2006), 123; Mearns and Mearns, *The Bird Collectors*, 288–89; Barrow, *A Passion for Birds*, 26.

133. Audubon, *Ornithological Biography*, 1:96, 326.

134. Mearns and Mearns, *The Bird Collectors*, 369.

135. George Bird Grinnell, "American Waterfowl and How to Take Them—XXVI," *FS*, 16 March 1901, 207.

136. E. W. Nelson, "Birds of North-Eastern Illinois," *Bulletin of the Essex Institute* 8 (December 1876): 140.

137. "Eggs and Birds: About Park Commissioner Reineke's Collection," *Buffalo Courier*, 25 December 1895, 7.

138. William Dutcher, "The Labrador Duck:—A Revised List of the Extant Species in North America with Some Historical Notes," *Auk* 8, no. 2 (April 1891): 201.

139. Amos Robbins to Thomas De Voe, Fulton Market 1859, De Voe Papers, New York Historical Society, quoted in Jeremy Fisher, "Feeding the Millions: Markets, Metabolism, and the Transformation of the Food System in New York City, 1800–1860," PhD diss., Pennsylvania State University, December 2012, 208.

140. B. B., "The Clapper Rail," *Ornithologist and Oologist*, May 1883, 40.

141. Katherine C. Grier, "Buying Your Friends: The Pet Business and American Consumer Culture," in *Commodifying Everything: Relationships of the Market*, ed. Susan Strasser (New York: Routledge, 2003), 45–49; Pearson, *Adventures in Bird Protection*, 125.

142. "Live Birds for Decoys and the Preserve," *Amateur Sportsman*, November 1909, 12.

143. Pigeon shooting originated in England in the late eighteenth century, and the first American club was founded in Cincinnati, Ohio, in 1825. The sport quickly expanded, with numerous clubs formed all over the country. Margaret H. Mitchell, *The Passenger Pigeon in Ontario* (Toronto: University of Toronto Press, 1935), 114.

144. Roosevelt, *Game-Birds of the Coasts and Lakes*, 288–301; Bogardus, *Field, Cover, and Trap Shooting*, 475–95. Bogardus's appended "Rules for Trap Shooting" is twenty pages long.

145. "Mr. Bergh on Pigeons and Snakes," *FS*, 16 June 1891, 388.

146. Davis, *Fred Kimble*, 35.

147. Schorger, *Passenger Pigeon*, 163–64; Dinsmore, *Country So Full of Game*, 97.

148. William Bruce Leffingwell, *The Art of Wing Shooting: A Practical Treatise on the Use of the Shot-Gun* (Chicago: Rand McNally & Company, 1895), 137.

149. For example, see the New York law in 1763 that established equivalent prices for domestic and wild meats. "New York, August 29," *Pennsylvania Gazette*, 8 September 1763, 1.

150. See Smalley, *Wild by Nature*, 26.

151. William T. Hornaday, *Taxidermy and Zoological Collecting*, 4th ed. (New York: Charles Scribner's Sons, 1894), 187, 234.

152. "Chitchat on Fashions for October," *Godey's Lady's Book*, October 1875, 390; "Chitchat on Fashions for June," *Godey's Lady's Book*, June 1869, 564.

153. Elizabeth Gillespie, ed., *National Cookery Book: Compiled from Original Receipts for the Women's Centennial Committee* (1876; repr., Bedford, MA: Applewood Books, 2005), 104. English customs for serving woodcock and snipe included removing the heads at the table and returning them to the kitchen, where they were smothered in mutton fat and seasoned with salt and pepper. Returned to each guest, the heads were grilled or burned over a lighted candle and then crunched, "after having well smothered it with cayenne pepper." See "Woodcock," *The Epicure*, October 1900, 333.

154. Peter Matheissen, *Wildlife in America* (New York: Viking Press, 1959), 166.

155. "John B. Drake's Game Dinner," *Chicago Legal News*, 29 November 1881, 189.

156. "The American Exhibition," *American Engineering*, 18 March 1887, 259.

157. Charles Augustus Goodrich, *The Family Tourist: A Visit to the Principal Cities of the Western Continent* (Hartford, CT: Case, Tiffany & Co., 1848), 225.

158. "English and American Marketing," *Harper's Bazaar*, 13 July 1878, 442.

159. "Our Cuisine," *CT*, 24 August 1874, 4.

160. Mark Twain, *A Tramp Abroad* (Hartford, CT: American Publishing Company, 1880), 574.

161. Elliot, *Wild Fowl of the United States and British Possessions*, 93.

162. "Bank Coffee House," *New York Evening Post*, 14 February 1823, 3.

163. Margaret Bailey Tinkcom, "Caviar along the Potomac: Sir Augustus John Foster's 'Notes on the United States,' 1804–1812," *William and Mary Quarterly* 8, no. 1 (January 1951): 77.

164. Robert Ridgeway, *The Ornithology of Illinois*, vol. 2, pt. 1 (1895; repr., Springfield, IL: H. W. Rokker, 1913), 160.

165. J. B. White, "Canvasback Ducks," *FS*, 14 November 1889, 324. The connection between avian diets and taste was common. The reason eggs of wild birds were "so highly esteemed," according the *Vermont Record*, was "owing to the flavor acquired from the food consumed." See "A Chapter on Egg-ology," *(Brandon) Vermont Record*, 5 April 1866, 7. Country folk eschewed blackbirds, the *New York Mail and Express* reported, because they were erroneously thought to feed on carrion. City dwellers in Philadelphia, however, discovered them to be a delicacy and attributed their taste to a selective diet of fruits, grains, and worms. See "A New Dainty: The Trade in Blackbirds for Culinary Purposes Is Greatly Increasing," *New York Mail and Express*, reprinted in *(Danville, KS) Harper County Express*, 11 June 1886, 1.

166. Bogardus, *Field, Cover, and Trap Shooting*, 169–71.

167. On the turn-of-the-century anxieties about modern civilization, see T. J. Jackson Lears, *No Place of Grace: Antimodernism and the Transformation of American Culture, 1880–1920* (Chicago: University of Chicago Press, 1981); Gail Bederman, *Manliness and Civilization: A Cultural History of Gender and Race in the United States, 1880–1917* (Chicago: University of Chicago Press, 1996); and David Shi, *The Simple Life: Plain Living and High Thinking in American Culture* (New York: Oxford University Press, 1985).

168. On wildness as an antidote to modern life, see Roderick Nash, *Wilderness and the American Mind*, 5th ed. (New Haven, CT: Yale University Press, 2014), 152–60.

169. Witmer Stone, *Bird Studies at Old Cape May: An Ornithology of Coastal New Jersey*, 2 vols. (1937; repr., New York: Dover, 1965), 2:248, 450.

170. Coues and Prentiss, "Aviafauna Columbiana" (1883), quoted in Cutright and Brodhead, *Elliott Coues*, 254.

171. De Voe, *Market Assistant*, 151.

172. "The Game Birds of Lassen County," *Pacific Rural Press*, 19 March 1892, 258; Lynn R. Meekins, "The Bird for the Thousandth Man," *Saturday Evening Post*, 1 March 1913, 24.

173. "Toothsome Eggs: A Epicure's Discourse upon the Rare Qualities of Some Wild Birds' Eggs," *Lawrence (KS) Daily Journal*, 6 November 1886, 1.

174. "Expensive Delicacies: The Delicious Food Eaten by a Backwoodsman," *Wellington (OH) Enterprise*, 7 August 1889, 5.

175. *Industries and Wealth of the Principal Points in Rhode Island* (New York: A. F. Parsons Publishing, 1892), 202; *Half Century's Progress of the City of Chicago: The City's Leading Manufacturers and Merchants, Part 1: History of Illinois* (Chicago: International Publishing Company, 1887), 273; "Birds as Hat Ornaments," *Friends' Intelligencer*, 5 July 1879, 317; "Bird Millinery," *Maine Farmer*, 14 June 1900, 7.

176. William Dutcher, "Some Notes on Some Rare Birds in the Collection of the Long Island Historical Society," *Auk*, July 1883, 277.

177. Henry Reed Taylor, comp., *Taylor's Standard American Egg Catalogue* (Alameda, CA: H. R. Taylor, 1904), 5, 38, 53.

178. On nature appreciation as a commodity, see, Roderick Nash, "The Export and Import of Nature," *Perspectives in American History* 12 (1979): 519–60; Dunlap, *Saving America's Wildlife*, 6–7.

179. Mearns and Mearns, *The Bird Collectors*, 95.

180. Society of American Taxidermists, *First Annual Report of the Society of American Taxidermists* (Rochester, NY: Daily Democrat and Chronicle Book and Job Print, 1881), 13.

181. Frederic S. Webster, "Taxidermy as a Decorative Art," in *Annual Report of the Society of American Taxidermists* (Washington, DC: Gibson Brothers, Printers, 1884), 61–62.

182. One naturalist catalogued at least thirteen natural history journals that focused wholly or partially on oology. See, Frank L. Burns, comp., "A Bibliography of Scarce or Out of Print North American Amateur Trade Periodicals Devoted More or Less to Ornithology," *The Oologist* 32, no. 7, suppl. (1915): 1–32.

183. Barrow, *A Passion for Birds*, 16, 38–39.

184. Barrow, "Specimen Dealer," 514–19; Samuel E. Cassino, ed., *The Naturalists' Directory, 1884* (Boston: S. E. Cassino & Co., 1884).

185. William Hubbell Fisher, "Papers on the Destruction of Native Birds: Sixth Paper," *The Journal of the Cincinnati Society of Natural History* 9 (1886–87): 215–17.

186. Dutcher, "Destruction of Bird-Life," 199.

187. John H. Krider, *Krider's Sporting Anecdotes, Illustrative of the Habits of Certain Varieties of Game* (1853), ed. Milnor Klapp (Salem, OR: Sand Lake Press, 1988), 276–77.

188. Advertisement, *Amateur Sportsman*, November 1909, 12, 27.

189. Alexander Hunter, "The Club Houses of Currituck Sound," *Harper's Weekly*, 12 March 1892, 254; Van Campen Heilner, *A Book on Duck Shooting* (1939; repr., New York: Alfred A. Knopf, 1945), 238.

190. Henry David Thoreau, "Walking," *Atlantic Monthly*, June 1862, 669.

CHAPTER FOUR: The Sportsman

1. Merritt, *Shadow of a Gun*, 116–17.

2. Merritt, *Shadow of a Gun*, 121–22, 192, 225.

3. Merritt, *Shadow of a Gun*, 10, 177–80, 183–84, 223, 291.

4. Merritt, *Shadow of a Gun*, 290–91.

5. Merritt, *Shadow of a Gun*, 233–36.

6. Herbert, *Frank Forester's Field Sports*, 1:17–19.

7. George Perkins Marsh, *Man and Nature, or Physical Geography as Modified by Human Action* (New York: Charles Scribner, 1864), 97.

8. Herbert, *Frank Forester's Field Sports*, 1:12.

9. Roosevelt, *Game Birds of the Coasts and Lakes*, 15.

10. A. W. S., untitled item, *FS*, 24 April 1884, 247

11. On sportsmen's role in initiating conservation, see Reiger, *American Sportsmen and the Origins of Conservation*.

12. Philip Dray, *The Fair Chase: The Epic Story of Hunting in America* (New York: Basic Books, 2018), 13, 15, 27–28.

13. "Game and Fish Guardians," *NYT*, 16 January 1892, 3.

14. "New York Game Protective Association," *FS*, 26 December 1889, 450; Committee on the Historical Sketch of the New York Association for the Protection of Game, "Historical Sketch of the New York Association for the Protection of Game," Box 1, Folder 12, pp. 17, 19, 20, NYAPG.

15. List of Members of the Sportsman's Club of California, 1 June 1881, Box 4, Folder: S. T., Misc., 1925–44, Natural Resources—Fish and Game Division, JSHP.

16. Daniel Justin Herman, "The Hunters' Aim: The Cultural Politics of American Sport Hunters, 1880–1910," *Journal of Leisure Research* 35, no. 4 (2003): 455.

17. Dray, *Fair Chase*, 92–93; Daniel Justin Herman, *Hunting and the American Imagination* (Washington, DC: Smithsonian Institution Press, 2001), 11–12.

18. Lewis, *Hints to Sportsmen*, 16, 52, 102; Thomas L. Altherr, "The American Hunter-Naturalist and the Development of the Code of Sportsmanship," *Journal of Sport History* 5, no. 1 (Spring 1978): 7–22.

19. Charles Hallock, "Missionary Work among the Gunners," *FS*, 11 February 1875, 8.

20. "Some Offhand Definitions," *FS*, 20 October 1894, 335.

21. R., letter 3, *FS*, 7 March 1889, 133.

22. Polk Burhans, "Southwestward Ho!," *FS*, 10 January 1884, 473.

23. "Snap Shots," *FS*, 7 December 1895, 1. See also "The Utica Association," *FS*, 22 April 1886, 245.

24. "Our Sportsmen," *CT*, 23 November 1872, 3.

25. "Our Candid Advice," *FS*, 8 November 1888, 281.

26. Emerson Hough, "Chicago and the South," *FS*, 9 February 1893, 116; Hough, "Chicago and the West," *FS*, 14 May 1891, 328; Hough, "Chicago and the West," *FS*, 8 May 1890, 308; Chicago (pseud.), "Italian Joe and 'De Plov,'" *FS*, 12 June 1890, 411.

27. L. I. Flower, "Vigorous Language on the Grouse Question," *FS*, 6 July 1882, 446.

28. Don F. Bartlett, "Opinions," *FS*, 2 May 1895, 166; W. W. Brown, letter, *FS*, 17 May 1894, 225.

29. See, for example, "Game Bag and Gun: Game in Season for November," *FS*, 21 November 1878, 329.

30. "Our Candid Advice," *FS*, 8 November 1883, 281.

31. "Central Illinois Association," *FS*, 16 October 1884, 226.

32. Mills (pseud.), "Sound Sense from Wisconsin, *FS*, 13 April 1882, 208.

33. Greenberg, *Feathered River across the Sky*, 143; Mershon, *Passenger Pigeon*, 77.

34. Roney, "Efforts to Check the Slaughter," 92.

35. Greenberg, *Feathered River across the Sky*, 117; Tober, *Who Owns the Wildlife?*, 236–37n64; "Bergh Comes to the Rescue of the Birds," *Illustrated Police News*, 11 January 1872, 6. Well-publicized cases brought against trapshooting include Paine v. Bergh, 1 City Ct. R. 160 (N.Y. 1874). The New York legislature subsequently passed an act revising anti-cruelty statutes to exclude shooting by "members of sportsmen's clubs or incorporated societies." David S. Garland and Lucius P. McGehee, eds., *The American and English Encyclopedia of Law*, 2nd ed., vol. 8 (Northport, Long Island, NY: Edward Thompson Co., 1898), 448, n. 3.

36. Untitled item, *St. Louis Globe-Democrat*, 10 October 1875, 4; State v. Bogardus, 4 Mo App 215, 219 (1877). Bogardus was cleared of the crime upon appeal, the court concluding that killing live pigeons was no more inhumane than any other of the "so-called 'sports'" that allowed "for man's enjoyment of his legitimate dominion over the brute creation." Daniel McKinley, *A Chronology and Bibliography of Wildlife in Missouri* (Columbia: University of Missouri Library, 1960), 22–23.

37. "The Anti–Pigeon Shooting Bill," *FS*, 28 July 1881, 514. On the elite sportsmen's opposition to pigeon-shooting bans, see Janet M. Davis, *The Gospel of Kindness: Animal Welfare and the Making of Modern America* (New York: Oxford University Press, 2016), 112.

38. "The Anti–Pigeon Shooting Bill," 514.

39. "Mr. Bergh's Anti–Pigeon Shooting Bill," *FS*, 14 July 1881, 468; "The Little Hellbender's Appeal," *FS*, 23 June 1881, 404; "A Puzzle for the Anti-cruelty Society," *Health Reformer*, July 1896, 85.

40. United States Patent Office, *Annual Report of the Commissioner of Patents for the Year 1877* (Washington, DC: Government Printing Office, 1878), 397.

41. M., "State Pigeon Tournaments," *FS*, 17 November 1881, 310.

42. T. S. Palmer, *Chronology and Index of the More Important Events in American Game Protection, 1776–1911*, Bulletin No. 41, US Department of Agriculture, Biological Survey (Washington, DC: Government Printing Office, 1912), 20, 23, 49.

43. Grinnell, Bryant, and Storer, *Game Birds of California*, 55; Palmer, *Chronology and Index*, 27–31.

44. Palmer, *Chronology and Index*, 17.

45. Lewis, *Hints to Sportsmen*, 104.

46. Palmer, *Chronology and Index*, 24, 27–31; Sullivan, *Waterfowling on the Chesapeake*, 18; Charles Ditto, interview, 23 August 1930, 471, FTP.

47. "New York Game Protective Association," *FS*, 26 December 1889, 450. Early members of the association included Herbert (Frank Forester) and Robert B. Roosevelt. See Committee on

the Historical Sketch of the New York Association for the Protection of Game, "Historical Sketch," pp. 17, 19, 20.

48. Palmer, *Chronology and Index*, 29, 57, 62.

49. "Game Protection in New Jersey," *Country*, 20 July 1878, 596; New Jersey Legislature, Senate, *Journal of the Thirty-Fourth Senate of the State of New Jersey*, vol. 102 (Morristown, NJ: Vogt Bros., 1878), 565.

50. T. S. Palmer, *Hunting Licenses: Their History, Objects, and Limitations* (Washington, DC: Government Printing Office, 1904), 12–13. Eastern Shore counties in Virginia adopted a similar nonresident license in 1894, requiring out-of-state wildfowl hunters to purchase memberships in the Eastern Shore Game Protective Association. Ibid., 13.

51. John W. Hamer to Royal Phelps, 31 May 1874, Box 1, Folder: Correspondence, 1874–1875, NYAPG (emphasis original).

52. B. A. Hoopes to Charles Hallock, 17 November 1874, Box 1, Folder: Correspondence, 1874–1875, NYAPG.

53. H. W. Leonard to Royal Phelps, 19 April 1875, Box 1, Folder: Correspondence, 1874–1875, NYAPG.

54. Seth Green to Royal Phelps, 25 January 1879, Box 1, Folder: Correspondence, 1874–1875, NYAPG.

55. John Y. Page to Royal Phelps, 17 December 1875, Box 1, Folder: Correspondence, 1874–1875, NYAPG.

56. Royal Phelps to John Y. Page, 23 December 1875, Box 1, Folder: Correspondence, 1874–1875, NYAPG.

57. New York State Legislature, *Laws of the State of New York Passed at the Ninety First Session of the Legislature, 1868*, vol. 2 (Albany: Van Benthuysen & Sons' Steam Printing House, 1868), 1763–64.

58. "Game and Fish Guardians," *NYT*, 16 January 1892, 3.

59. Charles Whitehead to Royal Phelps, 7 April 1870, Box 1, Folder: Court Cases: Phelps, Bangs, NYAPG.

60. "The New York Association for the Protection of Game," *FS*, 18 November 1873, 218; Tober, *Who Owns the Wildlife?*, 216.

61. Charles Hallock, *The International Association for Protecting Game and Fish: Report of Charles Hallock, Secretary* (New York: Forest and Stream Publishing Company, 1876), 3–4.

62. Hallock, *International Association*, 3–8.

63. Thomas Cuthbert, Plaintiff's Points on Demurs, Royal Phelps v. J. H. Racey, District Court in the City of New York for the Sixth Judicial District, 12 March 1874, Box 1, Folder: Cases: Phelps v. Joseph H. Racey, NYAPG.

64. New York State Legislature, *New York Statutes, 1871*, 2:2320; Charles P. Daly, Unpublished Decision, New York Court of Common Pleas, Royal Phelps v. J. H. Racey, 4 May 1874, Box 1, Folder: Cases: Phelps v. Joseph H. Racey, NYAPG; Phelps v. Racey 60 NY 10, 13 (1875). A manuscript copy of Daly's unpublished opinion in the records of the NYAPG has "The Forest & Stream" written on it, suggesting that the decision in this case was of great interest to the wider community of activist sportsmen and was to be communicated immediately to that magazine.

65. Daly, Unpublished Decision, NYAPG.

66. T. S. Palmer, "Lest We Forget," *Bulletin of the American Game Protective Association* 14, no. 1 (January 1925): 11–12; Tober, *Who Owns the Wildlife?*, 169n26; Phelps v. Racey, 13.

67. Magner v. The People of Illinois, 97 Ill 320, 331 (1881).

68. Here the Illinois case differed materially from the Kansas Supreme Court's decision in *State v. Saunders* (1877), which ruled unconstitutional the state's game law prohibiting out-of-state transportation of prairie chickens. The court concluded that the law allowed dead birds "to

become the subject of traffic and commerce" and interference with their transportation was therefore an illegal restraint on free trade. State v. Saunders, 19 Kan. 127, 128–30 (1877).

69. Magner v. People, 329, 336.

70. "The Game Law," *CT*, 25 January 1881, 6; "Notice to Dealers in Game," *CT*, 6 February 1881, 11.

71. Palmer, "Lest We Forget," 11–12; Tober, *Who Owns the Wildlife?*, 169n26.

72. "Snap Shots," *FS*, 22 December 1892, 1.

73. "Who Is Responsible?," *FS*, 5 February 1885, 21.

74. Gris (pseud.), "Nebraska Prairie Chickens," *FS*, 17 July 1890, 512.

75. W. F. A., "Prairie Chickens for Market," *FS*, 26 February 1891, 107.

76. Merchant, *Spare the Birds!*, x, 22, 25–26. Grinnell dissolved his Audubon Society in 1889 due to financial issues.

77. "A Review," *Audubon Magazine*, February 1887, 16; Merchant, *Spare the Birds!*, 26; "The Audubon Society," *FS*, 11 February 1886, 41.

78. Chapman, "Birds and Bonnets," 84; Sialia (pseud.), Letter 1 (no title), *FS*, 3 April 1884, 183.

79. "Bird Destruction," *FS*, 14 January 1886, 482.

80. "Snipe Decoration," *FS*, 4 November 1886, 281.

81. C. M. Stark, "Foxes and Grouse," *FS*, 5 March 1898, 184.

82. S. H., "Suggestions for Reform," *FS*, 2 January 1890, 472.

83. Hermit (pseud.), "'Market-Hunting Outrage!,'" *FS*, 27 November 1897, 428.

84. W. W. McCain, "Pennsylvania Game Law," *FS*, 1 May 1897, 349.

85. Mills (pseud.), "Sound Sense," 208.

86. Harry Hunter, "In Dakota Sloughs," *FS*, 13 November 1884, 307; Garganey (pseud.), "Diminution of the Ducks," *FS*, 17 January 1889, 516. See also "Made a Great Record," *FS*, 4 January 1896, 12; and B. A. Mayor, untitled item, *FS*, 6 July 1882, 447.

87. Untitled item, *FS*, 18 December 1884, 406.

88. Hoodoo (pseud.), "A Word about Market Hunters," *FS*, 8 August 1889, 46.

89. Hermit, "'Market-Hunting Outrage!'"

90. Nom de Plume, "Shooting for Fun and Profit," *FS*, 14 May 1889, 153.

91. "The Rational and the Insane," *FS*, 21 January 1899, 1.

92. Abel Crook, "The Proposed Amendments," *FS*, 3 February 1881, 9.

93. *Missouri Republican* article, quoted in "The St. Louis Convention," *FS*, 6 August 1885, 27.

94. K. (pseud.), "Slaughtering Game," *FS*, 21 June 1883, 405.

95. Untitled item, *NYT*, 23 July 1881, 4. The New York Supreme Court of Judicature articulated this principle in the foundational property law case of *Pierson v. Post* (1805), in which the court had to determine at what point a wild fox became the property of the pursuer. See Pierson v. Post, 3 Cai. R. 175, 2 Am. Dec. 264 (1805).

96. Nor'east (pseud.), "Maine Game Exportation," *FS*, 7 February 1889, 46.

97. Straight Hand (pseud.), untitled item, *FS*, 10 January 1884, 476.

98. Untitled item, *FS*, 12 November 1885, 306; Special (pseud.), "New England Grouse," *FS*, 24 October 1889, 267.

99. "Monmouth Shooting and Fishing Club," *FS*, 24 February 1887, 47.

100. "Possession in Close Season," *FS*, 12 November 1885, 301.

101. "Ruffed Grouse," *FS*, 19 October 1882, 221.

102. "The Utica Association," *FS*, 22 April 1886, 245.

103. Palmer, *Hunting Licenses*, 12–15. Oregon was the only state outside the South to enact a market hunting license.

104. Rex (pseud.), "Ruffed Grouse in Vermont," *FS*, 13 April 1882, 209; "The Zoological Garden," *FS*, 20 September 1888, 161.

105. Tober, *Who Owns the Wildlife?*, 215.

106. Hough, "Chicago and the West," *FS*, 20 August 1891, 85; Hough, "Chicago and the West," *FS*, 19 March 1891, 170–71.

107. "One-Man Power," *FS*, 14 November 1889, 321.

108. "Monmouth Shooting and Fishing Club."

109. "Notes and Comments," *FS*, 10 November 1887, 301.

110. Cohannet, "Taxing the Gun," *FS*, 24 November 1889, 268.

111. Hough, "Chicago and the West," *FS*, 1 October 1890, 269.

112. Mount Tom (pseud.), "Game Bag and Gun: Who Is Responsible?," *FS*, 24 July 1897, 67.

113. Brant (pseud.), "Madison County, New York," *FS*, 20 February 1882, 87.

114. Hough, "Chicago and the West," *FS*, 10 September 1891, 146.

115. Merritt Starr and Russell H. Curtis, eds., *Annotated Statutes of the State of Illinois in Force January 1, 1885, Supplement Embracing Session Laws, 1885–1892* (Chicago: Callaghan and Co., 1892), 663.

116. "The Game Question," *Ice and Refrigeration*, June 1894, 398.

117. William B. Mershon, "Michigan's Many Bills," *FS*, 5 April 1883, 189; Mershon, *The Passenger Pigeon*, 69.

118. "District Attorney Nicoll and Delmonico," *FS*, 23 February 1893, 155; "Delmonico Pays $450.00," *FS*, 6 April 1893, 291.

119. Hough, "Chicago and the West," *FS*, 1 October 1891, 221; Hough, "Chicago and the West," *FS*, 27 August 1891, 108; Hough, "Chicago and the West," *FS*, 20 August 1891, 85.

120. "Quail Is Expensive: It Costs $25 a Bird If Eaten at This Time," *CT*, 14 March 1894, 8; "Blow Is after Sebree and Hall: Saratoga Hotel President and Manager to Answer Charge of Violating the Game Laws," *CT*, 24 February 1897, 12.

121. "Must Have Been Trapped," *Sacramento Daily Union*, 7 October 1892, 3.

122. "The St. Louis Convention," *FS*, 6 August 1885, 27.

123. Tober, *Who Owns the Wildlife?*, 57.

124. "To Modify the Game Laws: Produce Dealers Hold a Meeting and Perfect an Organization," *New York Times*, 15 January 1885, 2; Tober, *Who Owns the Wildlife?*, 203.

125. "The Game Question," *Ice and Refrigeration*, June 1894, 398.

126. Untitled item, *Ice and Refrigeration*, August 1893, 104.

127. St. Lawrence (pseud.), "New York Fish and Game Interests," *FS*, 11 December 1890, 414; "Fish Dealer and Fish Commissioner," *FS*, 1 January 1891, 469.

128. Hough, "Chicago and the West," *FS*, 10 September 1891, 146.

129. Emerson Hough, "Chicago and the West," *FS*, 10 September 1891, 146.

130. "New York Fish and Game Interests," *FS*, 11 December 1890, 414.

131. "A Plank," *FS*, 3 February 1894, 1.

132. After *FS* published its "no-sale" plank in 1894, subsequent issues of the magazine contained nearly twice as many articles about market hunting than in any previous year since the magazine's founding in 1873. Most of the letters written were in support of the "stop the sale of game plank." See, for example, L.W. M., Letter 8—untitled item, *FS*, 21 July 1894, 50; E. M., Letter 3—untitled item, *FS*, 21 April 1894, 336. More support from contributors came in later years. See, for example, "Snap Shots," *FS*, 7 December 1895, 1; William Henry Atherton, "The 'Rich' and the 'Poor,'" *FS*, 17 April 1897, 306; "The 'Forest and Stream' Plank," *FS*, 22 October 1898, 328; Joseph Kalbfus, "The Pennsylvania Law," *FS*, 11 March 1899, 186.

133. Spike Horn, "Wisconsin Deer and Ducks," *FS*, 14 July 1894, 29.

134. John R. Cotter, "A Reformed Pot-Hunter," *Recreation*, August 1899, 116; "The 'Forest and Stream' Plank," *FS*, 22 October 1898, 328.

135. Edwin W. Sandys, "Rod and Gun: The Sale of Game," *Outing*, June 1894, 64.

136. "A New Audubon Society," *FS*, 18 April 1896, 314.

137. C. Frank Scott, "A Firm Stand on the Plank," *FS*, 11 November 1899, 390.

138. "Fish and Game Protection in New York," *FS*, 20 July 1895, 53.

139. K., "In Support of the Plank," *FS*, 19 May 1894, 423.

140. W. R. H., "The Problem of the West," *FS*, 8 September 1894, 203.

141. L. W. M., "Pennsylvania Game Interests," *FS*, 1 September 1894, 184.

142. "The Minnesota System," *FS*, 29 December 1894, 1. Similar provisions became part of other states' game codes in subsequent years.

143. American Express Company v. People, 133 Ill. 649, 649, 656 (1890).

144. Ex Parte Maier, 103 Cal. 476, 480 (1894).

145. Hough, "Chicago and the West," *FS*, 9 October 1890, 229. In a statement dripping with sarcasm, Hough described Brusewitz in 1890 as "a bright and living example of the heights of glory that the illustrious game warden system is capable of attaining." He also suggested that a "tin man" would be an improvement over Brusewitz. Hough, "Chicago and the West," *FS*, 23 October 1890, 270; Hough, "Chicago and the West," *FS*, 20 August 1891, 85; Maurice R. Bortree, *First Annual Report of the Game Warden for the City of Chicago, November 1, 1892* (Springfield, IL: H. W. Rokker, State Printer and Binder, 1893), 4–8.

146. In the first year after his appointment (August 1891–November 1892) he filed twenty-five complaints against Chicago merchants for possession and sale of out-of-season game. See Bortree, *First Annual Report*, 4–8.

147. "Says It Was Well for the Game," *Chicago Tribune*, 28 June 1893, 8.

148. Hough, "Chicago and the West," *FS*, 2 September 1893, 185.

149. "Quail Is Expensive," *Chicago Tribune*, 14 March 1894, 8.

150. "Change in Illinois Game Warden," *FS*, 17 May 1901, 366.

151. Hough, "Chicago and the West," *FS*, 9 November 1895, 403; "Chas. Blow Accused of Attempted Murder," *Chicago Tribune*, 4 December 1900, 8.

152. "Good Season For Hunters," *Chicago Tribune*, 16 August 1896, 30.

153. "Wild Game in Plenty," *Chicago Tribune*, 14 January 1894, 14.

154. Hough, "Chicago and the West: The New Facts—The Illinois Game Law Imbroglio," *FS*, 16 February 1895, 129–30.

155. "Says the Criticisms Are Unjust: Game Warden Blow Defends His Bill Now before the Legislature," *Chicago Tribune*, 3 February 1895, 5.

156. "They Fight over Game Laws: Commission Merchants and Sportsmen Cannot Come to an Agreement," *Chicago Tribune*, 7 February 1895, 8.

157. Hough, "Chicago and the West: The New Facts—The Illinois Game Law Imbroglio"; Hough, "Chicago and the West," *FS*, 23 March 1895, 230. See also "On a War Footing," *FS*, 23 February 1895, 141.

158. Hough, "An Unlucky Newspaper," *FS*, 9 February 1895, 111. Some fourteen articles specifically about the "Blow Bill" appeared in *FS* during the spring of 1895 and into 1896, even after the bill was defeated in the Illinois legislature.

159. Hough, "Chicago and the West: The New Facts—The Illinois Game Law Imbroglio"; "Says the Criticisms Are Unjust"; Hough, "Chicago and the West," *FS*, 9 February 1895, 111–12.

160. W. R. H., "Prairies and Market Stall," *FS*, 23 February 1895, 151.

161. W. R. H., "Nebraska Shooting," *FS*, 20 April 1895, 307.

162. Palmer, *Hunting Licenses*, 16.

163. "New York's 'Blow Bill,'" *FS*, 23 February 1895, 141; "Snap Shots," *FS*, 21 December 1895, 1. Cities that had longer sale periods for game, such as New York, Boston, and Washington, DC, were known as "dumping grounds" for illegally killed wildfowl. In 1896, the Massachusetts Fish and Game Protective Association estimated that as much as 95 percent

of the game sold in Boston came from outside the state. See Tober, *Who Owns the Wildlife?*, 238n84.

164. Hough, "Chicago and the West," *FS*, 23 February 1895, 150.

165. Hough, "Chicago and the West: Blow's Flop," *FS*, 2 March 1895, 169; "The Illinois Markets," *FS*, 9 March 1895, 181. A "sportsman's bill" (known as the Mott Bill) introduced by the Illinois State Sportsman's Association also failed to pass when it was vetoed by Governor Altgeld, much to the dismay of Hough and *Forest and Stream*. Hough, "Chicago and the West," *FS*, 23 March 1895, 230; Hough, "Chicago and the West," *FS*, 4 May 1895, 347.

166. Hough, "Chicago and the West," *FS*, 6 April 1895, 265.

167. "The Markets and the Game," *FS*, 16 February 1895, 121.

168. "The Work in Maine—1883," *FS*, 3 January 1884, 453.

169. Hough, "Chicago and the West," *FS*, 20 February 1897, 148.

170. "Snap Shots," *FS*, 22 December 1892, 529; "Snap Shots," *FS*, 29 December 1892, 551. The National Sportsman's Association was renamed the National Game, Bird, and Fish Protective Association in 1893. See "Notes and Queries," *American Angler*, December 1893, 103.

171. "Destruction of the Birds," *Chicago Tribune*, 11 August 1895, 12; Theodore W. Cart, "The Lacey Act: America's First Nationwide Wildlife Statute," *Forest History* 17 (October 1973): 6–7; Grinnell, *American Duck Shooting*, 576–82.

172. "Game Protection in Congress," *FS*, 5 February 1898, 107.

173. Merritt, *Shadow of a Gun*, 239–42.

CHAPTER FIVE: The Criminal

1. Merritt, *Shadow of a Gun*, 248, 308.

2. Merritt, *Shadow of a Gun*, 308–9.

3. "Game Warden Makes a Big Haul," *CT*, 24 July 1895, 2

4. "Raids a Game House," *CT*, 21 July 1895, 15; "Game Warden Makes a Big Haul," *CT*, 24 July 1895, 2; "Warden Blow Asks for Assistance," *CT*, 27 July 1895, 9; "Merritt Game Case Draws a Big Crowd," *CT*, 30 July 1895, 7.

5. Untitled item, *Kewanee Courier*, 24 July 1895, 1.

6. Merritt, *Shadow of a Gun*, 310.

7. "Warden Blow Asks for Assistance," 9.

8. "To Vigorously Prosecute Merritt," *CT*, 28 July 1895, 11.

9. "Warden Blow Asks for Assistance," 9.

10. Untitled item, *Kewanee Courier*, 24 July 1895, 1.

11. "The Merritt Case," *(Kewanee, IL) Independent*, 25 July 1895, 5.

12. "Merritt Game Case Draws a Crowd," 7; untitled item, *Kewanee Courier*, 31 July 1895, 1.

13. "State Wins Merritt Case," *CT*, 31 July 1895, 5.

14. "State Wins Merritt Case," 5.

15. "Illinois Game Dealer Convicted," *New York Times*, 31 July 1895, 1; "Contraband Game Gets H. C. Merritt of Kewanee into Trouble," *Macomb (IL) Journal*, 1 August 1895, 7; untitled item, *(Lexington, KY) Courier-Journal*, 31 July 1895, 1.

16. Hough, "Chicago and the West," *FS*, 3 August 1895, 96.

17. "Game Laws Are Being Enforced," *CT*, 8 August 1895, 3.

18. Hough, "Chicago and the West," *FS*, 10 August 1895, 118.

19. "Fine of $110,000 for Violating Game Laws," *Lowell (MI) Ledger*, 9 August 1895, 6.

20. "Hearing Merritt Illegal Game Case," *CT*, 17 August 1895, 9.

21. "Merritt Is Let Off with Light Fine," *CT*, 18 August 1895, 1; untitled item, *(Lexington, KY) Courier-Journal*, 31 July 1895, 1.

22. Hough, "Chicago and the West," *FS*, 24 August 1895, 171.

23. Dick of Connecticut (pseud.), "About Big Bags," *FS*, 4 January 1896, 11; Hermit (pseud.), "'Market Hunting Outrage!'," *FS*, 27 November 1897, 428; C. M. Stark, "Foxes and Grouse," *FS*, 5 March 1898, 184; Mount Tom (pseud.), "Who Is Responsible?" *FS*, 24 July 1897, 67.

24. "The Rational and the Insane," *FS*, 21 January 1899, 41.

25. W. W. Mc Cain, "Pennsylvania Game Law," *FS*, 1 May 1897, 349.

26. "The Rational and the Insane," 41.

27. Hermit, "'Market-Hunting Outrage!'," 428.

28. Palmer, *Hunting Licenses*, 9–19; Palmer, *Chronology and Index*, 10–11, 13–16.

29. George Shields, "Guns and Ammunition," *Recreation*, January 1904, 55–57.

30. Kelpie (pseud.), "A Gloomy Growl from Michigan," *FS*, 2 February 1896, 175.

31. J. M. R., "Arkansas Non-residents," *FS*, 23 July 1898, 67; "The Forest and Stream Plank (From the Report of the Vermont Fish and Game Commissioners)," *FS*, 22 October 1898, 328; C. Frank Scott, "A Firm Stand on the Plank," *FS*, 11 November 1899, 390.

32. Hough, "Chicago and the West: Movements of Western Sportsmen," *FS*, 26 August 1899, 166–67.

33. "The Forest and Stream Plank," *FS*, 22 October 1898, 328; Paper Shell (pseud.), "Pennsylvania Small Game," *FS*, 6 November 1897, 363; "The Illinois Game Outlook," *FS*, 4 August 1900, 86.

34. Atherton, "The 'Rich' and the 'Poor.'"

35. T. S. Ford, letter, *Recreation*, October 1899, 119.

36. Paper Shell, "Pennsylvania Small Game."

37. John B. Burnham, "Game and Fish Laws in New York," *FS*, 26 December 1908, 1016.

38. J. B. Fisher and L. T. Christian, Letter 5, *FS*, 19 September 1908, 455.

39. "Ducks Begin to Fall," *Baltimore Sun*, 2 November 1911, 12.

40. "Pete and His Cannon: The Veteran Hunter Takes a Shy at Ex–County Clerk Hamilton; He Then Proceeds to Tell Why He Is the Enemy of the 'Ballee' Duck," *Sacramento Daily Union*, 18 January 1891, 5.

41. "Where the Game Goes," *Sacramento Daily Union*, 19 March 1895, 4; "Relative to Game Protection: Some Should Be Saved for the Average Citizen," *Sacramento Daily Union*, 12 November 1894, 3.

42. "Destruction of Wild Game," *Sacramento Daily Union*, 26 November 1894, 4.

43. Elmer E. Pedlar to Board of Fish and Game Commissioners, State of California, 10 November 1910, Folder: Personnel, 1907–47, JSHP.

44. Sullivan, *Waterfowling on the Chesapeake*, 77; E. C. Hinshaw, interview, 23 February 1930, 174–79, FTP; B. V. Palmer, interview, 26 April 1930, 320, 377, FTP.

45. Charles Cristadoro, "Canvasbacks at $20 per Pair," *FS*, 12 December 1903, 467.

46. Robert S. Miller, interview, 22 February 1930, 193, FTP.

47. Sullivan, *Waterfowling on the Chesapeake*, 79.

48. Pearson, *Adventures in Bird Protection*, 119.

49. "Warden Loveday's Work," *FS*, 17 November 1900, 389.

50. Elmer E. Pedlar to Fish and Game Commission, 23 November 1910, Folder: Fish and Game Enforcement, 1907, 1910, 1926–28, JSHP.

51. Stuart B. McIver, *Death in the Everglades: The Murder of Guy Bradley, America's First Martyr to Environmentalism* (Gainesville: University Press of Florida, 2003), 160; Charles M. Brookfield, "The Guy Bradley Story," *Audubon Magazine*, July/August 1955, 170–84.

52. "Tragedy of the White Heron: Victim of Woman's Vanity," *New York Times*, 13 August 1905, SM4. See also "Dies to Save the Birds: Killed by Egret Hunters," *Muncie (IN) Morning Star*, 12 August 1905, 5; "Martyr in the Birds' Cause," *Washington Post*, 13 August 1905, 45.

53. E. W. S., "Save the Birds," *Outlook*, 27 February 1909, 508.

54. "News and Comment," *Colusa (CA) Daily Sun*, 2 December 1914, 1.

55. "The General Field of Sport," *San Francisco Call*, 25 January 1891, 10.

56. C. H. Ames, Letter 7, *FS*, 12 January 1907, 57.

57. Edgar F. Randolph, "Bloodless Sport," *FS*, 15 September 1906, 415.

58. "Game Law Involved," *(New Haven, CT) Morning Journal-Courier*, 23 November 1895, 1.

59. Geer v. Connecticut, 161 U.S. 519, 522–23, 534–35 (1896).

60. Bean, *Evolution of National Wildlife Law*, 18–19.

61. J. M. R., "Arkansas Non-residents."

62. "The New Game Law," *Scranton (PA) Tribune*, 31 May 1897, 4; Joseph Kalbfus, "The Pennsylvania Law," *FS*, 11 March 1899, 186.

63. J. M. R., "Arkansas Non-residents."

64. T. S. Palmer and H. W. Oldys, *Laws Regulating the Transportation and Sale of Game*, US Department of Agriculture Bulletin No. 14 (Washington, DC: Government Printing Office, 1900), 13; "Non-export Law Upheld," *FS*, 7 March 1896, 189. By 1901 only Montana, Kentucky, Louisiana, and Mississippi did not have nonexport laws. See Palmer and Oldys, *Laws Regulating the Transportation and Sale of Game*, 60–65.

65. Hough, "Chicago and the West," *FS*, 10 December 1898, 471.

66. "Acquitted By the Jury: Is the California Game Law a Dead Letter in This Section?" *San Francisco Call*, 5 November 1895, 9.

67. Palmer and Oldys, *Laws Regulating the Transportation and Sale of Game*, 40–41.

68. "New York's 'Blow Bill,'" *FS*, 23 February 1891, 141; Clarence F. Birdseye, *The Revised Statutes, Codes, and General Laws of the State of New York*, 2nd ed., vol. 2 (New York: Baker, Voorhis & Company, 1896), 1294.

69. "The Market Season for Game," *CT*, 18 August 1899, 6.

70. "Greater New York," *FS*, 3 April 1897, 261.

71. Hough, "Chicago and the West: A Relic of the Past," *FS*, 5 March 1898, 185. Suits against transportation companies for shipping illegally killed in-state game were easier to prosecute. See, for example, American Express Company v. People, 133 Ill. 649, 759–60 (1890).

72. 33 Cong. Rec. 4873 (1900).

73. 33 Cong. Rec. 4871–73 (1900).

74. Magner v. People, 336–37; Theodore W. Cart, "The Lacey Act: America's First Nationwide Wildlife Statute," *Forest History* 17, no. 3 (October 1973): 10; Betsy Mendelson, "Law and the Environment," in *The Cambridge History of Law in America*, ed. Michael Grossberg and Christopher Tomlins (New York: Cambridge University Press, 2008), 3:503.

75. W. H. Tallett, "Spring Shooting," *FS*, 26 January 1901, 66.

76. "Game and Fish Interests: Papers Read at the Meeting of the National Game and Fish Wardens and Commissioners at Mammoth Hot Springs, N.Y.," *FS*, 9 August 1902, 107.

77. State of Utah, *Message of the Governor of Utah to the Fourth Session of the Legislature . . . January 15, 1901* (Salt Lake City: Deseret News, 1901), 36–37.

78. F. M. Newbert to John P. Babcock, 8 November 1911, Folder: F. M. Newbert, 1911, JSHP; Casper Whitney, "The Sportsman's View-Point," *Outing*, July 1901, 462.

79. Hough, "Chicago and the West: Proposed Changes in Game Law," *FS*, 29 December 1900, 507.

80. People ex rel. Hill v. Hesterberg, 76 N.E. 1032, 1035 (N.Y. 1906).

81. Wells Fargo & Company Express v. State, 79 Ark. 349 (1906).

82. See, for example, Rupert v. United States, 181 F. 87 (8th Cir. 1910, Okla.); Eager v. Jonesboro, 103 Ark. 288 (1912); and Jonesboro, L. C. & E. R. Co. v. Adams, 117 Ark. 54 (1915).

83. Allen v. State, 11 Ga. App. 75, 77 (1912).

84. Ex parte Kenneke, 136 Cal. 527, 529 (1902).

85. State v. Shattuck, 96 Minn. 45, 104 N.W. 719, 720 (1905).

86. "Game Dealers Checked," *San Francisco Call*, 16 January 1896, 10.

87. Aberdeen (pseud.), "St Louis Notes," *FS*, 19 March 1898, 226.

88. "New Game Decisions," *New York Produce Review and American Creamery*, 1 March 1899, 30.

89. "Brooklyn Game Dealers Organize," *New York Produce Review and American Creamery*, 8 March 1899, 30.

90. "Poison for Vermin," *FS*, 14 January 1905, 25.

91. T. S. Palmer and H. W. Oldys, *Digest of Game Laws for 1901*, US Department of Agriculture Bulletin No. 16 (Washington, DC: Government Printing Office, 1901), 52. See, for example, Smith v. State, 58 N.E.1045 (Ind. 1901). Contracts with cold storage warehouses promising to keep frozen game during the closed season were also void. See Haggerty v. St. Louis Ice Co., 44 S.W. 1114 (Mo. 1898).

92. People v. Bootman, 95 App. Div. 469, 473 (N.Y. App. Div. 1904); "State Loses Game Law Case," *New York Times*, 7 December 1904, 8.

93. People v. Bootman, 72 N.E. 505, 508 (N.Y. 1904); Robert S. Anderson, "The Lacey Act: America's Premiere in the Fight Against Unlawful Wildlife Trafficking," *Public Land Law Review* 16 (1995): 41.

94. J. Warren Pond, "Report of the Chief Game Protector," in *Eighth and Ninth Reports of the State of New York Forest, Fish, and Game Commission, 1902–1903* (Albany, NY: J. B. Lyon Company, 1903), 87.

95. People v. Martin, 123 App. Div. 335, 337 (N.Y. App. Div. 1908).

96. Silz v. Hesterberg, 211 U.S. 31, 40, 44 (1908).

97. Dwight Williams Huntington, *Our Feathered Game: A Handbook of the North American Game Birds* (New York: Charles Scribner's Sons, 1903), 4.

98. "$250,000 Game Fine: New York State to Start Proceedings in Cold-Storage Case," *(St. Regis Falls, NY) Adirondack News*, 2 October 1909, 1.

99. F. C. Riehl, "The Illinois Game Outlook," *FS*, 4 August 1900, 86.

100. John C. Sharp, *Third Biennial Report of the State Fish and Game Commission . . . for the Years 1899–1900* (Salt Lake City: Deseret News, 1901), 36–38.

101. "A Platform Plank," *FS*, 3 February 1894, reprinted in *FS*, 5 January 1901, 2; W. H. Tallett, Letter, *FS*, 26 January 1901, 66; Hough, "Chicago and the West: Wars of the Wardens," *FS*, 24 August 1895, 161.

102. Hough, "Chicago and the West: Texas Edging toward the Plank," *FS*, 14 February 1903, 128.

103. George A. Irwin, "Sooners at Work in Florida," *FS*, 13 October 1906, 576; Herbert Brown, "Arizona Game Animals," *FS*, 18 December 1909, 979; "The Massachusetts Bill," *FS*, 28 April 1900, 321.

104. Theodore Roosevelt to John B. Burnham, 19 September 1911, reprinted in 67 Cong. Rec. 9616; Charles Bradford, *The Wild Fowlers, or Sporting Scenes and Characters of the Great Lagoon* (New York: G. P. Putnam's Sons, 1901), 156–57.

105. T. S. Palmer and R. W. Williams Jr., *Game Laws for 1906*, US Department of Agriculture, Farmers' Bulletin No. 265 (Washington, DC: Government Printing Office, 1906), 34.

106. "Snap Shots," *FS*, 30 June 1900, 501.

107. Hough, "Chicago and the West: The Black Duck," *FS*, 13 March 1897, 208; "The Plume Bill," *FS*, 10 February 1900, 101; "Bird News and Notes: A New Bird Law," *(Meadville, PA) Chautauquan*, June 1900, 300.

108. "Editorials," *Bird Lore*, February 1900, 31.

109. "Agreement between the members of the Millinery Merchants Protective Association of New York and Audubon Society of the State of New York," Manuscripts and Archives Division,

New York Public Library, New York Public Library Digital Collections, http://digitalcollections
.nypl.org/items/981bce80-0067-0136-7831-7f2c784ec770.

110. "Ethics and the Sportsman," *FS*, 21 June 1902, 484.

111. Merritt, *Shadow of a Gun*, 306

112. Merritt, *Shadow of a Gun*, 308–9, 313; "Contraband Game: Gets H. C. Merritt, of Kewanee, into Trouble—Can Be Fined over Half a Million Dollars," *Macomb (IL) Journal*, 1 August 1895, 7; Hough, "Game Bag and Gun: Chicago and the West," *FS*, 3 August 1895, 96.

113. "Sensational Charges: They Grow Out of the Kewanee Game Case," *Rock Island Argus*, 22 August 1895, 3.

114. Untitled item, *(Kewanee, IL) Independent*, 2 February 1896, 1; "Abbreviated Telegrams," *Rock Island Argus*, 1 July 1896, 1; Hough, "Chicago and the West: Wardens and the Markets," *FS*, 29 February 1896, 174.

115. Hough, "Chicago and the West: Wars of the Wardens," *FS*, 24 August 1895, 161.

116. Merritt v. People, 68 Ill. App. 273 (1896); Hough, "Chicago and the West: Game Wardens," *FS*, 29 February 1896, 173.

117. Hough, "Chicago and the West: The Wardens and the Markets," *FS*, 29 February 1896, 174.

118. Merritt v. People, 169 Ill. 218 (1897); Hough, "Chicago and the West: Illinois Game Law Declared Invalid," *FS*, 11 December 1897, 467–68.

119. "Game Warden Has Trouble," *CT*, 26 December 1897, 10; "State Will Appeal Game Case," *CT*, 17 June 1899, 4; "Birds Are Set Free," *CT*, 4 June 1898, 16.

120. Merritt, *Shadow of a Gun*, 318.

121. Merritt, *Shadow of a Gun*, 316.

122. Merritt, *Shadow of a Gun*, 318.

123. Hough, "Chicago and the West: Warden Wins First Round," *FS*, 11 June 1898, 463; Merritt, *Shadow of a Gun*, 318; "Merritt Is Acquitted," *Rock Island (IL) Argus*, 16 June 1899, 5; "State Will Appeal Game Case," *CT*, 17 June 1899, 4; People v. Merritt, 91 Ill. App. 620, 623 (1900).

124. Hough, "Chicago and the West: The Kewanee Storage Case," *FS*, 24 June 1899, 497.

125. Merritt, *Shadow of a Gun*, 316, 319, 322.

126. Hough, "The Interstate Wardens' Convention," *FS*, 19 February 1898, 146.

127. Merritt, *Shadow of a Gun*, 316–17.

128. Merritt, *Shadow of a Gun*, 316, 319, 322.

129. "Chas. Blow Accused of Attempted Murder," *CT*, 4 December 1900, 8.

130. Rueben H. Donnelley, comp., *The Lakeside Annual Directory of the City of Chicago, 1899* (Chicago: Chicago Directory Company, 1899), 264; "Chas. Blow Accused of Attempted Murder," *CT*, 4 December 1900, 8. See also "Unnatural: Crime of a Husband and Father," *Akron (OH) Daily Democrat*, 3 December 1900, 4; "Attempt Failed: Chicago Man Tried to Murder His Wife and Three Children," *Stark County (OH) Democrat*, 4 December 1900, 1.

131. "Chas. Blow Accused of Attempted Murder."

132. Hough, "Chicago and the West: The Clocked Stopped," *FS*, 8 January 1898, 28.

133. Hough, "Chicago and the West: The Wardens and the Market."

134. Hough, "Chicago and the West: The Clocked Stopped," 28.

135. R. C. Darr, interview, 24 and 26 February 1937, 1–2, FTP.

136. Buck Lufkin, interview, 1 March 1930, 142, FTP.

137. Missouri Game and Fish Department, *Second Annual Report of the State Game and Fish Commissioner, 1911* (Jefferson City, MO: Hugh Stevens Printing Company, 1911), 99.

138. Missouri Game and Fish Commission, *Annual Report of the State Game and Fish Commissioner, 1932* (Jefferson City, MO: Botz, Hugh Stephens Press, 1933), 122.

139. David Benac, *Conflict in the Ozarks: Hill Folk, Industrialists, and the Government in Missouri's Courtois Hills* (Kirksville, MO: Truman State University Press, 2010), 88; Kathy Love,

"Vantage Point: The Tilt toward Conservation," *Missouri Conservationist Magazine*, October 1995, https://mdc.mo.gov/conmag/1995/10/vantage-point.

140. Benac, *Conflict in the Ozarks*, 89–90; untitled item, *(Ironton, MO) Iron County Register*, 26 October 1905, 4.

141. "The Game Law: Plain Words Misconstrued to Suit Parties Interested," *Marshall (MO) Republican*, 4 August 1905, 1.

142. "Judge Overrules Warden's Construction of Game Law," *St. Louis Republic*, 21 October 1905, 13.

143. "Local Poultry Gossip," *New York Produce Review and American Creamery*, 10 January 1906, 458.

144. Missouri State Fish and Game Commission, *Annual Report of the State Fish and Game Warden of the State of Missouri for the Year Ending in 1905* (Jefferson City, MO: Hugh Stephens Printing Company, 1906), 42.

145. State v. Heger, 93 S.W. 252, 253–55 (Mo. 1906).

146. Missouri State Game and Fish Commission, *Annual Report, 1911*, 72.

147. Pearson, *Adventures in Bird Protection*, 147; T. S. Palmer, Henry Oldys, and Charles E. Brewster, *Game Laws for 1907*, US Department of Agriculture, Farmers' Bulletin No. 308 (Washington, DC: Government Printing Office, 1907), 36.

148. George Kennedy, "Poor Old Missouri," *FS*, 14 December 1907, 937.

149. Missouri State Fish and Game Commission, *Annual Report, 1911*, 74; T. S. Palmer, Henry Oldys, and C. E. Brewster, *Game Laws for 1909*, US Department of Agriculture, Farmers' Bulletin No. 376 (Washington, DC: Government Printing Office, 1909), 39.

150. Missouri State Game and Fish Commission, *Annual Report, 1911*, 34–35.

151. Missouri State Game and Fish Commission, *Annual Report, 1911*, 6–8, 14, 39, 73–74.

152. Missouri State Game and Fish Commission, *Annual Report, 1911*, 28.

153. Loch Laddie (pseud.), "Missouri Game Affairs," *FS*, 23 September 1911, 486.

154. Love, "Vantage Point."

155. Palmer, Oldys, and Brewster, *Game Laws for 1909*, 40.

156. "Local Poultry Gossip," *New York Produce Review and American Creamery*, 10 January 1906, 458.

157. "To Protect Butchers and Others," *National Provisioner*, 5 October 1907, 43.

158. William T. Hornaday, "The Bayne No-Sale-of-Game Law," *Bulletin of the New York Zoological Society*, May 1913, 981; Hornaday, *Thirty Years War for Wild Life* (New York: Charles Scribner's Sons, 1931), 155.

159. "A Great Victory for Wildlife," *Field and Stream*, August 1911, 350–51; "Restricts Sale of Game," *New York Times*, 27 June 1911, 3; Madison Grant, "Bayne-Blauvelt Bill," *New York Zoological Society Bulletin*, July 1911, 773.

160. "Bayne-Blauvelt Bill Becomes a Law," *Brooklyn Eagle*, 29 June 1911, 20.

161. "General Agricultural News: A Victory for Wild Life," *Cornell Countryman*, October 1911, 26.

162. Missouri State Game and Fish Commission, *Annual Report, 1911*, 138; G. Dan Morgan, "Bobwhite," *FS*, 16 December 1911, 872; "A Great Victory for Game Protection: New York State Takes a Long Step Forward," *Rod and Gun in Canada*, August 1911, 318, 320; "Our Medicine Bag," *Rod and Gun in Canada*, July 1912, 216.

163. Henry Oldys, C. E. Brewster, and Frank L. Earnshaw, *Game Laws for 1911*, US Department of Agriculture, Farmers' Bulletin No. 470 (Washington, DC: Government Printing Office, 1911), 36–37.

164. William T. Hornaday, "The Struggle for a Bayne Law in California," *New York Zoological Society Bulletin*, May 1913, 982.

165. T. S. Palmer and Henry Oldys, *Game Laws for 1908*, US Department of Agriculture, Farmers' Bulletin No. 336 (Washington, DC: Government Printing Office, 1908), 39.

166. Grinnell, Bryant, and Storer, *Game Birds of California*, 14.

167. Elmer E. Pedlar to Fish and Game Commission, 23 November 1910, Folder: Fish and Game Enforcement, 1907–10, 1926–28, JSHP.

168. Pedlar to Board, California Fish and Game Commission, 10 November 1910, Folder: Fish and Game Enforcement, 1907, 10, 1926–28, JSHP.

169. Pedlar to Board of Fish and Game Commissioners, 3 November 1910, Folder: Personnel, 1907–47, JSHP.

170. Charles A. Vogelsang to T. S. Palmer, 1 December 1909, Folder: O. P., Misc., 1909–11, 1925–46, JSHP.

171. John Babcock to Frank M. Newbert, 26 November 1911, Folder: Newbert, F. M., 1911, JSHP.

172. "From Friday's Daily," *Madera (CA) Mercury*, 25 February 1911, 3.

173. Elmer E. Pedlar to Fish and Game Commission, 25 April 1911, Folder: O. P., Misc., 1909–11, 1925–46, JSHP.

174. "Ducks Driven to Gun Club Grounds: Market Hunters Testify Game Warden Used State's Launch to Scare Up Game," *San Francisco Call*, 21 March 1911, 17.

175. "Chief Deputy's Office Is Abolished," *Sacramento Union*, 30 November 1911, 26.

176. Golden Gate (pseud.), "Present Day Needs in Game Conservation," *FS*, 4 January 1913, 13.

177. "Nineteen Reasons for the Passage of a Non-sale American-Killed Wild Game Law in California," *Western Wild Life Call*, 7 February 1913, 3–4; Walter Taylor, "Conserve the Wild Life the Duty of the Individual," *Western Wild Life Call*, 7 February 1913, 4; Joseph Grinnell, "The Eastern Passenger Pigeon Is Extinct: Shall the Wild Pigeon of California Be Forced to the Same Fate," *Western Wild Life Call*, 7 February 1913, 5–6.

178. "Beware of Compromise!! Pass the Flint-Cary Bill without Amendment!," *Western Wild Life Call*, 7 February 1913, 13, reprinted as "Pass the Flint-Cary Bill without Amendment," in *FS*, 29 March 1913, 407.

179. William Frederic Bade, "Tricks of Class Legislation Exposed," *FS*, 15 February 1913, 205.

180. "Money or Personal Effort? An Editorial from the Fresno Republican," *Western Wild Life Call*, 7 February 1913, 7.

181. "Game Belongs to the People," *Marin County (CA) Tocsin*, 13 June 1914, 4.

182. "Would Give Everybody a Chance to Eat Game," *Mariposa (CA) Gazette*, 2 May 1914, 2.

183. "The Initiative a Failure," *Western Wild Life Call*, 4 September 1914, 6; "California Fighting Game-Hogs," *Outing*, October 1914, 124.

184. Ernest Thompson Seton, "The Sale of Game Always Exterminates," *FS*, 31 October 1914, 562; Emerson Hough, "Market Hunter a Destroyer," *FS*, 31 October 1914, 562; Stewart Edward White, "Shall Use of Game Be Determined by the Pocket Book?" *FS*, 31 October 1914, 562–63.

185. William Finley, "Oregon Calls on California to Do Her Duty!" *Western Wild Life Call*, 7 February 1913, 12; William T. Hornaday, "Protect the Wild Game of California," *San Pedro (CA) Daily News*, 1 October 1914, 4; Bade, "Tricks of Class Legislation Exposed," 205.

186. Frank Newbert, *Swat the Market Hunter and Give the Boy a Chance* (Sacramento: n.p., 1914), n.p; "Catechism on the Non Sale of Game," *Western Wild Life Call*, 4 September 1914, 3; "Nineteen Reasons," 4.

187. "Nineteen Reasons," 4; "Catechism," 3.

188. "Vote Yes on the Non Sale of Game," *Western Wildlife Call*, 4 September 1914, 8

189. "Against Repeal of Game Law," *FS*, 17 January 1914, 73; Charles F. Holder, "This Law Means Extermination, Extinction, Annihilation," *FS*, 25 July 1914, 107

190. "The Initiative a Failure," 6.

191. "To Whom Will You Entrust the Making of Our Game Laws?" *Western Wild Life Call*, 4 September 1914, 7.

192. Hornaday, "Protect the Wild Game of California," 4.

193. "Woman's Suffrage and Game Protection," *FS*, 8 August 1914, 184; "California Fighting Game-Hogs," *Outing*, October 1914, 124.

194. "Nineteen Reasons," 4.

195. "Woman's Suffrage and Game Protection."

196. "Nineteen Reasons," 4.

197. Newbert, *Swat the Market Hunter*, n.p.

198. "Appeals to Voters to Stop Terrible Slaughter of Birds by Voting 'No' on Proposed Law," *Sacramento Daily Union*, 17 August 1914, 7.

199. "180 and More Reasons for Voting for the Sale of Game to People," *San Francisco Examiner*, 1 November 1914, 53.

200. Non-Sale of Game, California Proposition 31 (1914), UC Hastings Scholarship Repository, accessed 21 June 2021, https://repository.uchastings.edu/ca_ballot_props/64.

201. Frank M. Newbert, "Reason for Defeat of Non Sale of Game Law," *FS*, 5 December 1914, 730.

202. Grinnell, Bryant, and Storer, *Game Birds of California*, 13.

203. "Market Hunter Holds California," *FS*, 28 November 1914, 696.

204. Oldys, "Game Market of To-day," 243.

205. Edward Howe Forbush, "Sale of Game," *FS*, 18 June 1910, 977.

206. Herbert K. Job, "Market Hunter's Days Are Numbered," *FS*, 6 July 1912.

207. Merritt, *Shadow of a Gun*, 236.

CHAPTER SIX: The Conservationist

1. Hornaday, *Thirty Years War for Wild Life*, 160. Recent biographical treatments of Hornaday's contentious role in conservation include Gregory J. Dehler, *The Most Defiant Devil: William Temple Hornaday and His Controversial Crusade to Save American Wildlife* (Charlottesville: University of Virginia Press, 2013); and Stefan Bechtel, *Mr. Hornaday's War: How a Peculiar Victorian Zookeeper Waged a Lonely Crusade for Wildlife That Changed the World* (Boston: Beacon Press, 2012).

2. Hornaday, "Sportsmen as Game Savers," *NYT*, 14 October 1907, 8.

3. Hornaday, "The Destruction of the Birds and Mammals of the United States," *Nature*, 28 July 1898, 308.

4. Hornaday, "Wild Life Protection Work of the Zoological Society," *Zoological Society Bulletin*, May 1913, 481.

5. Hornaday, *The Statement of the Permanent Wildlife Protection Fund*, Vol. 1: *1913–1914* (New York: Permanent Wildlife Protection Fund, 1915), 60–61.

6. Hornaday, *Thirty Years War*, 160–61. Hornaday complained he had never received due credit for the passage of the Bayne-Blauvelt bill. State no-sale laws, Hornaday argued, were the original impetus for the abolition of the game market, not the later federal effort in the Migratory Bird Treaty Law, as many conservationists claimed. Wildlife conservationists, he quipped, were "lazy historians."

7. Hornaday, *Our Vanishing Wild Life: Its Extermination and Preservation* (New York: New York Zoological Society, 1913), 304.

8. Theodore Roosevelt to John Burnham, 19 September 1911, in *Theodore Roosevelt Papers: Series 3: Letters Sent, 1888–1919*, Subseries 3A: Carbon Copies of Letters Sent, 1894–1919, Vol. 27, 1911, Sept. 11–22, 4942, https://www.loc.gov/item/mss382990001/.

9. Roosevelt to Burnham, 4943.

10. Merritt, *Shadow of a Gun*, 182–83.

11. In the 1900 census, Merritt is identified as a "capitalist" instead of a produce dealer. 1900 US Census, Kewanee, Henry Co., Illinois, Enumeration District 0021, Henry Clay Merritt, microfilm 124036, 10, Ancestry.com.

12. Merritt, *Shadow of a Gun*, 9.

13. Merritt, *Shadow of a Gun*, 242.

14. Merritt, *Shadow of a Gun*, 239.

15. Merchant, *Spare the Birds!*, 25–26, 235.

16. Lloyd Willis, *Environmental Evasion: The Literary, Critical, and Cultural Politics of "Nature's Nation"* (Albany: State University of New York Press, 2011), 77–78.

17. Hornaday, *Our Vanishing Wild Life*, 138.

18. Henry Palmer, "Diary of Henry Palmer, from May 5 to August 18, 1891," reprinted in Walter Rothschild, *The Avifauna of Laysan and the Neighboring Islands* (London: R. H. Porter, 1893–1900), ix–x.

19. Walter K. Fisher, "Birds of the Laysan and Leeward Islands, Hawaiian Group," in *U.S. Fish Commission Bulletin, 1903* (Washington, DC: Government Printing Office, 1906), 770.

20. Mark J. Rauzon, *Isles of Refuge: Wildlife and History of the Northwestern Hawaiian Islands* (Honolulu: University of Hawaii Press, 2001), 106–7.

21. Homer R. Dill and William A. Bryan, *Report of an Expedition to Laysan Island in 1911* (Washington, DC: Government Printing Office, 1912), 12, 27; Hornaday, *Our Vanishing Wild Life*, 138–42.

22. Hornaday, *Our Vanishing Wild Life*, 142.

23. Grinnell, Bryant, and Storer, *Game Birds of California*, 261.

24. "Found a Colony of White Herons," *New York Times*, 30 October 1907, 8.

25. "Satirizes Women's Hats: Then Women Anti-vivisectionists Question Columbia Lecturer," *New York Times*, 18 February 1909, 2.

26. "Found a Colony of White Herons," *New York Times*, 30 October 1907, 8.

27. "Cruel Fate of the Egrets," *Boston Globe*, reprinted in *Life*, 4 July 1907, 40.

28. See, for example, "Cruelty and Extermination," *Outlook*, 15 June 1912, 324; "Exterminating the Birds," *Youth's Companion*, 3 March 1910, 116; "Murderous Millinery," *Living Age*, 8 December 1906, 636; Sara Thorp Thomas, "The Destruction of Birds," *Colman's Rural World*, 25 January 1900, 3; Rudolf Cronau, "A Continent Despoiled," *McClure's Magazine*, April 1909, 639; "The Slaughter of the Innocents," *Maine Farmer*, 15 March 1900, 4.

29. H. C. Martin, "The Passenger Pigeon," *Avicultural Magazine*, December 1901, 33.

30. Mershon, *Passenger Pigeon*, ix, xii, 141–62.

31. Hornaday, *Wild Life Conservation*, 9.

32. Hornaday, *Our Vanishing Wild Life*, 42–52.

33. John Bates Clark, "The Economics of Waste and Conservation," *Atlantic* 106 (September 1910): 326.

34. Robert Miller, interview, 22 February 1930, 195, FTP.

35. R. C. Darr, interview, 8 March 1937, 69, FTP.

36. Charles Willard, interview, 20 March 1930, 308, FTP.

37. Frank Nye, interview, 26 April 1930, 398; William Brown, interview, 19 August 1930, 511, 515, both in FTP.

38. L. E. Bowman, interview, 21 August 1930, 441, 445–46; William Burnett, interview, 32 January 1930, 4; Joe Kautzky to Fred Thompson, 29 January 1930, 579; A. C. Connor, interview, 22 September 1930, 551–54; Frank Dodd, interview, 16 September 1930, 536; Fred Carlson, interview, 15 January 1930, 282; Hoyt Elbert, interview, 19 January 1930, 61–62; [Billy] Finnicum, interview, 21 February 1930, 3–4, all in FTP.

39. Elbert, interview, 62–63, FTP. A marginal notation, presumably made by the interviewer, directed the transcriber to "cut the names of McCormick & Case" from the typescript.

40. William Brown, interview, 19 August 1930, 515, FTP.

41. "Pete and His Cannon," *Sacramento Daily Union*, 18 January 1891, 5; "Preserved Ducks," *San Francisco Call*, 26 November 1894, 5. See also "Destruction of Wild Game," *Sacramento Daily Union*, 26 November 1894, 4; "Game Preserves: A Club Will Be Formed to Oppose the System," *San Jose Herald*, 7 February 1889, 3.

42. Charles Ditto to F. O. Thompson, 26 August 1930, 493, FTP.

43. George Shields, "Guns and Ammunition," *Recreation*, January 1904, 55–57.

44. Buck Lufkin, interview, 1 March 1930, 140–41, FTP.

45. Emerson Hough, "Chicago and the West: Pheasants," *FS*, 21 January 1899, 52.

46. Loch Laddie (pseud.), "Missouri Game Affairs," *FS*, 23 September 1911, 486.

47. Oregon's early-twentieth-century limit on ducks was fifty birds a week. Alabama allowed hunters to take twenty-five ducks per day. Oregon State Game Commission, Oregon State Game Code, 1905 (Salem, OR: J. R. Whitney, 1905), 9; Alabama Department of Game and Fish, *First Biennial Report, Department of Game and Fish of the State of Alabama, February 23, 1907–September 30, 1908* (Montgomery, AL: Brown Printing Company, 1908), 34.

48. "Gunning from a Car," *FS*, March 1920, 120.

49. Leonidas Hubbard Jr., "Iowa's Lack of Sportsmanship: Where the Pot-Hunter Reigns Supreme," *Outing*, May 1901, 192.

50. Hough, "Chicago and the West: Pheasants."

51. "Game Protection in Congress," *FS*, 5 February 1898, 107.

52. Emerson Hough, "Chicago and the West: Left Out the Clam," *FS*, 20 February 1897, 147; Palmer and Oldys, *Laws Regulating the Transportation and Sale of Game*, 33.

53. C. H. Ames, Letter 7, *FS*, 12 January 1907, 57.

54. Annual compendiums of game laws that summarized provisions related to seasons, shipment, sale, and licenses were issued as Department of Agriculture bulletins under various titles between 1900 and the 1930s. Earlier game law compilations included *Dodge's Digest of Bird and Game Laws* (1865), *Suydam's Fur, Fin, and Feather* (1868), *Hallock's Bibliography for Sportsmen* (1880), *Smith's Law of Field Sports* (1886), and *Reynold's Book of the Game Laws* (1890). Palmer, *Chronology and Index*, 53, 59.

55. Roosevelt to Burnham, 4942; Hornaday, *Our Vanishing Wildlife*, 153, 260, 304.

56. Palmer, *Chronology and Index*, 12.

57. Mearns and Mearns, *The Bird Collectors*, 3.

58. Foster, *Working for Wildlife*, 128; T. J. Pearson, "The Government as a Game Warden," *Country Life in America*, January 1917, 41.

59. Grinnell, "Bird Life as a Community Asset"

60. Massachusetts State Board of Agriculture, *Forty-Eighth Annual Report*, xv.

61. Pleas, "The Turkey Buzzard"; "A Mosquito Brief". Commentators were divided on whether carrion species were useful or detrimental to health, with some arguing that these birds transmitted diseases such as anthrax to livestock. See Harry Morris, *Carrion Feeders as Disseminators of Anthrax or Charbon*, Louisiana Bulletin No. 136 (Baton Rouge, LA: Rameriz-Jones Printing Co., 1912), 4.

62. Barrow, *Passion for Birds*, 123–25, 133.

63. Wells W. Cooke, *Distribution and Migration of North American Ducks, Geese, and Swans* (Washington, DC: Government Printing Office, 1906), 9–10.

64. Trefethen, *Crusade for Wildlife*, 153; Henry M. Reeves, "FWS Operational Branches," in *Flyways*, 346–48.

65. George Bird Grinnell, "Preface to the Twelfth Thousand," in *American Duck Shooting* (New York: Forest and Stream Publishing, 1918), 2c.

66. Hornaday, *Thirty Years War*, 162.

67. Grinnell, "Preface to the Twelfth Thousand," 2c.

68. Dorsey, *Dawn of Conservation Diplomacy*, 183–84.

69. 49 Cong. Rec. 2725, 4330 (1913) (statement of Rep. Mondell).

70. 49 Cong. Rec. 4332, 4335–36 (1913).

71. 49 Cong. Rec. 4331 (1913) (statement of Rep. Moss).

72. Mark Weston Janis and Noam Weiner, "Treaties in US Law from the Founding to the Restatement (Third)," in *Supreme Law of the Land? Debating the Contemporary Effects of Treaties within the United States Legal System*, ed. Gregory H. Fox, Paul R. Dubinsky, and Brad R. Roth (New York: Cambridge University Press, 2017), 40; Hornaday, *Thirty Years War*, 164.

73. T. S. Palmer, W. F. Bancroft, and Frank L. Earnshaw, *Game Laws for 1914: A Summary of the Provisions Relating to Seasons, Export, Sale, Limits, and Licenses* (Washington, DC: US Department of Agriculture, 1914), 9.

74. Mendelson, "Law and the Environment," 3:503.

75. "A Long Step Forward," *FS*, 1 February 1913, 144.

76. T. Gilbert Pearson, "The Government as a Game Warden," *Country Life in America*, January 1917, 41; Clement Vose, "State against Nation: The Conservation Case of *Missouri v. Holland*," *Prologue* (Winter 1984): 233–47, reprinted in *A Nation of States: Federalism at the Bar of the Supreme Court*, ed. Kermit Hall (New York: Garland, 2000), 404.

77. Many sources claim that the Weeks-McLean bill prohibited commercial sale of migratory game, but this provision was not specifically a part of the act. See, for example, Trefethen, *Crusade for Wildlife*, 170; John K. Powell, "The Migratory Bird Treaty Act: On the Wings of an Executive Branch Reinterpretation," *Florida Bar Journal* 92, no. 9 (November 2018): 45.

78. Palmer, Bancroft, and Earnshaw, *Game Laws for 1914*, 36.

79. J. S. Hunter, "Estimates of Ducks Sold in the Markets, between 1911 and 1916," in Grinnell, Byrant, and Storer, *Game Birds of California*, 13.

80. 54 Cong. Rec. Appendix, 116 (1917) (statement of Rep. Doolittle).

81. The Department of Agriculture's 1914 annual digest of game repeated the same language about commercial sales from previous years, including the notation, first articulated in 1906, that laws regarding game sales were "passing through a transition stage." Palmer, Bancroft, and Earnshaw, *Game Laws for 1914*, 35.

82. Lewis, *Hints to Sportsmen*, 102; "Spring Duck Shooting," *FS*, 11 April 1903, 287; "New York Spring Duck Shooting," *FS*, 9 April 1904, 294. On the divisiveness of spring shooting, see Karen Bicha, "Spring Shooting: An Issue in the Mississippi Flyway, 1887–1913," *Journal of Sport History* 5 (1978): 65–74.

83. Cooke, *Distribution and Migration of North American Ducks, Geese, and Swans*, 9–10.

84. Bradford, *Wild Fowlers*, 151.

85. "Game Bag and Gun: North American Game Protective Association," *FS*, 1 February 1902, 90.

86. Allan Brooks, Letter 4, *FS*, 16 July 1910, 95.

87. Hornaday, *Our Vanishing Wildlife*, 387.

88. Dorsey, *Dawn of Conservation Diplomacy*, 200–201; Kimberly G. Smith, "One Hundred Years Ago in the American Ornithologists' Union," *Auk* 131, no. 4 (April 2017): 776.

89. Pearson, *Adventures in Bird Protection*, 154; Vose, "State against Nation," 404.

90. William T. Hornaday, *The Statement of the Permanent Wild Life Protection Fund, Vol. 2: 1915–1916* (New York: Permanent Wild Life Protection Fund, 1917), 141–43.

91. Ray P. Holland, "It Was Mostly Luck," unpublished manuscript, xiii–3, Box 10, RPHP.

92. Ray P. Holland, Field Diary, 1914, Box 15, RPHP. All field diaries are in Box 15 of RPHP. Field diaries are not paginated and exact dating is inconsistent. They are cited hereafter by Field Diary and year.

93. Holland, Field Diary, 1914.

94. Holland, Field Diary, 1914.

95. Holland, Field Diary, 1914.

96. Holland, Field Diary, 1915.

97. Holland, Field Diary, 1914.

98. Holland, Field Diary, 1915.

99. United States v. McCullagh, 221 F. 288, 292 (D. Kans. 1915).

100. Holland, Field Diary, 1915; "On the Firing Line: Weeks-McLean Law Developments," *Recreation and Outdoor World*, August 1914, 106.

101. Holland, Field Diary, 1916.

102. State v. Sawyer, 113 Me. 458 (1915); State v. McCullagh, 96 Kan. 786, 788–89 (Kans. 1915); Mark Cioc, *The Game of Conservation: International Treaties to Protect the World's Migratory Animals* (Athens: Ohio University Press, 2009), 72.

103. "Migratory Bird Law Unconstitutional," *Game Breeder and Sportsman*, April 1915, 8; "Conservation—or Conversation?" *FS*, March 1915, 160.

104. George J. Klein, "Wild Ducks Dying in Kansas," *FS*, May 1915, 294.

105. "Migratory Bird Law Unconstitutional"; "Conservation—or Conversation?," 160.

106. E. T. Grether, "Illinois Has Joined the 'No Sale of Game' States," *FS*, August 1915, 486; Palmer, Bancroft, and Earnshaw, *Game Laws for 1914*, 3.

107. This nonexport exception for Mississippi County in Arkansas was declared void in *Jonesboro, L. C. & E. R. Co. v. Adams* (117 Ark. 54 [1915]). The Arkansas Supreme Court decided, however, that, in this case, the shipment of game from Arkansas to Illinois was not a violation of the Lacey Act.

108. Grether, "Illinois Has Joined the 'No Sale of Game' States," 486–87.

109. Grether, "Illinois Has Joined the 'No Sale of Game' States," 486–87.

110. Louis Devincenzi, letter, 20 November 1916, Folder: Personnel, 1907–1947, JSHP.

111. J. W. Dobbins, "Game Bird Conditions in the Middle West," *FS*, May 1915, 293.

112. Holland, Field Diaries, 1915, 1914, 1916, 1917.

113. Holland, Field Diary, 1914.

114. Holland, Field Diary, 1917.

115. Holland, Field Diary, 1916.

116. Holland, Field Diary, 1914.

117. Holland, Field Diary, 1915.

118. E. C. Hinshaw, interview, 23 February 1930, 174–79, FTP.

119. Untitled item, *Des Moines Register*, 21 June 1914, 25.

120. "On the Firing Line: Are We Heckling Hinshaw?" *Recreation and Outdoor World*, August 1914, 110–11; "Hinshaw Explains the Law," *Sioux City (IA) Journal*, 13 April 1914, 3.

121. Holland, Field Diaries, 1918, 1917.

122. Holland, Field Diaries, 1915, 1916.

123. Pearson, "The Government as a Game Warden," 41.

124. Holland, Field Diary, 1916.

125. Holland, Field Diary, 1917.

126. Pearson, "The Government as a Game Warden," 40–41.

127. Later treaties on migratory birds were concluded with Mexico, the former Soviet Union, and Japan. Foster, *Working for Wildlife*, 145; Phillips, *American Game Mammals and Birds*, 9;

Albert M. Day, *North American Waterfowl* (Harrisburg, PA: Stackpole, 1949), 48; Trefethen, *American Crusade for Wildlife*, 148–50.

128. A. C. Connor, interview, 22 September 1930, 549–51, FTP.

129. "Keep Market Hunters Out," *FS*, February 1918, 96.

130. "The Migratory Bird Law," *FS*, August 1917, 360.

131. Grinnell, Bryant, and Storer, *The Game Birds of California*, 55–56.

132. 56 Cong. Rec. 7357 (1918).

133. 56 Cong. Rec. 7362–63 (1918) (statement of Rep. Cooper).

134. 54 Cong. Rec. Appendix, 116 (1917) (statement of Rep. Doolittle).

135. 54 Cong. Rec. Appendix, 308 (1917) (statement of Rep. Shackleford); Jessica Scott and Andrea Folds, "From Friend to Foe: The Complex and Evolving Relationship of the Federal Government and Migratory Birds It Is Bound to Protect," *Environmental Law* 49, no. 1 (2019): 193–94. Hornaday denied the charges of collusion, noting that his organization had, in fact, raised the ire of gun manufacturers because of its stance against automatic pump guns and repeating rifles. See 54 Cong. Rec. Appendix, 428–29 (1917) (statement of Rep. Hulbert).

136. "New Law Bars Pot Hunting," *NYT*, 18 August 1918, 14.

137. George A. Lawyer, *Federal Protection of Migratory Birds: Yearbook of the Department of Agriculture, 1918*, no. 785 (Washington DC: Government Printing Office, 1919), 5. Game farms could propagate birds for sale but only under federal permit.

138. Migratory Bird Treaty Act, 16 U.S.C. § 703. Exception was made for scientific collecting and propagation purposes.

139. Lawyer, *Federal Protection of Migratory Birds*, 7, 13; Day, *North American Waterfowl*, 49.

140. Charles Ditto, interview, 23 August 1930, 494, FTP.

141. Ditto, interview, 497–98, FTP.

142. Connor, interview, 550–51, FTP.

143. Ditto, interview, FTP, 500.

144. Connor, interview, 549–51, FTP.

145. Agriculture Appropriations Bill, 1920, H.R. Rep. No. 980, at 428–29 (1919) (statement of E. W. Nelson).

146. Idaho Fish and Game Department, *Biennial Report of the Fish and Game Warden of the State of Idaho, 1923–1924* (Boise, ID: Fish and Game Dept., 1939), 13.

147. *Migratory Bird Refuges and Public Shooting Grounds: Hearings Before the Committee on Agriculture, February 16 and 17, 1922, on H.R. 5823*, 67th Cong. 11 (1922) (statement of R. P. Holland).

148. Holland, Field Diary, 1919.

149. Holland, Field Diary, 1918.

150. Resolution of the Judiciary Committee, 17 February 1919, *Journal of the House of the State of Missouri*, vol. 1: 1919 (Jefferson City, MO: Hugh Stephens Company, 1919), 338.

151. "Migratory Bird Law Is Again under Fire," *St. Louis Post-Dispatch*, 22 February 1919, 8.

152. "Ray Holland Arrested: Missouri Is Waging a Duck War on the United States," *Atchison (KS) Daily Globe*, 7 March 1919, 1.

153. Statement of Ray Holland, 31 June 1965, RPHP, quoted in Vose, "State against Nation," 409.

154. "M'Allister Is Held on Duck Hunting Charge," *(Shelbyville, MO) Shelby County Herald*, 12 March 1919, 7.

155. "To Test the Validity of Federal Law on Shooting Ducks in Mo.," *Chillicothe (MO) Constitution*, 11 March 1919, 1.

156. "Ray Holland Arrested," 1.

157. Vose, "State against Nation," 409; "Ray Holland Arrested," 1.

158. "Migratory Bird Laws Conflict," *Chillicothe (MO) Constitution*, 27 March 1919, 1.

159. "Test Cases," *Bulletin of the American Game Protective Association*, April 1919, 14.

160. United States v. Thompson, 258 F. 257, 258, 268 (E.D. Ark. 1919).

161. "Migratory Bird Treaty Act Upheld," August 1919, typescript, Box 9, Series 3, American Game Protective Association, Folder: June–Dec. 1919, RPHP.

162. Jonathon Q. Holmes to Ray Holland, 1 June 1919, Box 15, Folder: Missouri v. Holland, correspondence & clippings, RPHP.

163. "Upholds Bird Treaty," *Kansas City Times*, 3 July 1919, 4.

164. H[arry] H[argus] to Ray Holland, 10 July 1919, Box 15, Folder: Missouri v. Holland, correspondence & clippings, RPHP.

165. Francis M. Wilson to Ray Holland, 11 July 1919, Box 15, Folder: Missouri v. Holland, correspondence & clippings, RPHP.

166. H[arry] H[argus] to Witmer Stone, 10 July 1919, Box 15, Folder: Missouri v. Holland, correspondence & clippings, RPHP.

167. "Test Cases," 14.

168. Missouri v. Holland, 252 U.S. 416, 434–35 (1920). In Canada, the companion Migratory Birds Act was similarly sustained by the Supreme Court of Prince Edward Island in a decision (*King v. Russell C. Clark*, 1920) having to do with illegal sale of game in closed season. "Canada Declares Migratory Bird Legislation Constitutional," *Bulletin of the American Game Protective Association*, January 1921, 11.

169. Vose, "State against Nation," 399. Vose notes that the opinion has been seen "as an exposition on the treaty power and expression of constitutional theory. Yet many find the Holmes opinion baffling." Vose argues it can "be characterized as a test case of state resistance to national intervention, the Midwest against the Eastern hunting establishment, provincial versus cosmopolitan values."

170. John Dickinson Sherman, "U.S. Supreme Court Saves the Birds," *Ironton County Register* (Ironton, MO), 8 July 1920, 3.

171. "Bird Hunters Must Obey the Law," *The Daily Gate-City and Constitution Democrat* (Keokuk, IA), 3 September 1920, 9.

172. Ray Holland to George A. Lawyer, 2 August 1920, reprinted in Vose, "State against Nation," 412.

173. Tom Marshall, "Trap Gun and Rod," *Shreveport Times*, 9 April 1920, 7.

174. J. S. Hunter, "Report on Game Status and Game Protection," in *Thirtieth Biennial Report of the California Department of Natural Resources, Division of Fish and Game, for the Years 1926–1928* (Sacramento: California State Printing Office, 1928), 99. An undated typescript version of the report calls the federal law "the greatest step forward." In the published version of the report, this is changed to "a great step forward." See untitled typescript, n.d., Box 3, Folder: Game Conservation, 1928, n.d., JSHP.

175. [Hunter], untitled typescript, ca. 1928, p. 2, Box 3, Folder: Legislation, 1907, 1913, 23–49, JSHP.

176. Hunter, "Report on Game Status and Game Protection," 99.

177. Archibald Rutledge, "A Winter Home for Wild-Fowl," *FS*, February 1920, 82.

178. Frederick A. Willitis, "A Manual of Wild-Fowl Shooting," *FS*, September 1919, 462.

179. Old Camper (pseud.), "Confessions of a Market Hunter: A Story of Game Slaughter and the Shipment of Tons of Birds," *FS*, March 1915, 147–48.

180. "Book from the Pen of H. Clay Merritt," *Kewanee Daily Star-Courier*, 28 October 1904, 1.

181. Merritt, *Shadow of a Gun*, 242.

182. "Effort Made to Depose Trustee: Sons of H. Clay Merritt Make Many Charges against Their Father," *Kewanee Star Courier*, 17 November 1906, 1

183. "H. Clay Merritt Suddenly Dies," *Kewanee Weekly Star*, 4 December 1907, 1; "Merritt Will Filed in Court: Kewanee Man Leaves All Property to Widow," *Kewanee Weekly Star*, 11 December 1907, 1.

Epilogue

1. James Fulton, "'Herbie' Merritt Tells about Days When Site of City Hall Was Duck Pond," *Kewanee Star Courier*, 25 July 1927, 15.

2. Fulton, "'Herbie' Merritt."

3. E. A. Lincoln, "Figure of H. Clay Merritt Looms Large in Kewanee's Late 19th Century History," *Kewanee Star Courier*, 7 December 1934, 4; William Dines, "Mysterious Inscription over Merritt Building Reveals Oddity Builder," *Kewanee Star Courier*, 16 February 1937, 3.

4. Lincoln, "Figure of H. Clay Merritt," 4.

5. Frank P. Johnson, "Civil War Soldiers Drained Swamps and Reclaimed the Land," *Kewanee Star Courier*, 14 June 1944, 10.

6. "Personals," *Des Moines Tribune*, 27 September 1927, 5; "Old-Time Market Hunters Relate Stories of Early Iowa Duck Shooting: Fred O. Thompson Furnishes Unique Collection of Yarns," *Iowa Conservationist*, 15 November 1944, 81–82, 86–88; Robert Hale, "Hunting Tales Now Recalled," *Sioux City (IA) Journal*, 5 December 1948, 51.

7. Sec Taylor, "'Wizard of Spirit Lake' Hears Call," *Des Moines Register*, 10 August 1927, 13.

8. Jimmy Robinson, "Rod and Gun with Jimmy Robinson," *Minneapolis Star*, 2 January 1932, 10.

9. US Fish and Wildlife Service, "Old Time Market Hunter Reminiscences: Says Protection Bringing Game Back," press release, P.N. 128794, 21 January 1941, 1–2; The Nomad (pseud.), "The Campfire," *(Davenport, IA) Quad City Times*, 9 March 1941, 37.

10. Robinson, "Rod and Gun," 10.

11. US Fish and Wildlife Service, "Old Time Market Hunter Reminiscences," 1.

12. "Miner Bird Sanctuary Is Great Conservation Aid," *Sheboygan (WI) Press*, 19 April 1937, 15. The famous market hunting Kleinman brothers quit the business at the end of the nineteenth century with remorse for their excesses, and George Kleinman became a deputy game warden. See Henry Kleinman, "Game Bag and Gun: Hints and Points on Ducks," *FS*, 19 June 1890, 431; Hough, "Chicago and the West: Movements of Western Sportsmen," *FS*, 26 August 1899, 167.

13. Wilson, *Seeking Refuge*, 67–68; US Fish and Wildlife Service, "Old Time Market Hunter Reminiscences," 2.

14. Aldo Leopold, "Preliminary Report on a Game Survey of Northern Iowa for the Iowa Conservation Plan, Oct.–Dec. 1931," Iowa State Fish and Game Commission and the State Conservation Board, 1 February 1932, 43, unpublished report, State Historical Society of Iowa, Des Moines.

15. Edgar B. Nixon, comp. and ed., *Franklin D. Roosevelt and Conservation, 1911–1941*, vol. 1 (Washington, DC: Government Printing Office, 1957), 393.

16. Connett, *Wildfowling in the Mississippi Flyway*, 235; "Forest and Stream: Forty Eighth Year," Part 1, *FS*, January 1920, 25; "Selling Game," *Montgomery Times*, 20 December 1924, 4.

17. Mason Dixon, "This Little World," *Shreveport Times*, 18 December 1923, 6; "Feds War on Duck 'Leggers,'" *Escanaba (MI) Daily Press*, 23 July 1939, 12; Connett, *Wildfowling in the Mississippi Flyway*, 235; Philip Garone, *The Fall and Rise of the Wetlands of California's Great Central Valley* (Berkeley: University of California Press, 2011), 145; Rose Certini, "'Dillinger of Duck Hunting' Reminisces," *(Santa Rosa, CA) Press-Democrat*, 24 October 1993, 21. Blewett was arrested in a federal raid in 1935 and spent thirteen months on a federal road crew for his crimes.

18. Henry M. Reeves, "Recollections of My Career, My Colleagues, and My Retirement," unpublished manuscript, 2014, pp. 22–25, Henry M. Reeves Collection USU_499, Box 1, Special Collections and Archives, Utah State University Merrill-Cazier Library, Logan, Utah; "G-Men Grab 38 in Hunting Raids," *Fort Worth Star-Telegram*, 19 April 1956, 16; "'Ducklegging' Raids," *Tyler (TX) Morning Telegraph*, 24 April 1956, 4A; "Widespread Racket of Killing Game Fowl Broken by Raid," *Kerrville (TX) Times*, 19 April 1956, 6.

19. Gordon Charles, "Outdoors with Gordie," *Traverse City (MI) Record-Eagle*, 24 December 1958, 12; Bill Burton, "Adventures of an Undercover Man," *Baltimore Sun, Sunday Sun Magazine*, 6 January 1963, 10–11.

20. Ben Black, "Rod, Reel, and Blunderbuss . . . ," *Santa Cruz Evening News*, 29 December 1939, 5.

21. "Market Hunters Are Low-Lifes," *(Marysville, CA) Appeal-Democrat*, 5 April 1940, 16.

22. Fox, *Conservation Movement*, 159–82.

23. "United States Pondering Bird Protection," *Great Falls Tribune*, 12 December 1926, 3.

24. Irving Brant, "Compromised Conservation: Can the Audubon Society Explain?" Pamphlet 8 (New York: Emergency Conservation Committee, 1930), 5–8.

25. Dan B. Starkey to Carlos Avery, American Game Protective Association, 15 May 1925, RPHP.

26. Manly S. Harris, "Ducks in California," October 1929, n.p., Box 4, JSHP; Emerson Hough, "Time to Call a Halt," *Izaak Walton League Monthly*, August 1922, 1.

27. T. Gilbert Pearson to Ray Holland, 8 January 1926, Box 16, RPHP.

28. Brant, "Compromised Conservation."

29. William T. Hornaday, "Scourge of Guns and Commercialism," *Plain Truth about Game Conservation*, 1 December 1931, 1.

30. Hornaday, "Scourge of Guns and Commercialism"; Brant, "Compromised Conservation."

31. Hough, "Time to Call a Halt."

32. Grinnell, "Bird Life as a Community Asset," 20.

33. Grant Allen, "Aesthetic Feeling in Birds," *Popular Science Monthly*, September 1880, 663.

34. Grinnell, "Bird Life as a Community Asset," 20.

35. "Bird Study: Its Educational Value and Methods," *Journal of Education*, 28 May 1908, 608. See also Albert F. Siepert, "Shop Notes and Problems," *Manual Training Magazine*, January 1917, 212–16.

36. Thomas H. Montgomery Jr., "The Protection of Our Native Birds," *Bulletin of the University of Texas*, No. 79, Scientific Series No. 8, 1 October 1906, 17.

37. Manly S. Harris, "Address to the United Duck Hunters of California," 6 January 1927, Box 3, Folder F3735:607, JSHP. Leopold and Loeb orchestrated the kidnapping and murder of Bobby Franks in 1924, which they believed was "the perfect crime." See Hal Higdon, *Leopold and Loeb: The Crime of the Century* (Urbana: University of Illinois Press, 1975), 341.

38. Harris, "Ducks in California."

39. Robert K. Gilbert to J. S. Hunter, 16 December 1930, Folder G, Box 3, JSHP.

40. Irving Brant, "Shotgun Conservation," Pamphlet 14 (New York: Emergency Conservation Committee, 1931), 1.

41. Hough, "Time to Call a Halt."

42. James Whorten, *Before* Silent Spring: *Pesticides and Public Health in Pre-DDT America* (Princeton, NJ: Princeton University Press, 1974), 68–94.

43. Craig R. Allen, *Socio-ecological Resilience and the Law* (New York: Columbia University Press, 2014), 43.

44. James L. Cummins, "Critical Legislative and Institutional Underpinnings of the North American Model," in *The North American Model of Wildlife Conservation*, ed. Valerius Geist and Shane P. Mahoney (Baltimore: Johns Hopkins University Press, 2019), 63–64.

45. Aldo Leopold, "The Conservation Ethic (1933)," in *Essential Readings in Wildlife Management and Conservation*, ed. Paul R. Krausman and Bruce D. Leopold (Baltimore: Johns Hopkins University Press, 2013), 19.

46. Will H. Dilg, "Victory!" *Outdoor America*, July 1924, 38.

47. Leopold, "Preliminary Report on a Game Survey of Northern Iowa," 45–46.

48. Paul A. Matthews, "Last of the Market Hunters: Old Bert Made His Living with His Guns in the Days When Grouse Were Plentiful," *(Elmira, NY) Star-Gazette*, 13 November 1955, 5.

49. Leo G. Windish, "Adventures of Henry County's Greatest Hunter," *Kewanee Star-Courier*, 22 February 1972, 10; Stan Smith, "Woods and Waters," *New York Daily News*, 7 September 1950, 67.

50. Smith, "Woods and Waters," 67.

51. Valerius Geist, "North American Policies of Wildlife Management," in *Wildlife Conservation Policy*, ed. Valerius Geist and I. McTaggert Cowan (Calgary, Alberta: Detselig, 1995), 77–129; Valerius Geist, Shane P. Mahoney, and John F. Organ, "Why Hunting Has Defined the North American Model of Wildlife Conservation," *Transactions of the 66th North American Wildlife and Natural Resources Conference* 66 (2001): 176–79; Valerius Geist, "The North American Model of Wildlife Conservation: A Means of Creating Wealth and Protecting Public Health While Generating Biodiversity," in *Gaining Ground: In Pursuit of Ecological Sustainability*, ed. David M. Lavigne (Limerick, Ireland: International Fund for Animal Welfare, University of Limerick, 2006), 285–93.

52. Shane P. Mahoney, Valerius Geist, and Paul R. Krausman, "The North American Model of Wildlife Conservation: Setting the Stage for Evaluation," in *North American Model of Wildlife Conservation*, 2–3.

53. Fox, *American Conservation*, 182.

54. Terry Lee Anderson and Donald R. Leal, *Enviro-capitalists: Doing Good While Doing Well* (Lanham, MD: Rowman & Littlefield, 1997), 65–67.

55. Michael P. Nelson, John A. Vucetich, Paul C. Paquet, and Joseph K. Bump, "An Inadequate Construct? The North American Model—What's Flawed, What's Missing, What's Needed," *Wildlife Professional* (Summer 2011): 60; M. Nils Peterson and Michael P. Nelson, "Why the North American Model of Wildlife Conservation Is Problematic for Modern Wildlife Management," *Human Dimensions of Wildlife* 22, no. 1 (2017): 48, doi: 10.1080/10871209.2016.123400948.

56. See, for example, Schaeffle, "Some Notes on the Non-sale of Game."

57. According to a 2016 US Fish and Wildlife Service census, only 4.4 percent of the adult population hunts, down from 7.3 percent in 1991. See Nathan Rott, "Decline in Hunters Threatens How U.S. Pays for Conservation," National Public Radio, 20 March 2018, https://www.npr.org/2018/03/20/593001800/decline-in-hunters-threatens-how-u-s-pays-for-conservation.

58. Arguments for the NAM may be found in Shane P. Mahoney, "The North American Model of Conservation: What Does It Really Mean?" *PERC Reports* 38, no. 1 (Summer 2019), https://www.perc.org/2019/06/19/the-north-american-model-of-wildlife-conservation/; Shane P. Mahoney and Valerius Geist, eds., *The North American Model of Wildlife Conservation* (Baltimore: Johns Hopkins University Press, 2019); Joanna Prukop and Ronald J. Regan, "The Value of the North American Model of Wildlife Conservation: An International Association of Fish and Wildlife Agencies Position," *Wildlife Society Bulletin* 33, no. 1 (Spring 2005): 374–77; Robert D. Brown, "The History of Wildlife Conservation in North America," in *Wildlife*

Management and Conservation: Contemporary Principles and Practices, ed. Paul R. Krausman, James W. Cain III, and James W. Cain (Baltimore: Johns Hopkins University Press, 2013), 6–23.

59. Critical appraisals of the NAM include Peterson and Nelson, "Why the North American Model of Wildlife Conservation is Problematic," 44–46; Lauren Eichler and David Baumeister, "Hunting for Justice: An Indigenous Critique of the North American Model of Wildlife Conservation," *Environment and Society* 9, no. 1 (2018): 75–90, https://doi.org/10.3167/ares.2018 .090106; Andrea M. Feldpausch-Parker, Israel D. Parker, and Elizabeth S. Vidon, "Privileging Consumptive Use: A Critique of Ideology, Power, and Discourse in the North American Model of Wildlife Conservation, *Conservation and Society* 15, no. 1 (2017): 33–40.

60. Nelson, Vucetich, Paquet, and Bump, "An Inadequate Construct?" 58.

61. Davis Quinn, "'Framing' the Birds of Prey," pamphlet (New York: Emergency Conservation Committee, 1929), 19.

62. Carl Talbot, "The Wilderness Narrative and the Cultural Logic of Capitalism" in *The Great New Wilderness Debate*, eds. J. Baird Callicot and Michael P. Nelson (Athens: University of Georgia Press, 1998), 327.

63. Or, to use the Marxist term, commodity fetishism. See David Harvey, "Between Space and Time: Reflections on the Geographical Imagination," *Annals of the Association of American Geographers* 80, no. 3 (1990): 422–23; Elaine R. Hartwick, "Towards a Geographical Politics of Consumption," *Environment and Planning A: Economy and Space* 32, no. 7 (July 2000), 1183; Noel Castree, "Commodifying What Nature?" *Progress in Human Geography* 27, no. 3 (2003): 279–83, 288, 294; and William Cronon, *Nature's Metropolis: Chicago and the Great West* (New York: W. W. Norton, 1992), 256–57, 264.

64. Merritt, *Shadow of a Gun*, 226.

65. Legal scholars trace the origins of the public trust doctrine in natural resource law to *Martin v. Waddell* (1842), which arose from a conflict over oyster fishery rights, but the application of that principle to wildlife (birds) first occurred in *Geer v. Connecticut*. See Michael J. Bean and Melanie J. Rowland, *The Evolution of National Wildlife Law*, 3rd ed. (Westport, CT: Praeger, 1997), 10–15; and John Organ and Shane Mahoney, "The Future of Public Trust: The Legal Status of the Public Trust Doctrine," *Wildlife Professional* (Summer 2007): 20.

66. Geist, Mahoney, and Organ, "Why Hunting Has Defined the North American Model," 177–79.

67. Markus J. Peterson, M. Nils Peterson, and Tarla Rai Peterson, "What Makes Wildlife Wild? How Identity May Shape the Public Trust versus Wildlife Privatization Debate," *Wildlife Society Bulletin* 40, no. 3 (2016): 431; Susan Clayton and Susan Opotow, "Introduction: Identity and the Natural Environment," in *Identity and the Natural Environment: The Psychological Significance of Nature*, ed. Susan Clayton and Susan Opotow (Cambridge: MIT Press, 2003), 1, 4–5.

68. William Cronon, "The Trouble with Wilderness; or, Getting Back to the Wrong Nature," in *Uncommon Ground: Rethinking the Human Place in Nature*, ed. William Cronon (New York: W. W. Norton, 1996), 69; Jack Turner, "In Wildness Is the Preservation of the World," in *The Great New Wilderness Debate: An Expansive Collection of Writings Defining Wilderness from John Muir to Gary Snyder*, ed. J. Baird Callicott and Michael P. Nelson (Athens: University of Georgia Press, 1998), 617.

69. Aldo Leopold, *A Sand County Almanac and Sketches Here and There* (1949; repr., New York: Oxford University Press, 2020), 90, 94.

70. Leopold, *Sand County Almanac*, 94–95.

71. "Herbert Merritt, 88, Well Known Resident Summoned by Death," *Kewanee Star Courier*, 18 November 1946, 5; "Herbert Merritt, Last of Well-Known Local Family, Nears 82nd Birthday Friday," *Kewanee Star Courier*, 23 October 1940, 7.

72. "Caught under Falling Wall: George Wiley Receives Injuries at Merritt's Old Ice House," *Kewanee Daily Star-Courier*, 4 June 1901, 1.

73. "Merritt Home: Old Landmark Torn Down," *Kewanee Star Courier*, 5 June 1937, 5.

74. "File Amended Complaint in Merritt Case," *Kewanee Star Courier*, 26 April 1948, 5; Jerry Moriarity, "Merritt Building Was Once a Game Storehouse," *Kewanee Star Courier*, 1 August 1956, 5; "Demolition of Hotel near Completion," *Kewanee Star Courier*, 22 September 1958, 5.

75. Merritt, *Shadow of a Gun*, 225.

76. Kiner, *History of Henry County*, 1:75.

77. "About Johnson-Sauk Trail SRA," Illinois Department of Natural Resources, accessed 25 June 2021, https://www2.illinois.gov/dnr/Parks/About/Pages/JohnsonSaukTrail.aspx.

78. "Kewanee Scrapbook," *Kewanee Star Courier*, 31 May 1989, 31.

79. Frank P. Johnson, "Passing Henry County as Hunter's Paradise Told in Merritt's Book," *Kewanee Star Courier*, 24 April 1944, 5; Jerry Mursener, "Waterfowl Shooting—an Era in Henry County History," *Kewanee Star Courier*, 8 February 1964, 5; Randy Rodgers, "Galva Author Has New Book," *Kewanee Star Courier*, 12 December 1985, 7; Jeff Lampe, "*The Shadow* Knows," (Peoria, IL) *Journal Star*, 6 December 2008, https://www.pjstar.com/article/20081206/news /312069941; James A. Tober, *Who Owns the Wildlife? The Political Economy of Conservation in Nineteenth Century America* (Westport, CT: Greenwood Press, 1981), 67; Jennifer Price, *Flight Maps: Adventures with Nature in Modern America* (New York: Basic Books, 1999), 47.

80. "Recommended Reading," *Outdoor News Bulletin* (Wildlife Management Institute) 62, no. 8 (August 2008), https://wildlifemanagement.institute/outdoor-news-bulletin/august-2008 /recommended-reading; Ernie Stites, "John Curl: Rockland Pioneer Once Was Market Hunter," (Pocatello) *Idaho State Journal*, 26 January 1964, 15; Paul A. Matthews, "Last of the Market Hunters," *(Elmira, NY) Star-Gazette*, 13 November 1955, 5; "E. M. Brubaker Dies at 96," *(Uniontown, PA) Evening Standard*, 27 December 1955, 7; Hamm with Bakke, *Last of the Market Hunters*; Walsh, *The Outlaw Gunner*; Kimball and Kimball, *The Market Hunter* (Minneapolis: Dillon Press, 1969), 78–79.

81. Compare, for example, Kimball and Kimball, *Market Hunter*, and Tober, *Who Owns the Wildlife?*

82. For example, Price, *Flight Maps*; Greg Mitman, *Reel Nature: America's Romance with Wildlife on Film* (Cambridge, MA: Harvard University Press, 1999); Susan G. Davis, *Spectacular Nature: Corporate Culture and the Sea World Experience* (Berkeley: University of California Press, 1997).

83. Mucha Mkono, "Outrage over Cecil the Lion Slaying Three Years Ago Left Little in Its Wake," *Conversation*, 29 July 2018, https://theconversation.com/outrage-over-cecil-the-lion -slaying-three-years-ago-left-little-in-its-wake-99163; Liam Stack, "Pedals the Walking Bear Is Dead, New Jersey Officials Believe," *New York Times*, 17 October 2016, https://www.nytimes.com /2016/10/18/nyregion/pedals-the-walking-bear-is-dead-new-jersey-officials-believe.html.

84. Nathan Rott, "Fewer Americans Are Hunting and That Threatens Conservation Funding," NPR, 20 March 2018, https://www.scpr.org/news/2018/03/20/81805/fewer-americans-are -hunting-and-that-threatens-con/.

85. James McCarthy, "Environmentalism, Wise Use, and the Nature of Accumulation in the Rural West," in *Remaking Reality: Nature at the Millenium*, ed. Bruce Braun and Noel Castree (London: Routledge, 1998), 125–48; Richard White, "Are You an Environmentalist or Do You Work for a Living? Work and Nature," in *Uncommon Ground: Rethinking the Human Place in Nature*, ed. William Cronon (New York: W. W. Norton, 1996), 171–85.

86. Valerius Geist, "How Markets in Wildlife Meat and Parts, and the Sale of Hunting Privileges Undermines Wildlife Conservation," *Conservation Biology* 2, no. 1 (March 1988): 16.

87. Daryl Fears and Dino Grandoni, "The Trump Administration Has Officially Clipped the Wings of the Migratory Bird Treaty Act," *Washington Post*, 13 April 2018, https://www .washingtonpost.com/news/energy-environment/wp/2018/04/13/the-trump-administration -officially-clipped-the-wings-of-the-migratory-bird-treaty-act/; "Trump Administration Continues Efforts to Strip Away Bird Protections," *Audubon*, 5 June 2020, https://www.audubon .org/news/trump-administration-continues-effort-strip-away-bird-protections. Trump-era changes to the MBTA were quickly revoked by the Biden administration in 2021. See US Department of the Interior, "Interior Department Takes Steps to Revoke Final Rule on Migratory Bird Treaty Act Incidental Take," press release, 6 May 2021, https://www.doi.gov/pressreleases /interior-department-takes-steps-revoke-final-rule-migratory-bird-treaty-act-incidental.

88. Ivan Brinkley, "Be Conservation Minded," *(Mattoon, IL) Journal Gazette*, 4 November 1964, 5.

89. Merritt, *Shadow of a Gun*, 247–48.

90. Many environmental writers, philosophers, and researchers have elaborated at length about the problematic implications of imaginatively severing nature from culture. See, for example, Cronon, "The Trouble with Wilderness."; Price, *Flight Maps*, xviii; J. Baird Callicott and Michael P. Nelson, "Introduction," in *The Great New Wilderness Debate*, ed. J. Baird Callicott and Michael P. Nelson (Athens: University of Georgia Press, 1998), 19–20; Kim Ward, "For Wilderness or Wildness? Decolonising Rewilding," in *Rewilding*, ed. Nathalie Pettorelli, Sarah M. Durant, and Johan T. du Toit (Cambridge: Cambridge University Press, 2019), 34–54. Here, I make a different point about the extent to which modern consumer capitalism overwhelms efforts to keep segments of the physical world outside its influence.

Index